Comprehensive Anatomy of Motor Functions

Pierre Rabischong

Comprehensive Anatomy of Motor Functions

Springer

Professor Pierre Rabischong
Faculty of Medicine of Montpellier,
Montpellier, France

Translated from the original French edition published under the title: Anatomie compréhensive des fonctions motrices © De Boeck Supérieur SA, 2013

Additional material to this book can be downloaded from http://extras.springer.com

ISBN 978-3-319-04168-1 ISBN 978-3-319-04169-8 (eBook)
DOI 10.1007/978-3-319-04169-8
Springer Cham Heidelberg New York Dordrecht London

Library of Congress Control Number: 2014932552

Printed on acid-free paper

Springer is part of Springer Science+Business Media (www.springer.com)

Extra Materials on http://extras.springer.com

List of Figures

Foreword

It is a special pleasure and honour for me to write a preface for the new book of Professor Pierre Rabischong—a man and scientist that I have always admired. He is undoubtedly one of the most outstanding medical scientists of the last 50 years. His research activities, which go far beyond the field of the classical anatomy and the originality of ideas, have influenced not only anatomy but the whole medicine. Professor Pierre Rabischong mastered the descriptive anatomy at a very young age but that was not sufficient for him. He wanted to understand how the body functions and how the separate organs and systems are interconnected. The next level of his research was biomechanics—for him the artificial boundaries between specialties did never exist. In each of these areas he made major and innovative contributions. Actually, Pierre Rabischong is not merely an anatomist—he is rather an encyclopaedic scientist with major contributions in areas such as anatomy, physiology, computer-aided surgery, neuroprosthetics and biomedical technologies.

Second main characteristic of Professor Rabischong's personality is his strive and enthusiasm to explain how the human body is organised and functions to all medical students and colleagues. He gave numerous lectures around the globe, many of which I have witnessed and enjoyed personally. In the current book he aimed to show "how to understand anatomy instead of simply describing structures". I am very happy that he decided to summarise his ideas in a new book on the functions of the human body. The book covers topics such as motor apparatus, spinal functions, respiration, locomotion, prehension, and even—speaking, swallowing, bladder and sexual functions. Therefore, the book is very useful for all medical doctors, irrelevant of their specialty or areas of interests. Reading this special book is a pleasure and an intellectual challenge.

Hannover, Germany

Madjid Samii, M.D., Ph.D.,
Professor of Neurosurgery
President of the International Neuroscience Institute (INI)

Preface

The main objective of this book is to show how to understand anatomy instead of simply describing structures, which can be found in conventional books of anatomy. The main principle is that man did not build man, which is an indisputable evidence. As a result, we have the solution but not the problem. In other words, we are in the situation of an engineer faced with a machine that he has not built, whose functioning he wants to understand as well as the intentions of the manufacturer. The human machine is very complicated, but the more we know it, the more we discover a technical intelligence at a very high level which leads us to talk about high technology (high tech).

To use this comprehensive approach of anatomy, the idea is first to observe a function, if possible with the eye of an engineer versed in the use of technology and knowing engineering terms. Then, in order to identify the technical problems to be solved to achieve this function, we will have to define a chart of the technical specifications that need to be found inside the machine. Finally, it is necessary to go for the anatomical solution that we know increasingly well through means of sophisticated investigation, including magnetic resonance imaging—the true revolution in modern medicine. We should consider within this approach a better understanding of the nature of the biological components chosen by the constructor and certain technical impossibilities associated with this choice. Of course, we must accept the idea that the *constructor is excellent* and that it is difficult to find design errors in the motor programmes studied.

Three important aspects seem to emerge from any study of the living. Firstly, one of the hallmarks of the construction of man is the automation of functions. Without it, complex living systems cannot be managed by pilots mostly totally ignorant of technical problems. This is the case of most humans, let alone of all animals, which therefore cannot in any way interfere actively in their own mode of construction. Secondly, all human beings are unique individuals; that is to say that there will never be an identical copy of them, in the whole history of life. Everyone has a particular genome with the characteristics of the species to which he belongs, but with individual specifications that scientists are now trying to decipher within the non-coding part of the genome for proteins. This search of individuality is a strong feature of the history of life. It can justify the non-reproducibility of neurons, thus keeping the same brain structure throughout life, that is to say the same instrument of thought, which guarantees the continuity of a personality, but with the various acquisitions and cultural variables that are unique to each individual.

Thirdly, the life span of each individual is limited over time. Organs and cells follow a time cycle that leads either to natural regeneration in a well-defined programme, or to them maintaining themselves. But it is difficult to prevent the erosion of time, which degrades living systems and leads to aging that lead to increased pathologies and the use of a variety of medicines.

Finally, we also wish this book to be the basic knowledge required to adequately address all the different treatments of motor deficiencies including manual therapies that are, undoubtedly, a large part of the future of medicine of the motor system. We have also added a significant amount of explanatory drawings and images to facilitate the desired understanding.

Montpellier, France Pierre Rabischong

Acknowledgements

My first acknowledgement is for Madjid Samii whom I have known for many years, one of my best friends and a real master neurosurgeon appreciated all over the world.

I also want to express my thanks to all the students I had the great pleasure to teach and to all my assistants at the laboratory of anatomy of the Faculty of Medicine in Montpellier, who helped me in this fascinating task.

Without my collaborators of the Unit 103 of the INSERM, I would not have been able to achieve biomechanical researches nor make all the contacts I had with the industry, mainly in car manufacturing using modern robotics, aeronautics for specific piloting systems and for many biomedical R&D applications, particularly in the domain of motor handicap.

The donors who gave their body after death allowed us to prepare the anatomical illustrations of this book, showing the fantastic organisation of the human body, and they have to be honoured for this generous and useful action.

The many clinicians I am friends with, particularly the neurosurgeons of the WFNS (World Federation of Neurosurgical Societies) and the surgeons of the international group of hand surgeons (GEM) have to be thanked as well for their precious participation in international teaching courses they offered me.

Finally, I am very grateful to Nathalie Huilleret for her editorial Springer support and to Anne Guilloux for her fine work in the preparation of this book.

Contents

1 Basic Notions on Mobility Function

Abstract

Bones, joints and muscles are the components of the motor apparatus, and their construction can be understood by referring to the basic principles of biomechanics. A bone mineralisation servomechanism explains the plasticity of the system in relation with the specific role of mechanical compression factors applied to the skeleton and generates a piezoelectric microcurrent that stimulates the osteocytes. This mechanism also acts as the technical support for bone regeneration in case of fracture. The joints can be classified in two categories: immobile, like cranial sutures, and mobile, like the intervertebral amphiarthrosis and all the diathroses with different types depending on the articular surface shape and synthesis means. Muscles are viscoelastic actuators, non-reversible and non-linear able to create forces by shortening their contractile part no more than the third of their length which explains the particular mode of construction of the diverse mobile body segments. Command and control of muscles are of great clinical interest. The motor part concerns the functional "motor unit" corresponding to the number of muscular fibres activated in on/off mode by a motoneuron with a variable recruitment explaining a possible proportional control. The sensitive part (proprioception) is given by two transducers: the muscle spindles measure the stiffness of a muscle in order to detect their three different states—relaxed, contracted, stretched—and the Golgi organs measure the force. The position and movement control require a goniometric information given by the skin, thanks to the Ruffini transducers. The conscious and voluntary mobility control applies to movements and not to muscles.

1.1 Introduction

Mobility is one of the four functions managed by the nervous system, the others being: communication, biological maintenance and "survival kit". It allows to move various body segments in space or to keep them in a controlled position (posture). This function occupies a large part of the body weight and shapes the motor apparatus, including bones assembled in the skeleton, joints

P. Rabischong, *Comprehensive Anatomy of Motor Functions*, DOI 10.1007/978-3-319-04169-8_1,

and muscles that mobilise or immobilise them. The nervous system of the motor command represents a large amount of the central nervous system (encephalon with brain, brainstem and cerebellum) and peripheral nervous system (roots, plexus and nerves). Some specific concepts in biomechanics are essential to understand the complexity and intelligent design of all these organs [1–4].

1.2 Components of the Motor Apparatus

1.2.1 Skeletal Levers

They are formed by two types of bone tissue.

1.2.1.1 Spongy and Compact Bones [5–14]

Bones are resistant structures with their mineral content consisting mainly of microcrystals of hydroxylapatite, phosphate and calcium carbonate, which is attached to a protein matrix forming the basic substance consisting of 90 % collagen and mucoproteins. Bones take different forms that can be classified as short, long and flat.

Their formation during embryogenesis occurs in two ways:

- either directly on a fibrous membrane—bones of the vault of the skull;
- or indirectly by means of a cartilage mould which is ossified secondarily after penetration of blood vessels.

There are two types of bone tissue: compact and spongy.

Compact bone tissue is composed of functional units, the osteons, with a microvessel in the centre, which provides osteoblasts locally from the circulating blood (Fig. 1.1).

These by their alkaline phosphatase begin the construction of a circular structure made of concentric blades in which are placed the osteocytes, the true bone cells, which provide the mineralisation of the bone and its retention. Osteocytes are placed in microcavities, the osteoplasts, which communicate with fine canaliculi, creating an interconnected osteocyte network. The assembly comprises a structure combining the fascicular osteonic minicylinders whose viscoelastic mechanical behaviour is a function of the slope of the collagen fibres of the protein matrix.

From a mechanical point of view, the stress represents the reaction of a material to mechanical forces. This cannot be measured directly but only by the deformation (strain) of material, which is the particular domain of mechanics called extensometry (Fig. 1.2). It is possible to use some little electric circuits stuck on the material. These strain gauges can precisely measure the deformation by the variation of their electrical resistance. The longitudinal and transversal deformations under load factor are called the "longitudinal elasticity module of Young" which is in the order of 20,000 MPa (2,000 daN/mm^2) for bone with many variations among the different types of bones. The "Poisson coefficient" characterises the rate between longitudinal (ε_1) and transversal (ε_2) deformations [7]. Therefore, the bone is a non-homogenous and anisotropic material. These biomechanical parameters are important to take into consideration when looking at an osteosynthesis on a bone fracture using plates and screws in metal having a Young modulus of 200,000 MPa (20,000 daN/mm^2), very different from the one of the bones.

The cancellous bone has a honeycomb structure with longitudinal trabeculae following the main lines of force and fine spans across, "**shrink**", ensuring the strength of the whole (Fig. 1.3). This orientation along the force lines is already present in the foetus before the loading of the skeleton. This structure can be compared to that of the reinforcement of concrete with thick stems and thin metal interlaced wires ensuring the cohesion of the frame. Mechanically, the spongy tissue is resistant due to its **geometry** and not to the quantity of bone material that is reduced, with the advantage of a significant weight gain. However, it is slightly elastic and retains **residual deformations** in case of excessive compression, as is the case for fractures of the vertebral bodies or tibial plateau. Within its meshes, the fragile **hematopoietic tissue** that

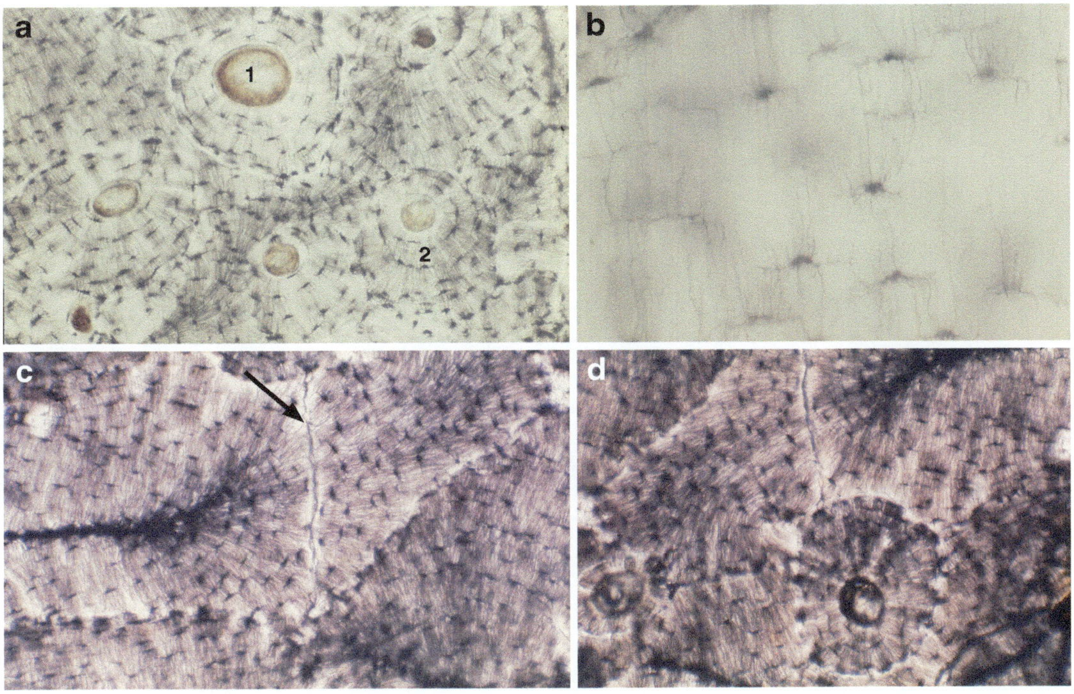

Fig. 1.1 Osteonic structure of compact bone seen on dry bone sections. (**a**) 1. Vascular microcanal of Havers; 2. Concentric circle of osteoplasts. (**b**) Detail of osteoplasts with their interconnected microcanalicular network creating the osteonic force transducer. (**c**) Osteoplast network with microcrack (*arrow*). (**d**) Osteon with concentric circle of osteoplasts

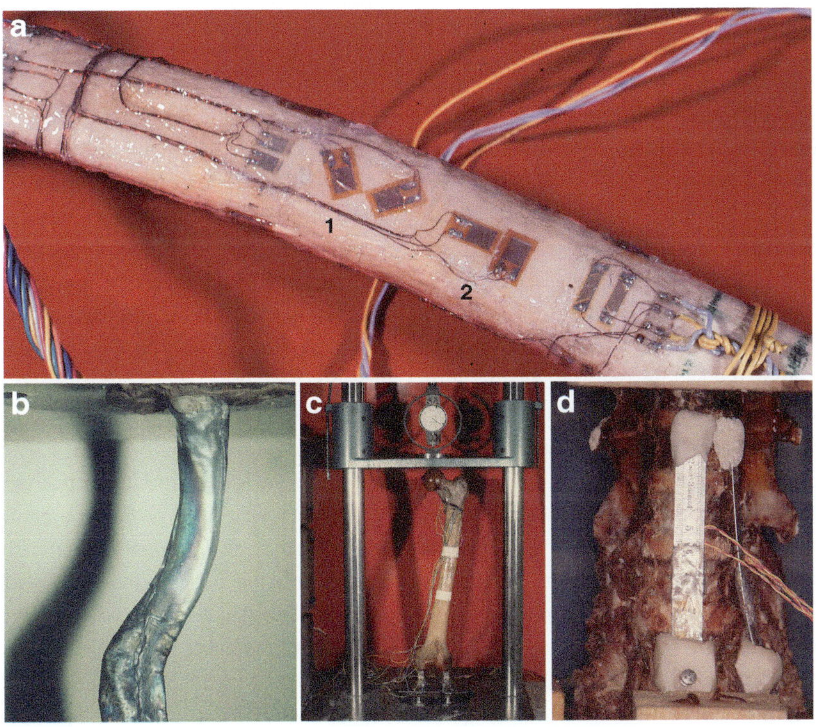

Fig. 1.2 Biomechanical experimental tests. (**a**) Strain gauge bridges on a femur: 1. Torsion (45°); 2. Traction/compression. (**b**) Photostress on a clavicle with blue colour for compression strain. (**c**) Femur equipped with strain gauges under compression machine test. (**d**) Two vertebras under compression measured with force transducer

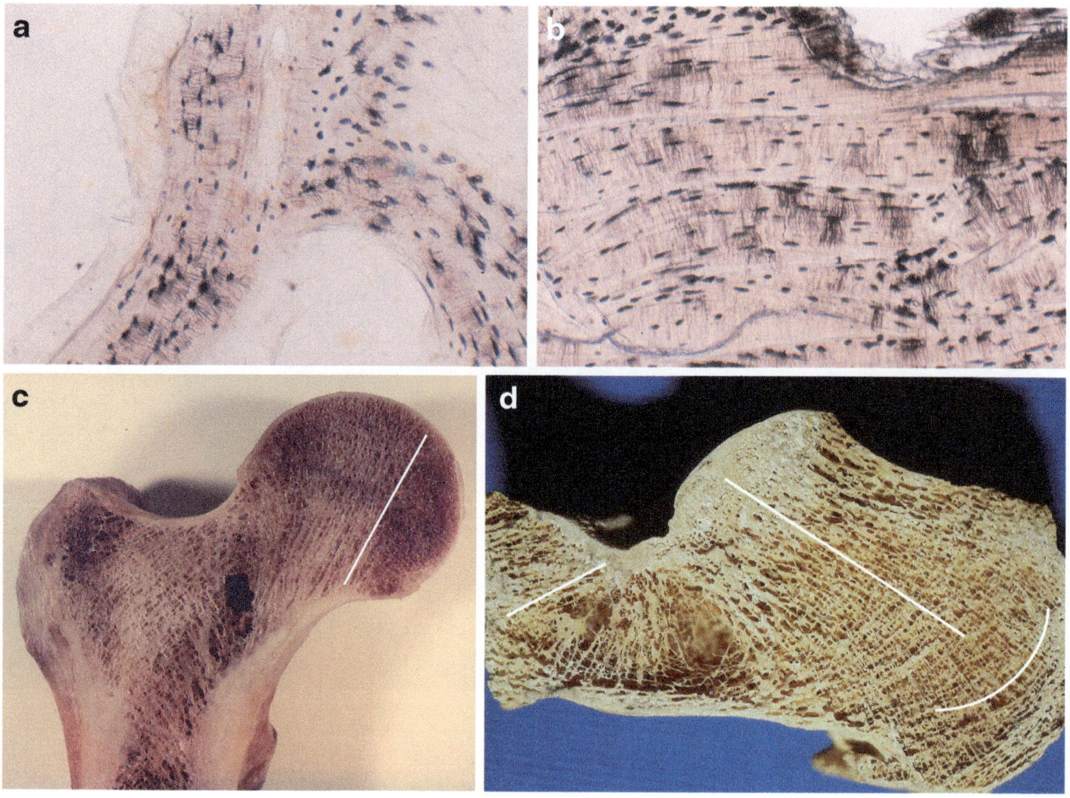

Fig. 1.3 Trabecular structure of spongious bone. (**a**) Microscopic view of trabecules with osteoplasts. (**b**) Lamellar ossification. (**c**) Femur superior part with trabecular orientation along the force lines. (**d**) Calcaneus: mechanical orientation of trabeculas

produces blood elements is protected. The wings of modern aircraft have this honeycomb structure in which kerosene flows.

1.2.1.2 Osteonic Force Sensor

There is a relationship between mechanical factors and bone, but it has not been possible to find force sensors in bone until now, except for few Pacinian corpuscles at the periosteum. These cannot account for the mechanical stress exerted on bone in three directions. Given the structure described above with its network of interconnected microcanaliculi, the whole bone becomes a force sensor sensitive to mechanical bone stress, mainly in compression.

1.2.1.3 Bone Mineralisation Servomechanism

Maintaining bone mineralisation is not a static and stable phenomenon. It depends on metabolic and endocrine factors. We should, however, emphasise the importance of mechanical factors that determine the existence of a true mineralisation servomechanism. Indeed, the mineral part of bone consists mainly of hydroxylapatite crystals that are formed in contact with the collagen fibres of the ground substance and bind along their longitudinal axis. It is known that compressing apatite crystals generate microcurrents corresponding to a piezoelectric effect, which can be either direct or inverse. Thus, putting bones in load creates microcurrents, which stimulate osteocytes enclosed in their osteoplasts and promote bone mineralisation. This hypothesis was verified experimentally, and it is possible to have a demonstration on patients with fractures immobilised by a cast in which a radiographic exam shows a clear decalcification, among paraplegics who do not verticalise regularly and have a dangerous osteoporosis, and finally, on astronauts living in space with reduced gravity, taking several months after return to Earth to recover normal mineralisation of their skeleton.

The **beneficial role of skeletal loading** activating the servo mineralisation also plays an important role in the repair of fractures. Bone has a preprogrammed tendency to repair in case of failure. Due to the action of **osteoclast cells** able to resorb bone, galleries are formed through the rupture zone, in which a cohort of osteoblasts provides bone reconstruction. However, for the success of this particular microscopic job, it is imperative that the two bone fragments are immobilised rigorously for a time. This justifies all orthopaedic and surgical methods: plaster, screws and nails. As clearly shown by the team of Professor Maurice Müller from Bern and Davos, compression correctly applied after fracture immobilisation significantly improves the progress of bone regeneration.

1.2.2 Joints

A joint is the junction of two skeletal parts. It is designed either to prevent movement by blocking the skeletal parts after the growth phase or to enable movement whose amplitude is variable, depending on the shape of the articular surfaces and the presence of their restraint capsules and ligaments.

1.2.2.1 Immobile Joints: The Cranial Sutures

The skull is composed of two parts: the neurocranium which houses the encephalon and the splanchnocranium which forms the skeleton of the facial cavities—lower orbit, nasal and oral cavities. The ultimate goal of skull growth is to form a bone box with a moving part, the mandible, and a junction with the cervical spine, the occipitocervical cardan.

The encephalon is formed during early development by different vesicles: telencephalic with a pair of vesicles forming the two cerebral hemispheres and an odd link forming the commissure area; and diencephalic and rhombencephalic forming the brainstem with its three parts: mesencephalic, pontine and medullary with behind, the cerebellum. In the embryo neural growth is fast, causing a characteristic phenomenon of winding. Then the encephalic part enclosed in a fibrous sac which will become later the dura expands by pushing the envelope. This membrane promotes on its surface some centres of direct ossification which gradually expand and coalesce to meet by suture lines, becoming as bone grows lines limiting the different bony pieces of the vault (Fig. 1.4).

The skull base that serves as a support for the brain is developed differently. The notochord, in relation with the central nervous system, plays a significant role by inducing growth of the vertebral bodies and the posterior neural arch. It extends within the skull base up to the posterior part of the sella, representing the central zone that houses the pituitary gland. Thus, the posterior fossa of the skull base has a central part that is of "spinal inspiration" and forms the clivus by the junction of the basilar process of the occipital bone and the posterior part of the sphenoid. In front of the sella is the prechordal part of the skull base with, in the centre, the olfactory cribriform plate of the ethmoid and, laterally, the surelevated orbital roofs. Between the prechordal part and the clivus, there is an angle that is in relation with the degree of development of the two morphogenetic components: the brain and the face. In humans, the angulation corresponds to a kyphotic aspect, but in animals to a lordotic aspect. This kyphosis of the skull base in humans has been named "foetalisation" by Bolk because it is similar to the foetal aspect. In addition, the nasal opening is always, whatever kyphotic or lordotic angulation of the base, parallel to the plane of the olfactory ethmoidal cribriform plate.

In fact, the important development of the telencephalon in humans generates a characteristic bulging of the frontal region as well as a rotation backward, responsible for the curved shape of internal structures such as the caudate nucleus and lateral ventricles. On the other hand, pushing forward on the frontal wall, it maintains the basal compass in its foetal shape, which is not the case for animals with lordotic skull base. Therefore,

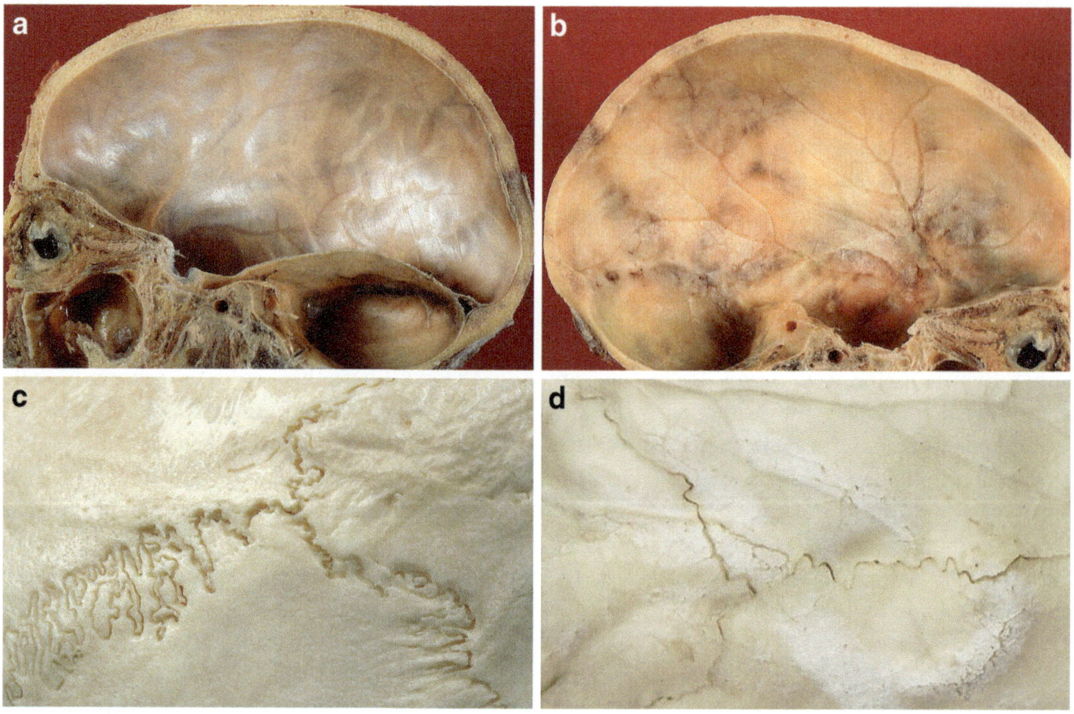

Fig. 1.4 Cranial sutures. (**a**) Endocranial view with dura mater. (**b**) Endocranial view without dura showing the complete suture ossification and the stamp of medial meningeal artery. (**c**) Exocrânial suture drawing like pitch line due to muscle traction avoiding any kind of movement. (**d**) Endocranial linear suture line drawing in relation with the progessive and regular telencephalic growing forces

the two fundamental morphogenetic factors are telencephalon and facial growth.

It should be noted that the posterior clivo-vertebral junction and the corresponding medulla oblongata/cervical spinal cord angulation present an angle almost identical in all vertebrates (Dabelow angle), which enables a dynamic representation of the variation of skulls from birds to humans.

Regarding the vault or calvarium, the plasticity of the cranial envelope changes gradually during growth, becoming less deformable due to ossification centres, which may be referred to as the neurocranial accommodation. Indeed, in the embryo, the telencephalon has a perfectly smooth superficial cortical part. The progressive implementation of the double neuronal cortical matrix, vertical and horizontal, results from the migration of neuroblasts from the periventricular zone towards the surface of the cortex by ascending along glial fibres. This generates a tangential growth, which tends to increase not only in volume but also in surface. This expansion during the end of the foetal period meet with the progressive lower plasticity of dura mater and skull, causing a phenomenon of folding of the cortex dominated by mechanical geometric factors, well studied by Brummelkamp. This creates sulci and gyri, which have many individual variations, but still keep the same anatomical trademarks in all individuals. This is the case of the central sulcus (Rolando) and lateral sulcus (Sylvius) with the fixer role in the backward rotation of the forebrain vesicles mentioned above, of the internal structures such as the insula and the basal ganglia. The corpus callosum, the largest telencephalic commissure, imposes on the interhemispheric cortex a parallel fissuration that does not exist in the case of agenesis of the corpus callosum.

It is important to consider that in this neurocranial accommodation, the growth of telencephalon is the motor of cranial expansion. Nothing can be opposed to its growing power. Therefore, the final shape of the skull is the expressiveness of the development of the telencephalon that is to say of the brain. In the range of phenotypic human variations, there are big brains and small brains. A "blowing occipital" can be considered as an indication of overdevelopment of the brain. It is logical to consider that given the functional value of neural components in mental speculation and understanding of problems, big brains should retain an advantage. In order to mitigate the difference, it may also be said that the same instrument may be played in many ways. But you cannot play the cello when you have a violin, which emphasises the importance of genetic factors in the construction of the brain instrument.

The thrust of telencephalic growth cannot be constrained by an abnormal sutural line growth. Cranial deformations (e.g. turricephaly, scaphocephaly) are known to be the hallmark of compensation growth found naturally in a different direction than the lacking suture. The same ethnic skull deformities as those surrounding the Mayas, who by binding the skull of newborns in a tight cap, push the telencephalic expansion in a vertical direction.

Some still believe in the theory put forward by Virchow who, while observing the skull of a dead microcephalic child with deep mental retardation, noted early synostosis of the cranial sutures and concluded that premature closure had stifled the development of the forebrain, explaining also mental retardation. The skull, however, can never prevent the brain from growing. It is illusory to believe that operating the premature synostosis of sutures causing some cranial deformities in young children can improve brain development.

A certain congenital malformation characterised by the agenesis of the telencephalon—called anencephaly—generates an absence of the cranial vault, but with a normal skull base and splanchnocranium. There is therefore no skull without a telencephalon. On the other hand, maintaining abnormal telencephalic thrust, as is the case of hydrocephalus by a disorder of the circulation of intracerebral cerebrospinal fluid, causes abnormal growth of the skull which can reach impressive proportions.

When brain growth is completed at about the age of 20 years, the sutures stop growing and close by synostosis starting with their endocranial side. In a suture, there are two parts separated by a thin layer of spongy bone, called the diploe, subject to different mechanical influences. The endocranial part has an almost linear suture, due to the gradual and steady nature of the expansive telencephalic thrust. The exocranial part, however, is subject to mechanical intermittent and variable tangential forces of traction exerted by the mandibular muscles, mainly the temporal muscle and the muscles of the neck fixed on the occipital bone and keeping the head in position. When these muscles have sufficient power to support the head and to allow chewing hard foods, they exert tangential traction on the exocranium, explaining the scalloped design of sutures. Due to mesh, it is difficult to separate the pieces of bone, and it becomes impossible when the endocranial synostosis is completed. It has been experimentally demonstrated in newborn dogs that the section of the temporal muscle leads to a suppression of the scalloped pattern of the lateral sutures. Post-traumatic hydrocephalus in young adults may lead to a dissociation of the sutures, but later, in 50-year-olds, it creates intracranial hypertension with a clinically discernible impact on vision by compression of the retina. The fundus of the eye in this case may lead to a diagnosis of an oedema of the papilla.

Therefore, the sutures in the foetus and the newborn are straight lines uniting the bony parts that keep for 1–2 years only some fibrous coalescence zones called fontanels: at the bregma which is the junction of the parietal and frontal bone, and at the lambda, junction of the parietal and occipital. In adults, the sutures close first on the endocranial, leaving the visible exocranial suture drawing, but without any possibility of moving it.

The so-called cranial osteopathy is based on the perception of a particular rhythm through the scalp and an attempt to mobilise sutures. However, this is not possible at a certain age as indicated above.

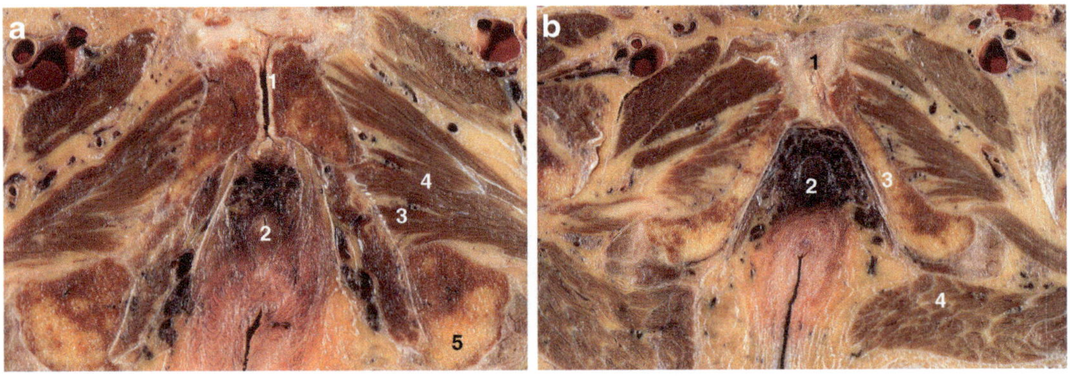

Fig. 1.5 Axial sections of pubic symphysis (elderly person). (**a**) Superior section: 1. Complete mechanical fissuration of the discus interpubicus; 2. Prostatic urethra; 3. Membrana obturatoria; 4. External obturatorius muscle; 5. Iliac bone. (**b**) Inferior section: 1. Inferior arcuate pubic ligament; 2. Prostatic venous plexus; 3. Ischiopubic branch; 4. Gluteus maximus muscle

The true clinical effectiveness of this therapeutic technique should be found, not in a possible mobilisation of bone pieces of the skull, but in the action on the skin covering the skull, which is very rich in cutaneous receptors of any kind and neurovascular complexes as demonstrated in acupuncture points. It seems important that osteopaths change their explanations which, given what we know about this type of therapy, will surely take much time.

1.2.2.2 Mobile Joints

They allow the movement of skeletal segments of two types:

- either with **no direct contact** between the skeletal parts (amphiarthroses) and interposition of fibrous tissue (synfibroses) or cartilage (synchondroses);
- or with **direct contact** between the bones, which requires a viscoelastic interposed joint: the articular cartilage (synovial joints).

Amphiarthroses

They are found in the sagittal plane, anterior and posterior.

- ***The Pubic Symphysis*** (Fig. 1.5)

It fits into the sacroiliac joint triangulation that breaks the pelvic ring with two sacroiliac diarthrosic joints in the back and the pubic amphiarthrosis in the front. This particular feature increases substantially the mechanical resistance of the pelvis. The pubic junction is made by very strong transverse fibres moored on a cartilaginous plate. It is subjected to vertical shearing forces that lead eventually to the creation of a crack. This can be the cause of pubic pain, especially in footballers related to the powerful sway of the lower limb with shock on the ball.

- ***The Intervertebral Amphiarthroses***

They form the intervertebral disc [15, 16] whose role is to maintain an incompressible but deformable space between the vertebrae to allow the amortisation of mainly vertical forces during locomotion (heel strike during walking, running, jumping) and the movements of the spine in the three directions of space. The only incompressible medium is liquid, but it is difficult to imagine a water bag between the vertebrae that will not resist high pressure. The original solution is to trap water in a hydrophilic material constituting a "fibrillar armed gel" made by proteoglycan fibres (65 %) associated with collagen fibres (25 % type II) located in the core (nucleus pulposus) of the disc. It is maintained in place by the peripheral part (60 % of collagen types I and II, 20 % proteoglycan) consisting of thick fibrous concentric strips oriented in three directions, one circular and two obliques [17].

The mooring of the disc is on a cartilage plate, an avascular tissue that is transected in its peripheral portion by the fibres of the peripheral zone

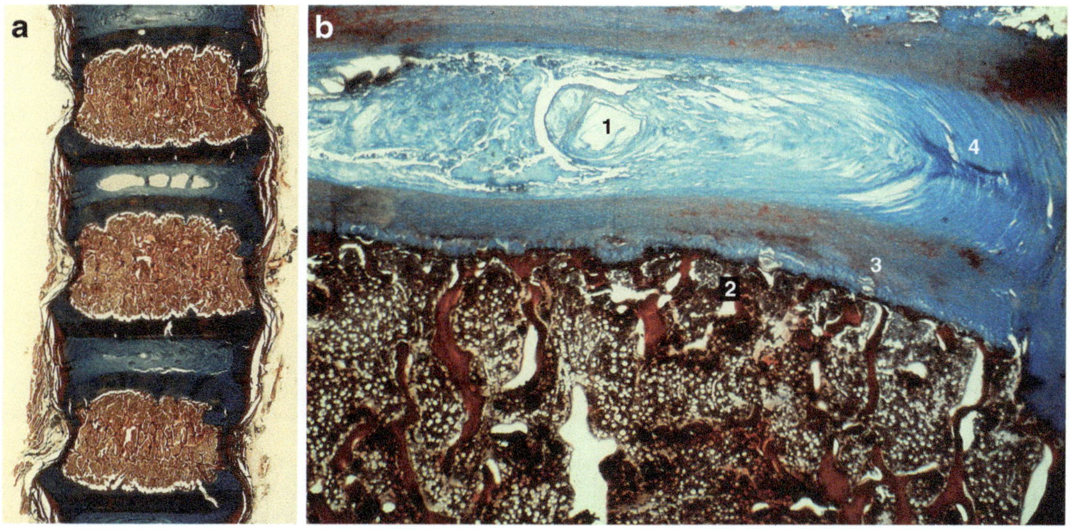

Fig. 1.6 Coronal histological sections of lumbar spine of a 3-year -old young boy. (**a**) General view. (**b**) Details of the disco-somatic complex: 1. Discal central zone (nucleus pulposus); 2. Subchondral venous plexus; 3. Cartilage plate; 4. Discal peripheral zone (annulus fibrosus)

which will bind to the bone of the vertebral endplate. In the centre, the hydrophilic material is in contact with the cartilage plate, which is semipermeable (Fig. 1.6).

There are therefore underload factor fluid exchanges between the disc and the subchondral plexus of the vertebral body. They are reversed during rest periods when lying. Thereby the alternate compression/decompression creates a discal pumping effect. The functional advantage of this pump is to create a gradient pressure inside the bone marrow of the vertebral body. This promotes the release of blood elements elaborated within the hematopoietic tissue into the large vertebral posterior vein connected to spinal epidural venous plexus.

The validation of this concept lies in the pathology. In case of the collapse of the disc, the pump does not work anymore and a more or less significant fatty degeneration of the vertebral bone marrow occurs along the disc border clearly visible on MRI and representing the Modic sign (Fig. 1.7).

The second important aspect of the disc is its mechanical fissuration due to normal compression and shearing forces related to local biomechanical conditions. This mechanical process

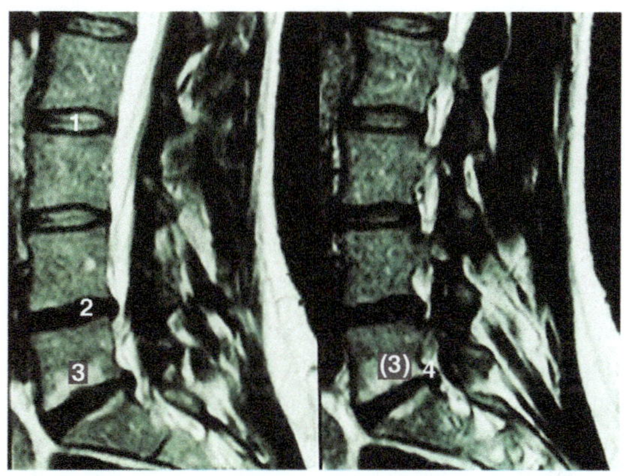

Fig. 1.7 Anachronesis of the lumbar spine on a MRI sagittal section. 1. Fissurated disc; 2. Black disc; 3. Fat involution of bone marrow (Modic sign); 4. Discal hernia

is characteristic of all connective spaces subjected to compression and shear (as, e.g. the pre-clavicular connective bursae acquired by prolonged friction of the rifle on the shoulder of soldiers or prepatellar bursae in "kneeling" professionals). It should be noted that the posterior part of the disc is the place of maximum compressive stress during flexion/extension of the spine, explaining the possible rupture of the annulus at this level. This cracking extending

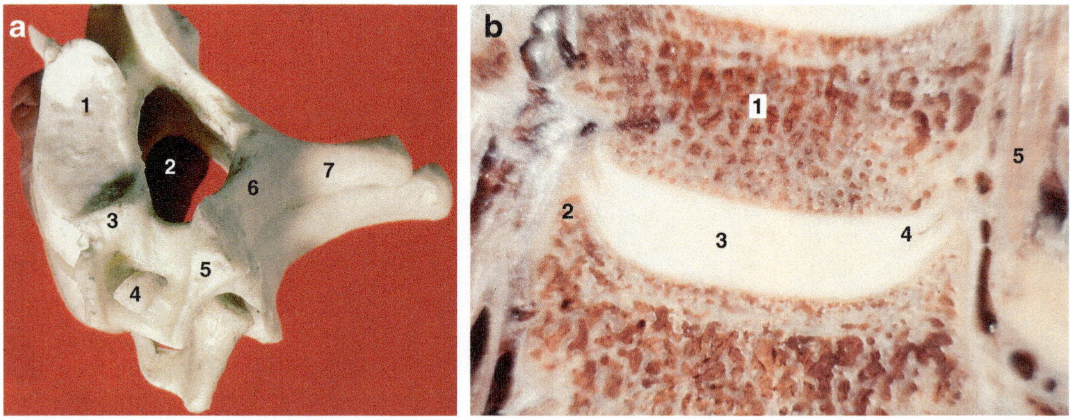

Fig. 1.8 The pseudo-uncovertebral joint. (**a**) Cervical vertebra: 1. Vertebral body; 2. Cervical vertebral canal; 3. Uncus; 4. Transverse foramen; 5. Superior articular process; 6. Cervical lamina; 7. Spinous process. (**b**) Coronal section of cervical spine of a young person (20 years old): 1. Vertebral body; 2. Uncus; 3. Vertebral disc; 4. Mechanical fissuration of the disc (false image of unco-vertebral joint); 5. Vertebral artery

back breaks the peripheral part of the disc, resulting in herniation of the central armated gel mainly in the young, with adequate hydration of the disc. This can lead to compression of roots and the spinal ganglion generating pain. The evolution over time of the disc (anatochronesis) tends to dehydration and involution of proteoglycans. This explains why full disc cracking without herniation can be observed in some elderly patients.

The disc starts cracking very early at the cervical junction of the uncus with the cervical vertebra endplate. This has led to a misinterpretation by former anatomists who have mentioned an uncovertebral joint corresponding to a particular type of joint between amphiarthroses/diathroses, which in fact does not exist (Fig. 1.8). The role of the uncus may therefore be interpreted more as a wall of protection for the vertebral artery, avoiding kinking of the disc as a form of rail, maintaining the vertebral bodies laterally. The arthrosic cervical spine pathology named uncarthrosis may of course alter the original construction scheme and compress the vertebral artery.

Multidirectional mechanical disc solicitations are obviously limited and channelled through the posterior facet joints, creating an articular intervertebral triangulation. Mechanically it is not a tripod, giving balanced support on three points called orthostatic or isostatic, but a hyperstatic support by more than three points which is not possible to calculate geometrically. The shape of the posterior articular processes varies according to the spine level and is strongly involved in limiting and orienting intervertebral movements:

- at the cervical level of great mobility, the posterior articular surfaces are planar and circular, facing backwards and forming a cylindrical column;
- at the thoracic level, they are also planar but more inclined backwards and outwards, in relation with the double costovertebral joint (rib head articulated on the body and rib tuberosity on the transverse process). This system greatly limits the axial movement of rotation leaving some amplitude in flexion/extension;
- at the lumbar level, the mechanical problem is to limit the horizontal shear while allowing flexion, extension, lateral and axial rotation. Thus the articular process takes the form of a vertical cylinder, forming a congruent joint of a trochoid type with a mainly anterior support zone. The two articular spaces in a horizontal section are on a circle that enables axial rotation. L5 bearing obliquely on S1 has facets with 45° orientation in a horizontal section, which increases the radius of curvature of the circle described above allowing a slight axial

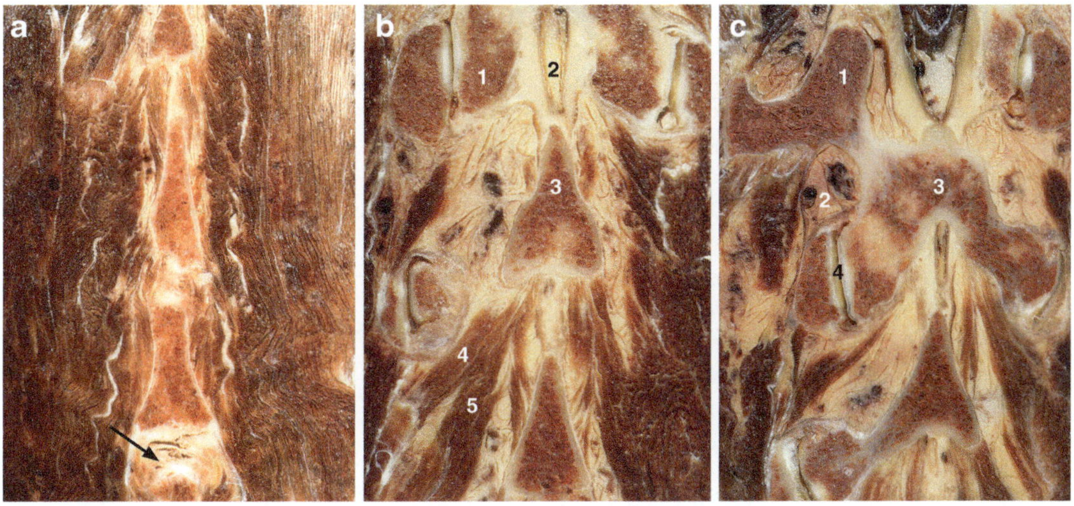

Fig. 1.9 Successive coronal sections of the spinous process (from back to front). (**a**) Mechanical delamination of the interspinous fibrous tissue by friction in relation with disc collapse (*black arrow*). (**b**) 1. Lamina; 2. Fat pad; 3. Spinous process; 4. Short spinous part of the multifidus; 5. Long spinous part of the multifidus. (**c**) 1. Superior articular process; 2. Posterior neurovascular pedicle; 3. Lamina (*little dog image*); 4. Posterior facet joint

rotation, but also the anteroposterior blocking slide. However, it may be observed in the form of a pathology, as an anterior or posterior listhesis. For completeness, we should add that the movements of the lumbar spine are channelled through the posterior facet and limited by the resistant articular capsule fixed closest to the joint line. One can also see on the edge of the synovial joint a false image of a meniscus, made only of synovial tissue without fibrous reinforcement or cartilage. When the dehydrated and cracked intervertebral disc is collapsed, the limitation of facet joint movements may generate specific pain by capsular distension, which can be attenuated by infiltration. Similarly, in disc collapse, the spinous processes may come into contact, which can lead to cracking by mechanical friction of the interposed fibrous tissue and can create a particular pain detected by palpation of spinous processes through the skin (Fig. 1.9).

Diarthroses

These are characterised by direct contact between the skeletal parts that require a special interface allowing slippage. Articular cartilage tissue responds well to this function; it has no blood vessels and is formed by a fundamental substance resistant and viscoelastic within which are chondrocytes. This is the ideal gasket spreading constraints and alignment of isostatic lines of force. Normal or accelerated wear of this joint is the main component of articular pathology. The absence of vessels requires nutrition by liquid with specific viscosity that favours the slip joint surfaces. Movement is essential to the dynamic penetration of the liquid within the surface layer of the cartilage. The synovial fluid formed by hyaluronate protein complexes is secreted from the blood by the synovial membrane with folds and villi [18–21].

One problem is the semantic notion of sliding and rolling. Sliding is comparable to the sudden braking of a car that covers the same contact area on a strip of land. Rolling, on the contrary, opposes point-by-point moving part on the other. There is frequently a combination of these two kinds of movement depending on the type of joint as well as some articular joint dysfunction.

Diarthroses mobility is conditioned by two factors:

- firstly, by the shape of the articular surfaces and their mutual relationship:
  - the junction of two spherical segments can have 3 DF (degree of freedom) as enarthroses: shoulder or hip;
  - a two-cylinder joint or two-cylinder joint segments could have 1–2 DF as trochoidal radioulnar or atlantoaxial joints;
  - two curved cylinders in a condylar joint such as the radiocarpal have 2 DF;
  - two pulleys with deep central cavities as in the humeroulnar trochlea have only 1 DF;
  - two flat articular segments as in the cervical facet arthrodies have no materialised DF;
  - less compatible surfaces, with a small area on one side and a larger on the other, as in the glenohumeral joint, increase mobility by decreasing congruence.
- secondly, the means of synthesis determine the range of motion:
  - the joint capsule is a fibrous more or less tight pocket that attaches to the edge of the articular cartilage and encloses the synovial membrane. It contains mechanoreceptors that do not have a great role in the perception of motion;
  - ligaments are fibrous non-extensible strips that limit movement and are placed either on the capsule in the strengthening (intracapsular ligaments) or outside the capsule and separated by a space (extracapsular ligaments). Some may be extrasynovial and intracapsular ligaments such as the cruciate ligaments within the knee or the round ligament of the femoral head. Only one ligament is elastic, the yellow ligament which links the vertebral lamina and provides a viscoelastic amortisation of spine flexion.

In addition to these passive restraints, the importance of some muscles which are true periarticular active ligaments should be emphasised, such as the piriformis muscle in the hip, the popliteal muscle of the knee or the anconeus muscle of the elbow. The significant role of mechanical periarticular tendons which pass over a joint and can stabilise it when contracted should also be emphasised, as is the case of the wrist or ankle. All these structures may be stretched beyond their normal limits of functioning, which can create a sprain with its succession of pathognomonic signs: pain and swelling depending on the neural richness in mechanoreceptors and vegetative fibres of the ligaments.

1.2.3 Muscles

Muscles are silent, viscoelastic, non-reversible and non-linear actuators that have no equivalent in the industry. There are nearly 600 in the body. They represent about 45 % of body weight. They provide a capacity to develop force by active shortening only in one way (not reversible) and without exceeding one-third of their length. This clearly determines the whole organisation of muscle systems.

1.2.3.1 The Functional Unit: The Myocyte

The myocytes are formed during the embryonic period by the merge of several myoblasts. After this merge, the myocyte is no longer reproduced by mitosis, which explains that the growth of the muscles is done more by myocyte hypertrophy than hyperplasia. Some myoblasts become satellite cells that retain their ability to divide and may possibly be involved in the regeneration of myocytes. These include myofibrils (Fig. 1.10) with a complex structure of sarcomeres consisting of two bands, dark and light (Fig. 1.11), justifying the name of striated muscle fibre. They are grouped in bundles. Muscle proteins are numerous and can be divided into three groups: the contractile proteins, actin (thin myofilaments) and myosin (thick myofilaments); the regulatory proteins acting as a switch for muscle contraction (troponin and tropomyosin); and the structural proteins guaranteeing the stability and elasticity of myofibrils.

There are different types of muscular fibres in relation with their use:

- **Type I (red) SO** (slow oxidative), high aerobic capacity, high mitochondrial density

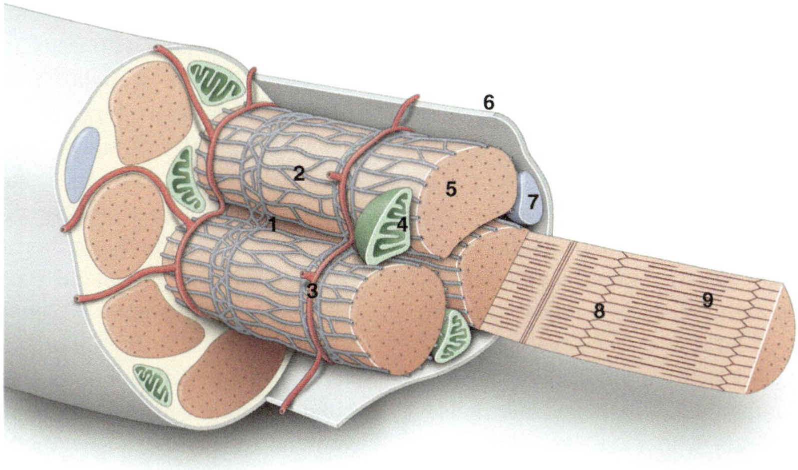

Fig. 1.10 Drawing of a myocyte (between 10 and 100 μm) made by the fusion of about a hundred myoblasts. 1. Terminal cisterns of the cytoplasmic network (Ca ions storage); 2. Cytoplasmic reticulum; 3. Tubule full of interstitial liquid and surrounded by the cisterns creating a triad; 4. Mitochondrion (production of TPA); 5. Myofibril; 6. Sarcolemma around the sarcoplasma containing glycogen and myoglobin; 7. Myocyte nucleus; 8. Z line of the sarcomere; 9. Overlapping of thick and thin myofilaments

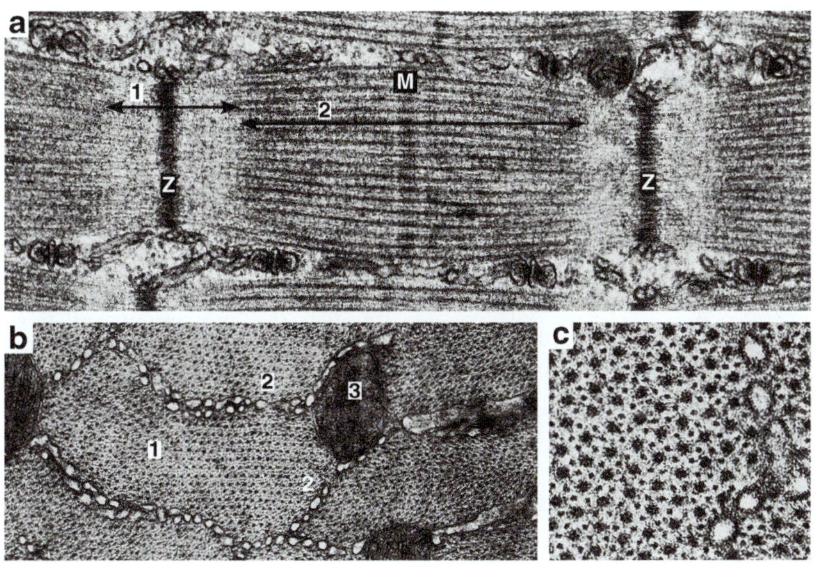

Fig. 1.11 Ultrastructure of sarcomere (after Tortora). (**a**) View of a sarcomere between two Z lines: 1. Light band; 2. Dark band with two types of filaments: thick (myosin) and thin (actin). The M line is formed by structural proteins, mainly titin and myomesin. (**b**) Transverse section of myofibrils: 1. Myofibril; 2. Tubular sarcoplasmic reticulum; 3. Mitochondrion. (c) Magnified view (×88,000): hexagonal shape of thick and thin myofilaments

- **Type IIa (white) FOG** (fast oxidative glycolytic), medium anaerobic and aerobic capacity, high mitochondrial density
- **Type IIb (white) FG** (fast glycolytic), high anaerobic capacity, low mitochondrial density

Muscle fibres are in packets embedded within a fibrous membrane, the epimysium. These packets can be isolated by dissection and are grouped to form a specific muscle, surrounded by its aponeurosis (fascia) that gives the anatomical reality [22]. They are terminated by a tendon of variable morphology in terms of its role and its topographic position. The tendinous fibres are strong and non-extensible, with very low elasticity. The attachment of the muscle fibres on tendons to ensure sufficient strength is a mechanically complex problem. The size of the tendons is partly genetically determined, usually with a sufficient safety factor. Some tendon ruptures, particularly of the Achilles' tendon, and may occur if a muscle overdevelopment is associated with a small-tendon genetic profile. This may be compared to the dental interincisal diastema of subjects who have inherited on a parental side a big jaw and on the other side small teeth. The reverse can also be seen with large teeth and a small jaw causing dental overlapping and rotations.

1.2.3.2 Internal Functional Muscle Architecture [23]

Muscle, from its operating principle of active shortening, may adopt different architectures that will allow some flexibility.

There are generally tendon ends (heads) that define fusiform muscles with two (biceps), three (triceps) or four (quadriceps) heads. Instead of a cylindrical tendon, one can find a large flat tendon, covered by a fibrous fascia, like the large muscles of the abdomen (e.g. the external oblique muscle with its fascia of Cooper). A muscle can be triangular with a concentration of muscle fibres on a tendon penetrating more or less within the muscle mass, like the deltoid muscle. Tendinous fibres can interrupt the continuity of a muscle creating then a digastric or a polygastric muscle. Muscles can activate a single joint (monoarticular muscles) or several (polyarticular muscles), which is important for the control of muscle synergies.

1.2.3.3 Functional Building Principles of Muscle Groups

In terms of mechanics, there exist only two possible movements: translation or rotation. Joint movements are rare in pure translation (spinal cervical or dorsal arthrodies).

Rotations are the more frequent, which however can be associated with translations, as is the case for moving the scapula on the chest wall. There are therefore two types of rotation:

- the first type of rotation is centered and balanced by the action of a muscular torque composed of two components acting in the opposite direction, for example, the horizontal rotation of the head by heterolateral synergistic action of the sternocleidomastoid on one side and splenius capitis on the other side, but with a vertical compression component due to the obliquity of the muscles explaining the frequent arthrosis of cervical spine. Another example is the prono-supination using the rotation of the radius bone by the muscular torque made by the biceps and the supinator muscles;
- the second possible type of rotation is done by a single muscle acting on a curved articular surface absorbing the compression stresses, such as in the flexion of the forearm on the arm. Some joints have a condylar articular surface with two radii of curvature and therefore two rotations along two perpendicular axes with a third combined rotation of both, which is said with instantaneous axis of rotation. This is also the case of the occipito-cervical cardan having two materialised axes of rotation and a combined rotation.

Regarding segmental movements, we can distinguish the movements of flexion/extension, abduction/adduction and axial rotation, with for each a logical construction plan given by the characteristics of the actuators.

Flexion Movements of the Elbow, Hip and Knee

It is interesting to note that the technical constraint limiting the active shortening of a muscle to one-third of its length obliges to not use the muscle under its best mechanical conditions. This is for a lever arm with maximum power and insertion of the active muscle at the distal end of the segment to bend: on the wrist for the flexion of the upper limb or on the knee and ankle for the flexion of the lower limb. It is therefore mandatory in order to bring closer to the joint axis of rotation the distal insertion of muscle, thereby shortening gain but losing power (which also has incidentally the advantage of not building a wing). In doing so, a resistance exerted on the end of the lever has a lever arm of maximum power, which puts the muscular action in a difficult position. The original solution is to associate to a first muscle component, combining polyarticular and monoarticular muscles, a second muscle component giving the extra force required and automatically shrink to a reflex against resistance respecting a principle of economy.

- **For elbow flexion**, the first component is formed by the biceps (polyarticular) and the brachial (monoarticular) and the second component by the humero-stylo-radial or brachioradialis symmetrical to the first component and activated only in case of resistance applied to the forearm. That can be easily controlled on oneself.
- **For hip flexion**, the first component is formed by the iliopsoas (mono- and polyarticular) and the second by the rectus femoris and sartorius.
- **For knee flexion**, the first component, polyarticular and monoarticular by hamstrings and short biceps, the second by the polyarticular part of the triceps surae (twins).

This contraction reflex against resistance of the second bending component has never been studied in details. It is completed at the spinal cord level and disappears in the case of lesion of the corresponding segment of the spinal cord. It is the mark of a great intelligence in muscle organisation and a search for optimisation in relation to technical constraints.

Extension Movements

They are coupled with bending and often less powerful. At the elbow and knee, they are placed under the control of two powerful muscles: the triceps (mono- and polyarticular) for the upper limb and the quadriceps for the lower limb whose problem in locomotion is to exert its force along the axis of the leg. This justifies the presence of the patella (the sesamoid of the quadriceps) guided in its movement by the femoral trochlea and restrained laterally by fibrous fins. Therefore, the misalignment of the patella can be a source of dysfunction of the knee.

Movements of Abduction/Adduction

They are located in a frontal plane. The most demonstrative is the abduction of the arm. It is done by the deltoid muscle. It remains the problem of the limited shortening of muscles. Indeed, the deltoid action corresponding to the third of its length can only give abduction of the arm horizontally. The mechanical solution to go beyond and get hands up above the head is to tilt the scapula which bears the humerus. This implies the "liberation" of the scapula that is attached to the axial skeleton only by the sternoclavicular joint. The flip of the scapula is provided first by the trapezius (specific muscle of the shoulder and not of the head) fixed to the clavicle and the spine of the scapula, which plays the role of lever, and second by the serratus pushing forward the medial border of the scapula. Paralysis of the trapezius prevents full arm abduction due to the swinging of the upper limb and the lack of tilt. The opposite of rocking motion is the "contratilt" of the scapula executed by bringing the scapula medially using the actions of the rhomboid muscle on the back and the pectoralis minor muscle on the front, which lowers the shoulder girdle.

To this has to be added the active adduction of the upper limb assured by a very powerful grouping strap supplying the pectoralis major, teres major and latissimus dorsi (the big three). One may wonder why a strap with so powerful actuators includes a muscle attaching to the lumbo-pelvic region. The answer lies in the necessity also for humans to climb which involves lifting the whole body from a handgrip above the head. That has not been considered as a reminder of the arboreal life of our alleged simian cousins.

For the lower limbs, abduction is interesting functionally by the ability to open the compass of both lower limbs to reach the perineum. But the most interesting action is the mechanical lateral stabilisation of the pelvis and trunk during station on foot in the leg stance phase of walking. This can be achieved by the combined action of a group of three muscles: medial, minimal glutei and tensor of fascia lata, which create an internal rotation of the lower limb when free foot support or pelvic rotation screw mounting responsible for the pelvic step. The adduction is the inverse function of protection and preservation of the perineum (custodes virginitatis) and also medial stabilisation of the pelvis.

Axial Rotations

They are, as seen above, effected in two ways: firstly, centred and balanced rotation torque as in pronation and supination of the hand, thanks to the double inverted trochoid joint radius on the ulna, enabled by the torque action of the biceps (fixed radius) and supinator muscle (attached to the ulna); secondly, unbalanced axial rotation of the humerus by a muscle couple: anterior component by subscapularis and posterior by infraspinatus and latissimus dorsi or in heterolateral synergic rotation of the head by sternocleidomastoid (SCM)/splenius capitis and cervicis or of the trunk by external one side and internal oblique rotation on the other side. Another example is, in case of bone support on the pelvic acetabulum, the rotation of the lower limb on the internal side by medial and minimal glutei and on the external side by all these muscles: piriformis, obturator internus, twins and quadratus femoris.

To sum, we can talk for these muscles group constructions about synergies, which combine several different muscles for further action, either on the same side with different muscles called ipsilateral such as actions of right SCM, splenius, scalene and elevator scapulae for right lateral tilt of the head; either on both sides named heterolateral such as SCM on the right side and splenius of the left side for left rotation of the head; and either on both sides by the same muscles as, for example, in supine position while lifting the trunk with synergy of both sides of the SCM and large muscles of the abdomen (obliques and transverse).

It should be added the muscle synergies that result in the fixation of a segment by cocontraction of agonists and antagonists to lock it in position or to push it across the whole body with manual support against resistance. The term synergies seem more correct than "muscle chains" which represent often a mental construction without serious physiological basis. Indeed, these synergies are fully automated and without voluntary analytical command. Their intimate mechanism and pathways in the nervous system are not yet fully clear and will require further study. What is certain is that the pilot guiding the human machine drives movements and not muscles. In addition, he feels his muscles only when they are painful. He also does not need, firstly, to know that he has muscles, secondly which muscle he has to activate in a particular movement, and thirdly to remember the name of his 600 muscles to perform complex motor tasks.

1.3 Command and Control

Motor control is dominated by two key concepts. Firstly, the motor system is highly complex considering the high level of performance observed. Secondly, considering most of the time the almost total technical ignorance of users, a simplified control is mandatory as well as is a very complicated execution of the process.

Therefore, a high level of process automation is the keyword of brain organisation. In addition, given the apparent simplicity of motor tasks control, the tendency is to underestimate the difficulty of achieving it, and in case of malfunctions, it is not easy for a patient to understand why recovery of a function may not be always possible. This is the case especially for paralysis.

Seen with the eye of an engineer, we can enter the necessary technical automation with specific servomechanisms which can be described in different levels of neural integration. That is made throughout the long conduction pathways: ascending from the periphery to the encephalic centres and from the centres to the periphery to activate the multiple actuators to be selected precisely for a specific task. However, we cannot avoid our reductionist tendency to expose the different operations, remembering, however, that several very different neural systems operate simultaneously in the brain, thanks to the parallel architecture type and non-sequential characterised brain activity. Finally, it is impossible to have a holistic perception of the vast interconnected neural network.

On the other hand, the control of motor functions is of heterarchical type, that is to say, with multiple interactions between different levels without a centralised hierarchical centre. This innovation involves a huge amount of specific transducers with intelligent wiring, creating a selective orientation flow through a discontinuous network having accurate synaptic doors with biochemical keys. Therefore, this gives a considerable value to neurotransmitters.

The synaptic connections are well known at the present time thanks to electron microscopy and modern neurochemistry. Two fundamental processes allow neural management: facilitation and inhibition. It is very hard for us to perceive or even imagine the billions of neurons working together with the technical success that we are enjoying constantly.

A number of technical problems must be explained: proportional control, motor control and central control.

1.3.1 Proportional Control

There are two types of control: all or nothing and proportional. The first type is, for example turning on the light in a room. Muscle fibres obey this principle: they contract or not. The second type is the possibility of proportionality in the order of force according to the needs. This is the case of the muscle, which has a number of muscle fibres that can be activated setting on minimum or maximum.

Motor nerve fibres arising from motoneurons by their endplates producing a neuromuscular junction take under their control a certain number of muscle fibres, representing the motor unit. These can be "recruited" to accomplish a determined effort. They represent the specific increments of force for proportional control.

The intelligence of the system is to not give the same functional value to all motor units. For example, the increment of force used in the control of an extra ocular striated muscle is not the same as in the gluteus maximus [24]. In the first case, the motor unit has something like 25 muscle fibres and, in the other, about 6,000. This explains also the possible variation in the sharpness of muscular control. It is therefore clear that it is mandatory for the management of motor units to find within the muscle actuator two types of transducers: one to indicate the level of force needed to be produced and second to give precise indication about the real supplied effort.

1.3.2 Motor Control

In the command and in the control, two possible types of regulation are distinguished.

The first type is called open-loop control, and in this case, the control is independent of the effector response as well as of local environmental conditions. For example, if a pedestrian is crossing the street without looking at a moving vehicle, the driver of the vehicle has to brake hard without thinking about proportionality.

The second type is when control is directly related to the effector implementation and has to be adapted according to external or internal factors and is called closed loop control. This implies that some specific transducers have to be integrated within the effector in order to transfer to the command system some pertinent information allowing to adjust it in an adaptive mode. Most motor functions of the human body work in closed loop, including fine regulations of the endocrine glands. A servomechanism (from *servus*: the slave) involves the binding of two technical parameters, one being secured to the other. It follows very specific rules and one of them, anticipation, has a great functional value.

Applying these principles to the basics of motor functions, it is necessary to find some specific features related to their control mode [25]. The first point is that the muscles are viscoelastic, non-reversible and non-linear actuators. For a particular action, a minimum of two actuators for each DF, called agonist and antagonist, is required. These viscoelastic body muscles pass through various states of mechanical stiffness corresponding to their three identifiable states for control: relaxed, contracted or stretched (passive or active).

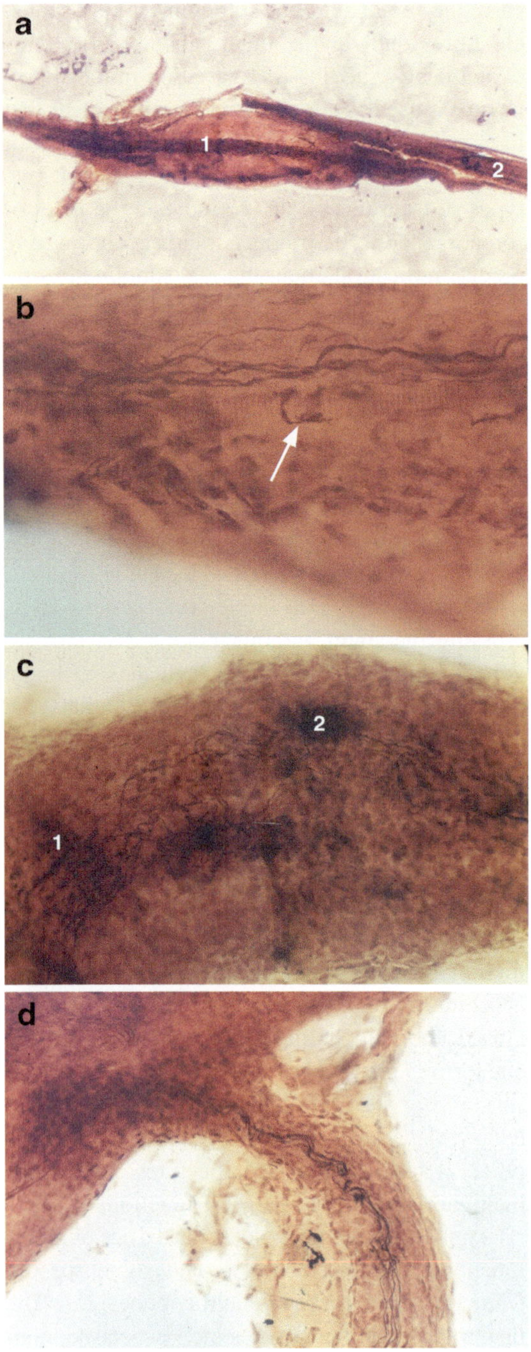

Fig. 1.12 Microdissection of a muscle spindle (1 cm long) with argentic impregnation showing the intrafusal nervous fibres. (**a**) 1. Intrafusal striated muscular fibres; 2. Extrafusal striated muscular fibres. (**b**) View of a motor plate of a gamma motor fibre (*arrow*). (**c**) 1. Gamma motor nervous fibres; 2. Annulo-spiral nervous fibres. (**d**) Nervous pedicle of the spindle

It is therefore logical to use a stiff transducer to characterise their three functional states. This is the muscle spindle [26]. It is about 1 cm long and is made of a fibrous sac, moored to the muscle fibres (Fig. 1.12). There is a bundle of fine striated muscle fibres inside, on which two types of nerve endings terminate: first, on distal ends, motor fibres from gamma motoneurons located in the anterior horn of the spinal cord and, second, in the middle equatorial part, sensory fibre type set in clusters or in spiral-coiled fibres.

This is a stretch transducer which triggers a signal that goes as quickly as possible by a monosynaptic junction to the alpha motor neurons activating the extrafusal muscle fibres of the corresponding muscle. This induced muscle contraction is a true reflex, the stretch reflex well described by Sherrington and also called myotatic reflex. The viscoelastic substrate of the sensor more or less contracted by the action of gamma fibres regulates its sensitivity and thus its gain. The more intrafusal fibres are "tense", the more the reflex is sensitive. The transducer is also sensitive to stretching speed, which was described as tonic and phasic system. Therefore, a hyperactivity gamma can create a muscular hypertonia.

It seems interesting to understand the purpose of this loop integrated into what we may call the fusorial servomechanism. In fact, the muscle spindle can adjust the level of muscle contraction needed for a specific task using its servomechanism. It behaves as a "recruiter" involving additional motor units until the stretch force due to mechanical resistance applied to a bone segment is stopping over.

We also understand that the muscle spindle placed in the active muscle mass may not indicate the level of force exerted by the muscle that we must know for the control. This is the reason for the existence of the Golgi organ, the force transducer placed at the junction between the tendon and the muscle (Fig. 1.13). It is formed by a microtendon surrounded by a fibrous capsule attached to the mother tendon and in which are sensory nerve fibres [27]. It is therefore a force sensor that provides information on the effort done and protects the muscle by triggering

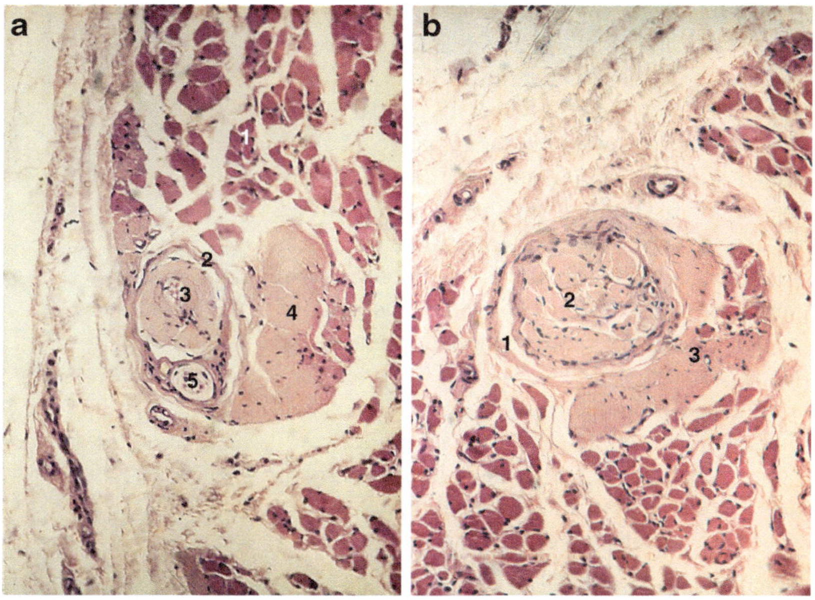

Fig. 1.13 Microscopic sections of Golgi organ. (**a**) Superior part: 1. Muscular fibres; 2. Fibrous capsule; 3. Neuro-tendinous organ with internal nervous fibres; 4. Tendon of the muscle; 5. Sensitive nerve of the Golgi organ. (**b**) Inferior part: 1. Fibrous capsule; 2. Microtendons of the Golgi organ; 3. Tendon of the muscle

inverse myotatic reflex inhibiting the alpha motor neurons and avoiding overdose contraction. In order to better understand, one can imagine a very large group of people wanting to raise a big load with many cables held by each of them. Individually they can feel some resistance and react by pulling, but they do not know what happens at the harness attached to the load. Therefore, only someone placed close to the load to move (the tendon fixed to a bone) can monitor the effectiveness of the multiple individual pulls [28].

The stretch reflex is expressed when pulling a living muscle, which triggers a contraction. This can be done by hitting a tendon with a reflex hammer. The answer may be live or attenuated and can test the responsiveness of the nervous system. Of course, this reflex response of the muscle must be controlled by a supraspinal level; otherwise it would not be possible to perform movements successively combining agonist and antagonist. It is in fact possible due to the reciprocal innervation of muscles, contracting the agonist and relaxing the antagonist, which is given by the spinal cord and brainstem control.

Neuropathology may allow a better understanding. In paralysed patients such as paraplegics, quadriplegics or hemiplegics, the fusorial servomechanism is no longer under central control and the active mobilisation of body segments causes abnormal contractions of the antagonists, creating an abnormal viscosity of the joints. This is called **spasticity**. It can be functionally useful if it is done in extension of the lower limb allowing the patient to stand up without any help. But when it maintains the upper limb in flexion, it creates a severe handicap by progressive anoxia of muscles with later fibrous retractions. Many drug treatments of spasticity are now available, but in certain irreducible cases, a surgical treatment by the posterior root neurotomy (rhizotomy) interrupting the signal from the sensitive fibres can be indicated.

Almost all muscles of the body are equipped with muscle spindles and Golgi organs with a variable number in relation to the control fineness. The lumbrical muscles of the hand, the only muscles of the body that have no bony insertion, are very rich in proprioceptors and represents the material of choice for studying spindles and Golgi organs as shown above [26, 27].

Anyway, there are some exceptions. Firstly, the vocal cord muscle, with variable tension and stiffness depending on the pitch, vibrates at very high frequencies beyond its normal bandwidth and is regulated by the ear. This explains why congenital deaf cannot speak. Secondly, the muscles of the face which are included inside the skin and whose regulation is done by facial skin sensitivity given by the trigeminal nerve (V), explaining why the facial nerve (VII)—which gives it motor innervation—is purely motor when entering within the facial mask. It can thus allow performing surgical anastomoses with another purely motor nerve, the hypoglossal nerve (XII), with some positive clinical results for facial reanimation in case of facial palsy.

We have proposed in a robotics conference in Singapore in 1992 to talk about robionics, combining robotics, electronics and biology, to try to explain the human motor coordination. Some American scientists created the term bionics in 1960.

This approach allows a double transfer of knowledge:

- firstly, from living to artificial system to make copies of some highly specialised biotransducers or by producing footed locomotion systems inspired from animals (biomimetics);
- secondly from artificial to living system in order to better understand human motor coordination according to the obvious similarity of technical problems between industrial robots and upper and lower human manipulators (Fig. 1.14). Therefore, to write equations for defining control algorithms for robot control, engineers know the mandatory need to identify two parameters: the state of the motors (electric or hydraulic) and the angles of the various segments. They use linear or angular potentiometers to measure angles.

It is the same technical requirement as for human motor control. The human brain must know the state of muscular actuators that is given by muscle spindles and Golgi organs as we have explained above.

Fig. 1.14 Painting robot (Renault factory). 1. Terminal organ activated by polymuscular actuators (elephant trump prototype); 2. One DF intermediate joint (equivalent to elbow joint); 3. Base with two rotation axes (2DF)

For angles of body segments, it is not possible to use potentiometers which do not exist in the list of available biocomponents. Muscles cannot provide this angular information, firstly because they are viscoelastic and redundant for the same joint movement and secondly because there is no library in the brain containing specifications of distances between muscle bony attachments and the joint axis, as well as the length of the muscles to make a complicated geometric computation angle.

To clarify this point, two important remarks have to be made. First, physiologists have described a special rate tension/length from the muscle spindle measurement, but we know clearly that they measure stiffness and not length, which allows the brain to identify for control the three possible states of muscles. Second, we commonly use a vectorial language with arrows representing muscle activity in direction and intensity. But it is a typical human language that the brain cannot understand and use. Ligaments rich in mechanoreceptors cannot give angle information and should be considered as indicators of joint excursion limit. Going over their mechanical stretching limit can create a painful sprain.

The intelligent solution of the constructor is to use the clothing-skin sensor covering all body segments as goniometer. Thanks to the Ruffini stretch transducers sensitive to mechanical deformation of the skin around the joints during movement, it is possible for the brain to identify positions and movement of all body segments without exception, including of course spine and limbs [29–31]. Vision improves skin goniometric accuracy. The use of a mirror is very useful for dancers or therapists to collect information about

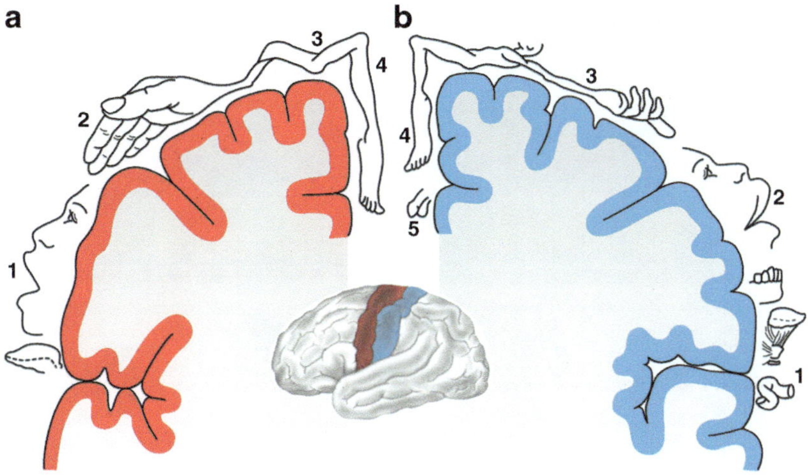

Fig. 1.15 Schematic representation of the motor key board (A) and sensitive key board (B) from the Homonculus of Penfield. (**a**) Precentral gyrus (area 4): 1. Swallowing, phonation, facial mimic and mastication; 2. Grasping; 3. Spine muscles and postures; 4. Locomotion (reduced surface within the interhemispheric scissure). (**b**) Postcentral gyrus (areas 1,2,3,5): 1. Abdominal visceras; 2. Head and neck skin and mucosa; 3. Conscious perception of upper limb with large surface for the hand; 4. Lower limb cutaneous sensibility; 5. Genital organs sensibility

the hidden parts of the body. So we can say that we live in a three-dimensional space with vision and skin, without the contribution of muscles that we do feel only when they are painful, due to unmyelinated C fibres placed between the muscle fibres and sensitive to chemical factors like lactic acid (metaboreceptors). This basic concept is still poorly understood by many therapists. It also explains very clearly the specific organisation of the cortex.

Indeed, it is logical to find in the central area of the cortex two juxtaposed keyboards linked together by short fibres creating a closed loop regulation system as shown by modern tractography (Fig. 1.15).

- First, a motor keyboard: area 4 of the precentralis gyrus of the frontal lobe, which groups all the muscles of the body represented by a particular topographic disproportional code. In 1920, a Canadian neurosurgeon William Penfield has well defined the different zones corresponding to muscle representation by intraoperative electrostimulation on more than a thousand of awake patients. At the base of area 4, a large face in relation to phonation and swallowing, a large hand in relation with prehension and a small lower limb in the interhemispheric cortex due to its mainly semi-automatic working mode in locomotion.
- Second, a sensitive keyboard: areas 1, 2, 3, 5 of the post centralis gyrus of the parietal lobe, where all skin sectors are included with the same type of disproportional representation. We describe a stato-kinesthetic loop that connects the two keyboards, allowing us to know at any time where we are and where we go with all our body parts, thanks to this position and movement in particular sense. Therefore, mobile and immobile body schemes are essentially due to the skin (Fig. 1.16).

This can explain why an insensitive segment cannot be coordinated by the nervous system. For example, a collateral sensory nerve damage of a finger still allows active mobilisation of the finger since there is no paralysis. But in the execution of a task with the corresponding hand, we can observe that the insensitive finger is excluded because it is not participating in the coordination. Hemineglect [32, 33], well described clinically in hemiplegia, relates to the alteration of the skin goniometer. We also must know that scars on the skin like at the midline cutaneous reference of spinous processes or at the knee need to be handled carefully in order

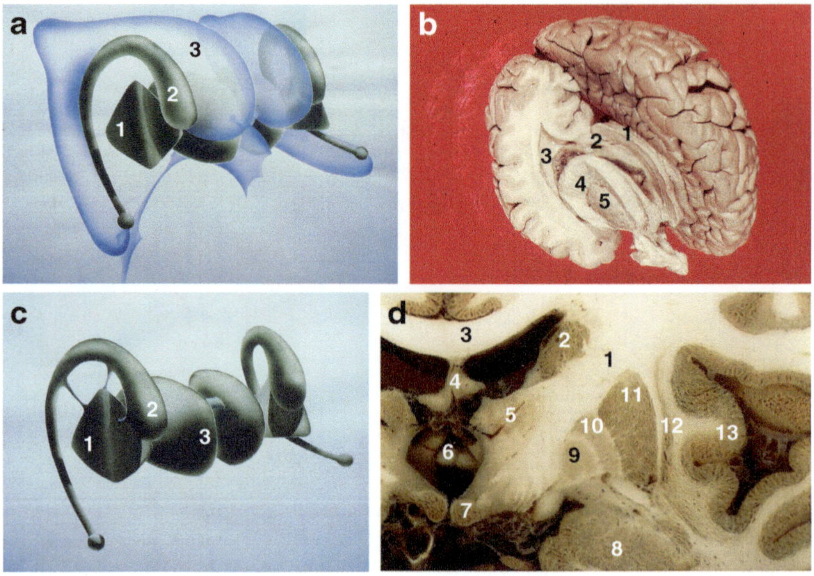

Fig. 1.16 Basal ganglia. (**a**) Schematic representation of the spiral growth of telencephalon around the central brain structures: 1. Lenticular nucleus; 2. Caudate nucleus with head, body and tail; 3. Lateral ventricle with the three frontal, occipital and temporal horns. (**b**) Dissection of the right hemisphere: 1. Corpus callosum; 2. Posterior pillar of fornix; 3. Occipital horn of the lateral ventricle; 4. Internal capsule (corona radiata); 5. Lenticular nucleus. (**c**) Drawing without lateral ventricles: 1. Lenticular nucleus; 2. Caudate nucleus; 3. Thalamus. (**d**) Coronal section of the brain (right side seen from behind): 1. Internal capsule; 2. Head of caudate nucleus (striatum); 3. Corpus callosum; 4. Fornix; 5. Thalamus; 6. Anterior commissure; 7. Mamillary body; 8. Amygdala; 9. Globus pallidus medialis (GPM); 10. Globus pallidus lateralis (GPL); 11. Putamen (striatum); 12. Claustrum; 13. Insula

to "realign" *(reset)* the skin goniometer and to avoid functional disruptions in the musculoskeletal associated systems.

1.3.3 Central Control of Mobility

Without going into the details of the description of the pathways and mobility centres to [34–44], several points are worth explaining.

In a didactic perspective, we always tend to simplify complex motor pathways by drawing schemes. But we must remember that everything works together: ascending and descending pathways, calls for motor recorded programmes stored after learning in the motor library of the basal ganglia and that the pilot of the machine most of the time has a rudimentary or no knowledge at all about technical problems. Therefore, most of the circuits work according to a preprogrammed automatic mode and can operate on two levels of control.

1.3.3.1 The Conscious Decision Level

It uses a global, simple and functional language: walking, taking an object, running, jumping, etc., therefore commanding movements and not muscles.

1.3.3.2 The Unconscious Execution Level

It processes the impressive amount of inputs and outputs, addressing three main issues:

- the first issue is the right choice of actuators for a given programme that cannot be consciously done, because the pilot is not supposed to know he has muscles and which to choose for a particular action. Therefore, a special circuit to properly use the motor keyboard goes from the decision zone in the premotor area to the motor cortex, to the lateral cerebellum and back to the corticospinal tract (pyramidal tract) in order to reach the appropriate metameric levels according to the polyradicular innervation of most of the muscles. At the same time, the skin sends information to the goniometric

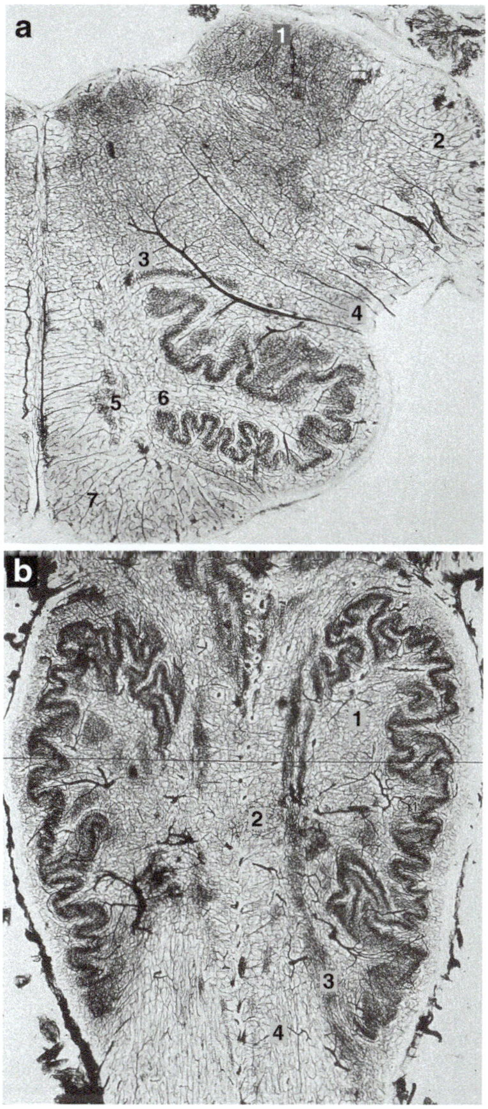

Fig. 1.17 Vascular microinjection of inferior olivary nucleus in medulla oblongata (H. Duvernoy). (**a**) Axial section of medulla oblongata: 1. Medial vestibular nucleus; 2. Inferior cerebellar peduncle; 3. Lateral accessory olivary nucleus; 4. Spinothalamic bundle; 5. Medial accessory olivary nucleus; 6. Hilum of the inferior olivary nucleus; 7. Corticospinal tract. (**b**) Coronal section: 1. Hilum; 2. Median lemniscus; 3. Lateral accessory olivary nucleus; 4. Corticospinal tract

cerebellar centres (dorsal and ventral spinocerebellar tracts) and cortical (sensitive key board of postcentral gyrus of the parietal lobe) for coordination;

- the second problem is the adjustment agonist/antagonist which is done by the identification of the states of the muscles by muscle spindle input in the intermediate cerebellum which interacts with the spinal cord. A lesion of this part of the cerebellum causes incoordination with dysmetria, hypermetria and intention tremor;
- finally, the third problem is the need for a possible detection in real time of errors in programme peripheral execution, which is done by a neural structure that is now well known: the inferior olive located within the medulla oblongata. It is a neuronal structure scalloped like a little cortex of 3 cm in diameter that has connections to the spinal cord (spino-olivary fibres) and with the cerebellum (olivo-cerebellar fibres) as well as a specific cortical control (cortico-olivo-spinal pathway) (Fig. 1.17).

Conclusions

Identifying technical problems behind of a particular function is the major characteristic of the comprehensive approach that we propose. But neuroanatomy has a long history and has gradually increased its knowledge capital so that the terminology has changed over time and do not always correspond to the immutable functional reality of the nervous system which has not changed since the beginning. It is obviously difficult to go against tradition and change what is learned and transmitted over generations.

For example, we have seen that the conscious decision in a determined action such as walking or looking at an object in space cannot be located in the central cortical areas which work in the broad field of unconscious execution. Therefore, we must locate it in an area located forward that we can call premotor area, which does not correspond to the commonly accepted nomenclature that speaks of a supplementary motor area. In this decision area, we can place the conscious control of gaze by the foveal fixation, which is undoubtedly one of the most powerful functions allowing to fix an object in space and to move the whole body around the fixation point. This is just an example that wants to draw attention to the prevalence of the possible and the necessary understanding of structures over the simple memorisation of details which is tedious and volatile.

References

1. Rabischong P (2003) Le programme homme. PUF, Paris
2. Kapandji AI (2011) Qu'est-ce que la BIOMECANIQUE? Sauramps Medical, Montpellier
3. Kapandji IA (2009) Physiologie articulaire : Tome I Membre supérieur Tome II Membre inférieur Tome III Tête et rachis. 6e Edit. Maloine S.A.
4. Dufour M, Pillu M (2006) Biomécanique Fonctionnelle: membres, tête, tronc. Elsevier Masson S.A.S.
5. Martin RB (1991) Determinants of mechanical properties of bones. J Biomech 24(Suppl 1):79–88
6. Cheng X (1997) Assessment of human vertebral strength: relationships to bone mass and trabecular microstructure in vitro. Acta Biomedica Lovaniensia 147, Leuven University Press, Leuven
7. Rabischong P, Konirsch G, Avril J (1964) Etude biomécanique du tissu osseux compact des os longs en fonction de leur structure. Bull Ass Anat:1:1359–1387.
8. Goldstein SA (1987) The mechanical properties of trabecular bone: dependence on anatomic location and function. J Biomech 20:1055–1061
9. Galante J, Rostoker W, Ray RD (1970) Physical properties of trabecular bone. Calcif Tissue Res 5:327–330
10. Ashman RB, Corin JD, Turner CH (1987) Elastic properties of cancellous bone: measurement by an ultrasonic technique. J Biomech 20:979–986
11. Gibson LJ (1985) The mechanical behaviour of cancellous bone. J Biomech 18:317–328
12. Evans JA, Takakoli MB (1990) Ultrasonic attenuation and velocity in bone. Phys Med Biol 35:1387–1396
13. Evans FG (1958) Relations between the microscopic structure and tensile strength of human bone. Acta Anat 35:285–301
14. Amling M, Herden S, Pösl M et al (1996) Heterogeneity of the skeleton: comparison of the trabecular microarchitecture of the spine, the iliac crest, the femur, and the calcaneus. J Bone Miner Res 11:36–45
15. Mizrahi J, Silva MJ, Keaveny TM et al (1993) Finite-element stress analysis of the normal and osteoporotic lumbar vertebral body. Spine 18:2088–2096
16. Rabischong P, Louis R, Vignaud J et al (1978) Le disque intervertébral. Anat Clin 1:55–64
17. Shirazi-Adl SA, Shrivastava SC, Ahmed AM (1984) Stress analysis of the lumbar disc-body unit in compression. Spine 9:120–134
18. White RK (1963) The rheology of synovial fluid. J Bone Joint Surg Am 45-A:1084–1090
19. Barnett CH, Cobbold AF (1962) Lubrication within living joints. J Bone Joint Surg 44-B:662–674
20. Clarke IC (1971) Surface characteristics of human articular cartilage. A scanning electron microscope study. J Anat 108:23–30
21. Bayourthe VP, Anklewicz J (1972) La lubrification articulaire. Rhumat Tome XXIV 8:55–74
22. Guimberteau JC (2004) Strolling under the skin. Elsevier S.A.S.
23. Bonnel F, Marc T, et al (2009) Le muscle, nouveaux concepts: anatomie, biomécanique, chirurgie, rééducation. Sauramps Medical
24. Winckler G (1964) Les aspects morphologiques de l'innervation des muscles striés. Arch Biol 75:943–973
25. Julia M, Hirt D, Perrey S, et al (2012) La proprioception. AMPR, Sauramps Medical
26. Rabischong P (1962) L'innervation proprioceptive des muscles lombricaux de la main chez l'homme. Rev Chir Orthop 48(3):234–244
27. Rabischong P (1961) Recherches sur la morphologie et la répartition des organes neurotendineux et des récepteurs épitendineux et intramusculaires encapsulés dans les muscles lombricaux de l'homme. Arch d'Anat, d'Histo et d'Embryo XLIV 5/8:329–349
28. Roll JP, Vedel JP (1982) Kinaesthetic role of muscle afferents in man, studied by tendon vibration and microneurography. Exp Brain Res 47:177–190
29. Halata Z (1988) Ruffini corpuscle – a stretch receptor in the connective tissue of the skin and locomotor apparatus. Prog Brain Res 74:221–229
30. Halata Z (1977) The ultrastructure of the sensory nerve endings in the articular capsule of the knee joint of the domestic cat (Ruffini corpuscles and Pacinian corpuscles). J Anat 124:717–729
31. Iggo A (1982) Cutaneous sensory mechanism. In: Barlow HB, Mollon JD (eds) The Senses. Cambridge University Press, Cambridge, pp 369–406
32. Vallar G (1998) Spatial hemineglect in humans. Trends Cogn Sci 2(3):87–97
33. Verdon V, Schwartz S, Lovblad KO et al (2010) Neuroanatomy of hemispatial neglect and its functional components: a study using voxel-based lesion-symptom mapping. Brain 133(3):880–894
34. Bossy J (1990) Anatomie clinique, neuroanatomie. Springer, Paris
35. Martin JH (2012) Neuroanatomy text and atlas. McGraw Hill, New York
36. Heimert L (1983) The human brain and spinal cord. Springer, New York
37. Nieuwenhuys R, Voogd J, van Huijzen C (2008) Human central nervous system, 4th edn. Springer, Heidelberg
38. Rabischong P (2004) Comprehensive anatomy of the spinal cord. Riv Neuroradiol 17:268–276

39. Duvernoy H (1999) The human brain. Surface, blood supply and three-dimensional sectional anatomy, 2nd edn. Springer, Wien NewYork
40. Marieb EN (1992) Anatomie et physiologie humaine. DeBoeck, Canada
41. Catani M, Thiebaut de Schotten M (2012) Atlas of human brain connections. Oxford University Press, Oxford
42. Tortora GJ, Derrickson B (2006) Principles of anatomy and physiology. Wiley, Hoboken, NJ
43. Moore KL, Dalley AF, Agur AMR (2010) Clinically oriented anatomy. Wolters Kluwer/Lippincott Williams and Wilkins, Philadelphia, PA
44. Rabischong P (2012) The skin neural interface. In: Ajeena IM (ed) Advances in clinical neurophysiology. InTech

2 Spinal Functions

Abstract

The spine is a complex biomechanical structure made by articulated vertebras in fixed position (postures) and moved in different directions by muscles. The spinal stabilisation is automatically regulated by a spinal servomechanism with three inputs: muscle transducers for the spinal balance, skin on the axial line of the spinal processes for global control of position and movement and vestibular utricular macula for the horizontal stabilisation of the head (inverse pendulum). A first group of muscles goes from spine to superior and inferior limbs and a second group is fixed directly on the spine (spinal muscles), and their complex architecture is now better known. Each segment has its own functional feature. The cervical spine is in close relation with vision, justifying the cardan joint for the occipitocervical junction and the specific protection system of the vertebral artery, the artery of life. The dorsal spine is linked to the respiratory function. The lumbar spine is articulated with the broken pelvic ring and its deformable joints: two posterior sacroiliacs and one anterior pubic in order to reinforce mechanical resistance of it. Rachidian pains have to be well clinically identified, and a particular emphasis has to be put on the important role of the spinal ganglion.

2.1 Introduction

The spine is an axial rod made of overlapping articulated bone pieces mobilised by muscles. It supports the cephalic mass, is leaning on the pelvic base and gives attachment to the upper and lower limbs. It has several segments with specific rules requiring to be studied separately, even if the spine is a very interactive system finding its unity in the complex function of spinal stabilisation.

In mechanical terms, the spinal beam is subjected to large stresses in compression that corresponds to the problem of compressed linear vertical beams generating buckling. At a maximum level of compression and even measuring strain, the direction of the rupture becomes unpredictable. It depends essentially on the degree of the homogeneity of the material. To avoid buckling, the most efficient system is to anticipate the direction of the

P. Rabischong, *Comprehensive Anatomy of Motor Functions*, DOI 10.1007/978-3-319-04169-8_2,

deformation until rupture by bending the beam which makes it predictable and allows to avoid it by using correctly positioned shrouds. This is the case of the spine which is formed by a succession of reversed curves, with bracing muscle placed in the concave part of the curves, except at the thoracic level due to mechanical transverse continuity given by the rib cage. The newborn, which has not yet walked (except some intrauterine trampling reassuring the mother), has not acquired the reverse spinal curves at birth. The chimpanzee that is not using a permanent bipedal posture has a spine with a single curvature.

2.2 Spinal Stabilisation [1–5]

This is the function that helps maintain spinal posture in three positions: flexion, extension, lateral left and right. It is dependent on the morphology of the vertebral triangle segmental support and especially the organisation of the muscles. This factor is predominant in all spine pathology and is not sufficiently taken into account, while modern imaging with new modalities (CT and MRI) allows for very precise anatomical images closer to the reality than conventional vector models. This function, for the same reason related to the ignorance of the pilot, is highly automated and integrated into what has to be called the spinal servomechanism. Like any servo, it has inputs that transmit imbalance information in order to induce stabilisation reactions by impulsions.

2.2.1 Spinal Servomechanism Inputs

They are three: the proprioceptive muscle sensors of the spinal balance, the cervicothoracoabdominal skin and the gravity sensor utricular macula of the vestibule.

2.2.1.1 Spinal Balance

It is defined as a system of two active right/left pendulums formed by spinal muscles attached directly to the spine like the shrouds and capable of exerting a pulling force on the spinal axis. The myotatic reflex we have described from stretching the muscle spindle will be the key for the spinal balance. Apparently, it determines paradoxical reactions that Isch has defined a long time ago by electromyography. Bending the spine on the right creates a concave curvature, and it is found that the muscle activity is on the side of the convexity to the left in relation to the stretch of the muscle spindles from this side. This is the same phenomenon which can be observed by placing both thumbs on the lumbar paravertebral mass and asking to walk in place. The lifting of the lower limb creates a fusorial stretching on the opposite side with an automatic contralateral muscular contraction. It is clear that the overall contraction of the spinal muscles is managed by spinal cord but also by supraspinal levels including the cerebellum.

2.2.1.2 Axial Goniometric Skin Reference Line

As we explained above, the skin is the goniometer of all the body segments, which allows isolating within the spine an axial skin baseline attached to the spinous processes. This axial reference line contributes largely to the cutaneous perception of spine movements. It can be disrupted in axial postoperative scars. In addition, it is possible to define a supraspinal fibrous complex (Fig. 2.1) which links the vertebrae together and is made by five anatomical structures: the skin (epidermis and dermis), the flat tendon of latissimus dorsi muscle, the thoraco-lumbosacral aponeurosis around the spinal muscles, the lumbosacral tendon and the supraspinal ligament which is not the strongest structure.

2.2.1.3 The Gravity Utricular Transducer

The cephalic mass based on the spinal rod is an inverted pendulum which plays an important role in stabilising the spine. It is desirable to achieve a proper balance of the system that the head is placed in horizontal position. This cannot be given by the visual system, because we have two eyes, but we only see one image while two right and left sensors are needed for good horizontal head stabilisation. This

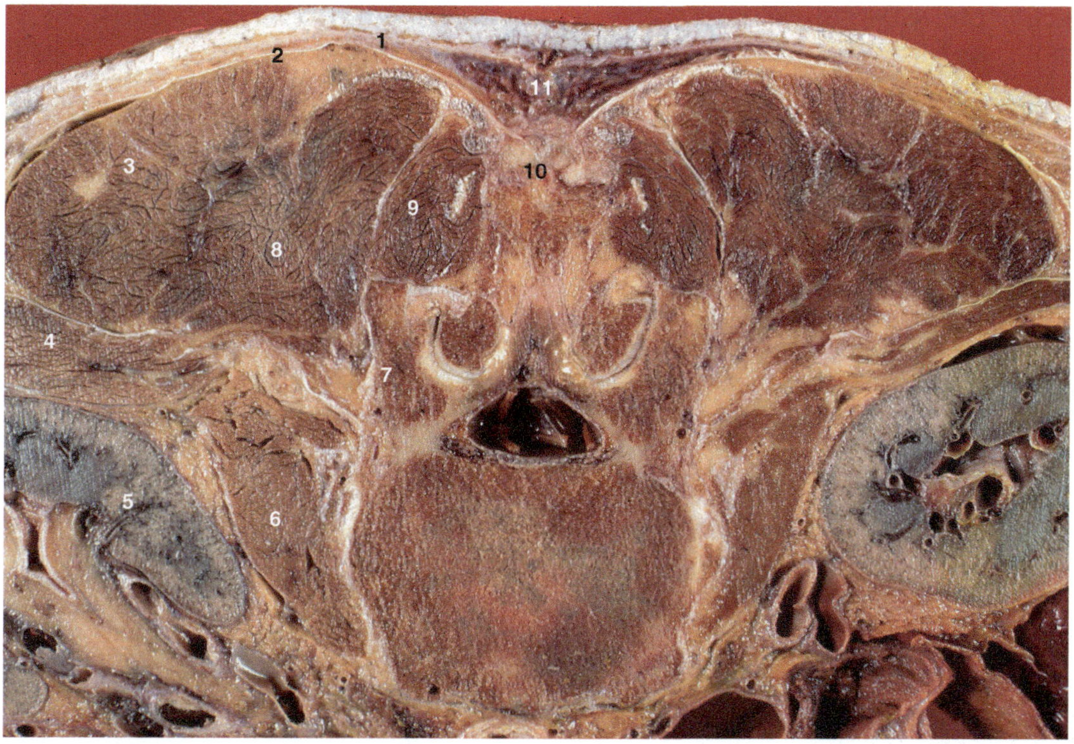

Fig. 2.1 Axial section of L3. 1. Tendon of the latissimus dorsi; 2. Thoraco-lumbo-sacral aponeurosis; 3. Ilio-costal muscle; 4. Quadratus lumborum; 5. Kidney; 6. Psoas muscle; 7. Facet joint; 8. Long dorsal muscle; 9. Transverso-spinal muscles (multifidus); 10. Spinal process; 11. Supraspinal fibrous complex

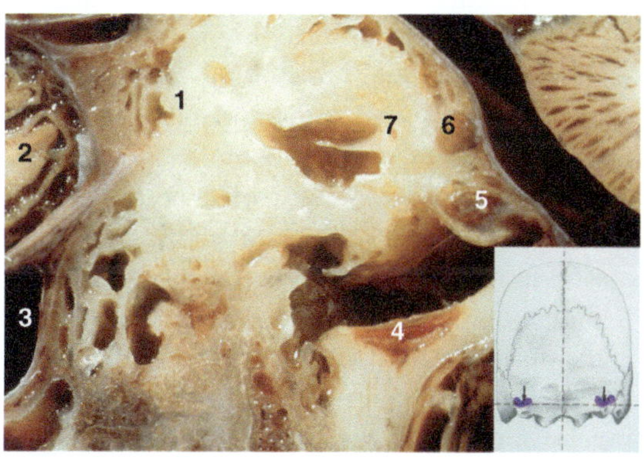

Fig. 2.2 Utricular macula on a sagittal section of temporal bone. 1. Posterior semi-circular canal; 2. Cerebellum; 3. Lateral sinus; 4. Tympanic membrane; 5. Tensor tympani muscle in its bony canal; 6. Geniculate ganglion (VII); 7. Utricular macula

justifies the existence in the vestibule of the inner ear of a gravity transducer, sensitive to vertical forces (Fig. 2.2). The utricular macula plays this role. It is formed by a thin bone lamella which is placed horizontally and covert by a mucous membrane reinforced with crystals. This changes the inertia of the system and creates what is called a flyweight in mechanics. It is not exactly the same system as fish or shellfish also have, a sensor formed by otolith, a calcareous concretion based on sensory cells, which allows the animal to know where the top and bottom are, allowing good navigation. An upside fish in water is nearing its biological end. In humans, it is called otolithic membrane to characterise the mucosa overlying bone lamella and sensory cells it contains. This horizontal stabilisation of the head intervenes significantly in the spinal stabilisation, and now we come to think that some adolescent idiopathic scoliosis, without apparent cause, may come from a functional deficiency of the utricular macula.

2.2.2 Muscular Corrections

Each vertebra has medial tendon attachments on spinous processes and lateral on lamina and transverse processes, as well as at the thoracic level on the posterior angle of the ribs. An economic spinal posture corresponds to total muscle relaxation obtained by a back support that neutralises the effects of muscle spindles stretching, but has the disadvantage to distend the posterior facet joint capsule, which can generate pain. The upright posture involves minimum muscle contractions for incomplete straightening. Posture leaning forward, however, requires significant contractions quickly painful in untrained subjects.

Muscles [6–8] acting on the spinal pelvic axis comprise two groups.

2.2.2.1 First Group: Spino-appendicular Muscles

The first group of muscles is attached to the spinous processes and activates either the upper limb: trapezius, rhomboid, serratus anterior, and latissimusdorsi; or the lower limb with a spinal action: iliopsoas; or the lower limb only: maximus medius and minimus glutei, tensor fascia lata, rectus femoris, adductors, medial and lateral obturators (Fig. 2.3).

2.2.2.2 Second Group: Spinal Muscles

The second group, the spinal muscles, has a specific action on the spine. Bonneau [9] in his thesis gave a comprehensive description. We must remember the following points (Fig. 2.4):

- there is on the back under the skin a fibrotendinous plan consisting, from the surface to the depth, of three formations: first the flat tendon of the latissimus dorsi from T6, second the thoraco-lumbosacral aponeurosis surrounding spinal muscles and encompassing within its deep surface the two mechanical tensors—superior and inferior serratus minor—and third the extremely resistant lumbosacral tendon forming a real posterior fibrous wall. This tendon is strongly attached to the lumbar spinous processes and the banks of the iliosacral groove till the coccyx. It represents the tendon for laterally iliocostal muscle and medially long dorsal muscle that fit with multiple chevrons on posterior angles of the ribs and transverse processes. Note that on the deep surface of the lumbosacral tendon are fixed a very thick oblique muscle fibres package that provide much of the force required to erect spine [10];
- below the fibrous plan, spinal muscles [11, 12] are arranged in three columns:
  - lateral, with the long dorsal muscle which has lumbar, thoracic cervical (neck transverse T1–T5/C3–C6) and cephalic portions (head transverse T3–C3/mastoid process);
  - intermediate with the iliocostal muscle having lumbar, thoracic and cervical portions (back angle T4–T8/transverse processes C4–C7);
  - medial with the transverso-spinal muscles called multifidus which is formed by 23 beams overlapped in the craniocaudal direction like tiles on a roof. Each beam has four different parts: the short lamellar (crossing one vertebral segment), long lamellar (two segments), short spiny (three segments) and long spiny (four to six segments).

It must be put in place after the muscles with a special place: the yellow ligament (Fig. 2.5). It is responsible for a resilient damping of the spine movement of flexion. So called because of its yellow colour due to a great number of dense elastin fibres it contains. It is the only true elastic ligament of the body. It is attached to the superior and inferior border of the laminae and gives a viscoelastic amortising return in the spine bending movement. It is present at all spinal levels and its morphology on the images of the spine is clearly related to the state of the disc [13, 14]. It appears relaxed and expanded when the disc collapse (pseudo-hypertrophy). It can also reverse its normal curvature by protruding into the spinal canal and causing an acquired narrow canal *(soft stenosis)* which can create neurological disorders that may require surgical recalibration of the canal by a minimally invasive approach. At the

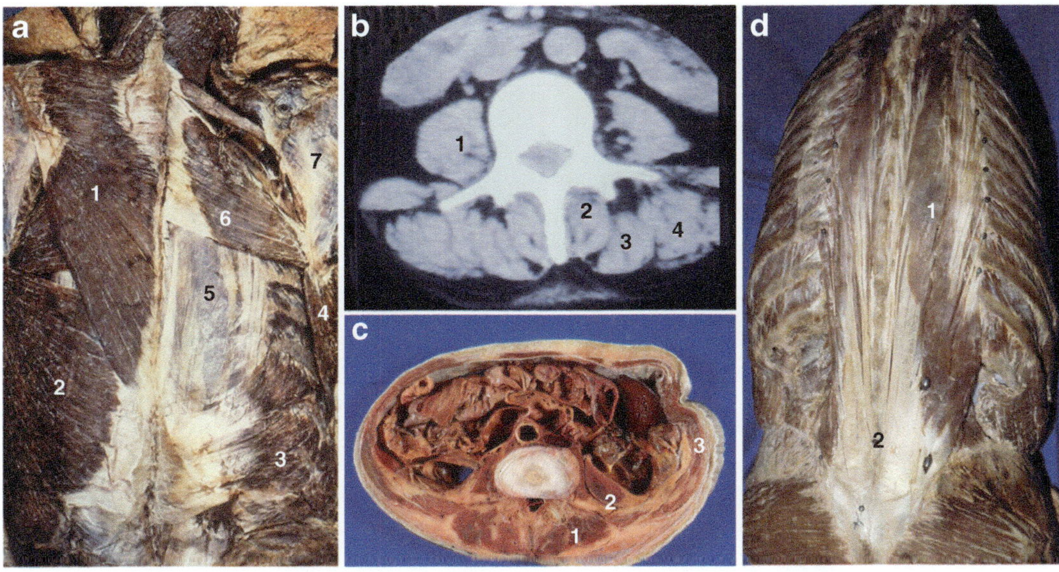

Fig. 2.3 The spine muscles. (**a**) Posterior view of the trunk: 1. Trapezius muscle; 2. Latissimus dorsi muscle; 3. Posterior and inferior serratus; 4. Serratus anterior; 5. Thoraco-lumbo-sacral aponeurosis; 6. Rhomboid muscle; 7. Scapula reclined outside. (**b**) Lumbar scanner of an axial section: 1. Psoas muscle; 2. Multifidus; 3. Long dorsal muscle; 4. Iliocostal muscle. (**c**) Axial section of the abdomen: 1. Spinal muscles; 2. Psoas muscle; 3. Abdominal wall large muscles. (**d**) Spinal muscles: 1. Long dorsal muscle; 2. Lumbosacral tendon

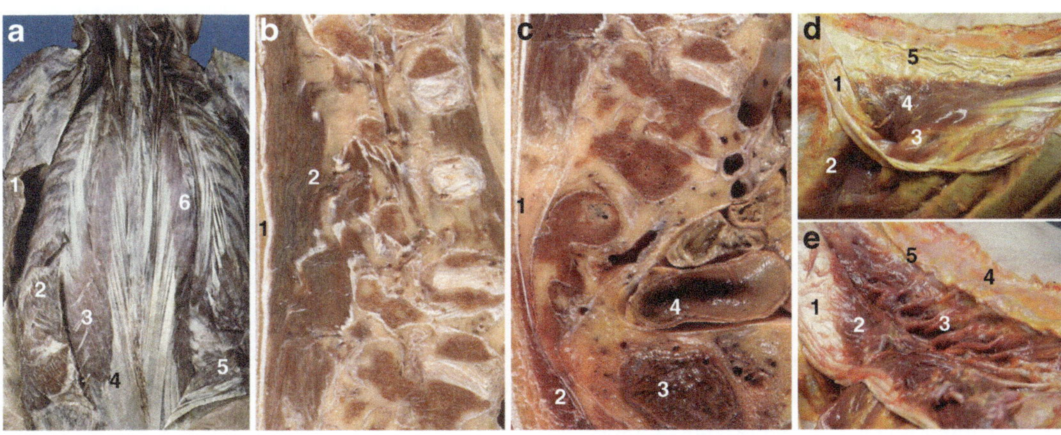

Fig. 2.4 Dissection of spine muscles. (**a**) Posterior view: 1. Latissimus dorsi; 2. Posterior and inferior serratus minor; 3. Transmuscular cutaneous nerves; 4. Lumbosacral tendon; 5. Ilio costal muscle; 6. Long dorsal muscle. (**b**) Sagittal section of lumbar spine: 1. Skin, aponeurosis and lumbosacral tendon; 2. Deep oblique muscular fibres. (**c**) Sagittal section of sacral spine: 1. Lumbosacral tendon; 2. Coccyx; 3. Rectum 4. Sigmoid colon. (**d**) Lateral view of lumbar spine after reclining the aponeurosis: 1. Thoraco-lumbo-sacral aponeurosis; 2. Iliac crest; 3. Posterior and inferior serratus minor. 4. Iliocostal muscle; 5. Lumbosacral tendon. (**e**) Lateral view: 1. Reclined lumbosacral tendon; 2. Deep muscular fibres of long dorsal; 3. Transverso-spinal muscle (multifidus); 4. Skin gliding plan. 5. Spinous processes

lumbar level, it is clearly visible on an axial section of the spine that it is shaped like a V covering posterior facet joint processes allowing to “smooth” the dorsal part of the foramen. Like all fibrous structures, it tends to involute with age, resulting in a decrease of elastic fibres and fibroblast proliferation, creating an image of true hypertrophy (Fig. 2.6).

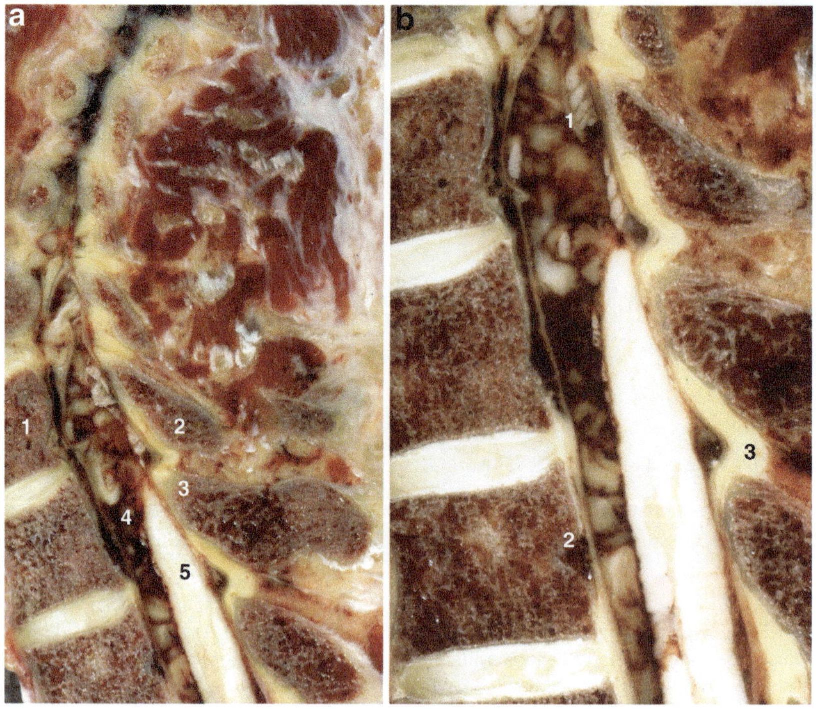

Fig. 2.5 Cervical yellow ligament. (**a**) Oblique sagittal section: 1. Vertebral body; 2. Spinal process; 3. Superior border of lamina; 4. Dura mater; 5. Spinal cord. (**b**) Sagittal section: 1. Cervical roots; 2. Posterior opening of the vertebral body vein covert by the posterior common vertebral ligament; 3. Yellow ligament

The enormous complexity of these muscles responds to the complex problem of stabilising the multi-segmented spinal rod and managing its movements. The human pilot of the machine does not feel this complexity unless its muscles become painful, sometimes creating an attempted involuntary automatic active spine immobilisation called lumbago.

Finally, to conclude this section on spinal stabilisation, some comments have to be given about pathomechanics. It seems not appropriate to talk about instability and a fortiori about microinstability to characterise abnormal stabilisation of one or more vertebral segments. In fact, in the official terminology of mechanics, the definition of an unstable structure is a structure that cannot be stabilized. That obviously is not the case of pathological spine. This is actually an abnormal stabilisation with hypo- or hypermobility that we must call a distability. This change in terminology will surely take a lot of time, but hopefully at least the readers of this book will agree to adopt this new terminology more consistently with the rigour of mechanics.

2.3 Cervical Spine and the Eye/Head Servomechanism

The cervical spine is the "head holder" and its mobility is directly related to that of the cephalic mass. A simple functional analysis helps to understand that the head movements are mainly related to the visual exploration of space in two main directions: vertical and horizontal. A blind person who lost this visual exploration is recognised by its cephalic immobility. This implies movements in flexion/extension and horizontal rotation of the cervical spine [15–18].

2.3.1 The Visual Programme

To understand the functional nature of these movements, it is essential to analyse the visual

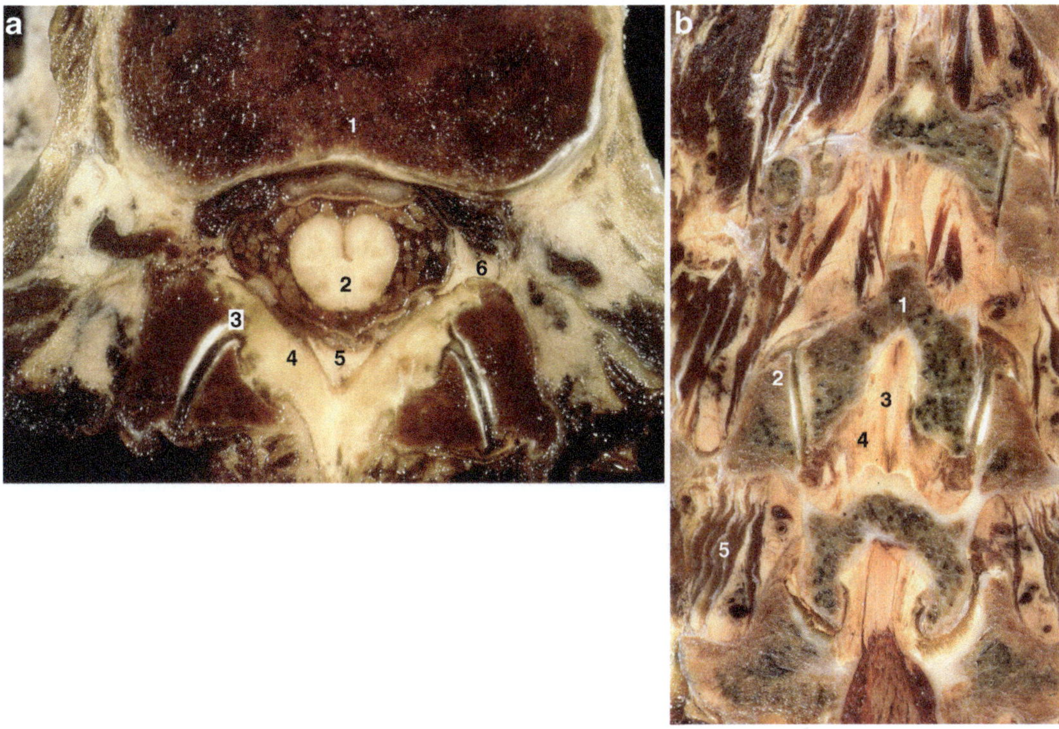

Fig. 2.6 Lumbar yellow ligament. (**a**) Axial section of L1: 1. Vertebral body L1; 2. Spinal cord; 3. Facet joint with pseudomeniscal synovial fold; 4. Yellow ligament. 5. Fat pad; 6. Vertebral foramen. (**b**) Coronal section of the posterior vertebral arch: 1. Lamina; 2. Facet joint; 3. Fat pad; 4. Yellow ligament; 5. Intertransverse muscle

program. Vision allows perceiving shape, colour and movement of objects in all directions. A further refinement, the third dimension (X, Y, Z), can be collected by stereoscopy (3D vision). This is possible, thanks to the existence of a gap between the two eyes placed, in humans, in a frontal position. The major technical problem is the fusion of the two images pixel by pixel. That can be done by the optic chiasm, which crosses at 50 % the retinal fibres and sends to the area of primary visual cortical projection of each occipital lobe two—left and right—half images. But the perfect fusion can only be made with a high resolution point placed in the retina of the eye on the optical axis. The macula gives 80 % of the vision and represents an extraordinary concentration of visual cones (135,000 mm^2). Peripheral vision is in fact less accurate.

At birth, it may happen that the two eyeballs are not perfectly centred, making impossible to merge the two retinal images for stereoscopy. Therefore, an amblyopia is detected in children, with a normal eye appearance but with a reduced acuity on one side. This is due to a macular active inhibition of this eye by cortico-retinal fibres described by Mawas. In other words, this means that the constructor has not only organised the complex establishment of vision, but he also predicted the failure. The treatment consists in blocking the good eye in order to disinhibit the other and then trying a realignment of the two visual axes by orthoptics or surgery, with rather mixed results.

In comparative anatomy, punctiform macula exists only in animals with stereoscopy: owls, raptors, cats, etc.; the others are using a panoramic vision with a macula in band as for ruminants which have low binocular vision field and two different retinal images which cannot be merged to give stereoscopy. Only the chameleon combines both types of vision: first panoramic when he looks at his prey using completely separate both eye vision with small pupillary opening

Fig. 2.7 The chameleon. (**a**) Independent movable eyes (panoramic vision). (**b**) Convergence of the two eyes to catch a pray with tongue projection (stereoscopic vision)

for precision and second stereoscopic by converging the two eyes and merging the two images, in order to guarantee the success of its quick tongue lick on its prey. It has a nice punctiform macula (Fig. 2.7).

In humans, the two eyeballs are moving along two axes: vertical and horizontal, but beyond 30 degrees of horizontal rotation, stereoscopy is impossible because of the image of the nose and optical parallax problems. We can deal with this problem by finding the solution of moving at the same time the head and the eyes, which is the eye/head servomechanism. But it is obviously necessary to coordinate and adjust the two movements, which means for the brain to know accurately the speed and direction of the movements of the head.

This explains why the vestibule, which has a three-dimensional angular accelerometer with three semicircular canals positioned orthogonally in the three planes of space and is able to measure very precisely direction and speed of the mobile head, was placed within the skull base. It could indeed be placed in the body since it can also quantify body movements in 3D, as, for example, when walking without visual control at night.

In addition, the vestibule and the cochlea have two different functions (stato-dynamic equilibration of the head and hearing) but are linked topographically in the same organ, the ear. The reason for that could be presumably found in their common technical problem: the precise osmotic regulation of their two perilymphatic and endolymphatic liquids. In fact it seems more intelligent and economical to put them together. Vertigo in Meniere's disease is the pathological illustration of this successful joint programme.

This also explains the important role of the 3D space reference of the vestibule that requires to be fixed very early in a dense bone area of the skull base. This is why semicircular canals and attached cochlea are the first organs with full growth at birth (Fig. 2.8e). It remains the problem of left-right symmetry in the adult semicircular canals. Caix from Limoges showed a low percentage of subjects with symmetrical canals and that could be a factor influencing susceptibility to vertigo.

Finally, the eye/head servomechanism has to coordinate movements of two different anatomical systems. It needs a common guide, which is the image perceived by visual pathways. It is interesting to note that with the artificial vision of robotic systems, it is the same problem and the same solution: give to images the crucial role of guiding servomechanisms.

This focuses on the very rich interactions between vision and vestibule illustrated by many reflexes that have been described. Eye is helping vestibule, which is a liquid sensor with a remanence (excitation maintained after stopping motion) by readjusting it (reset). Classical dancers, although they are not physiologists, know that to perform a "pirouette" perfectly (vertical rotation of the body) they must take within the environment a fixation point at each rotation (or the light from the balcony or the shiny bald head of the conductor...), explaining the separate head and body rotation.

The waltz, which has still escaped the trend of zoological modern dances, theoretically requires the same type of rotation system. This excludes talking to the partner by staring into each other's eyes with otherwise the risk of ending in so inglorious manner in a typical vestibular vertigo accompanied by nystagmus. That is characterised by saccadic eye movements induced by the vestibule. On a train trip, watching the traveller sitting in front of you that sees the passing scenery allows to observe a form of physiological nystagmus with a quick recall of the eyeballs not perceived by the subject. Therefore, when looking at scrolling objects, it is not possible for the eyes to follow them by going out of their orbital cavities. Aircraft pilots who do aerobatics also know that it is important to always keep the head within the direction of

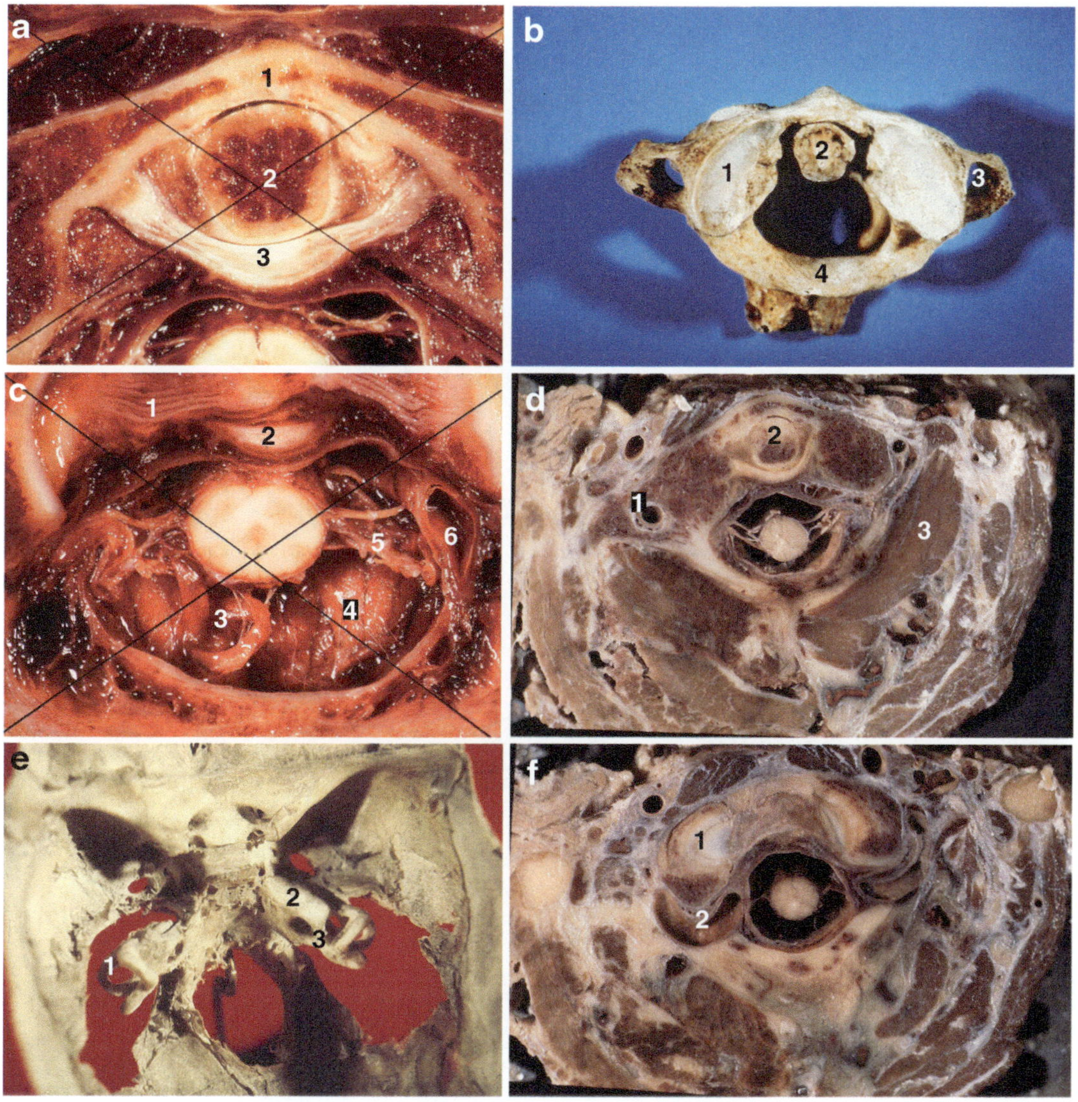

Fig. 2.8 Axial sections of the cardan joint. (**a**) Section of the atlanto-axoidal joint: 1. Anterior arch of atlas; 2. Odontoid process of C2; 3. Transverse ligament of C1 (transverse part of the cruciform atlantis ligament). (**b**) Superior view of atlas/axis: 1. Superior articular surface of atlas; 2. Odontoid process of C2; 3. Transverse foramen of C1; 4. Posterior arch of C1. (**c**) Axial section at the foramen magnum: 1. Alar ligaments; 2. Longitudinal fascicles of the cruciform ligament; 3. Posterior inferior cerebellar artery (PICA); 4. Cerebellar tonsils; 5. Roots of C1 cervical nerve; 6. Vertebral artery. (**d**) Axial section of the C1/C2 junction: 1. Vertebral artery; 2. Odontoid process of C2; 3. Inferior oblique muscle. (**e**) Dissection of a skull base of new born: 1. Lateral semicircular canal; 2. Cochlea; 3. Three semicircular canals (size of the adult). (**f**) Inferior axial section of the cardan: 1. Atlanto-axoidal joint; 2. Horizontal loop of the vertebral artery behind the lateral mass of C1

the velocity vector in order to avoid dizziness which may have catastrophic consequences. Nystagmography and vestibulometry are techniques that have made significant progresses, bringing to clinicians valuable data used also in posturology as we shall see later.

2.3.2 The Craniocervical Junction

The strong functional connection between visual exploration and cervical spine dynamics fully justify at the junction of the head and cervical spine a joint system with two degrees

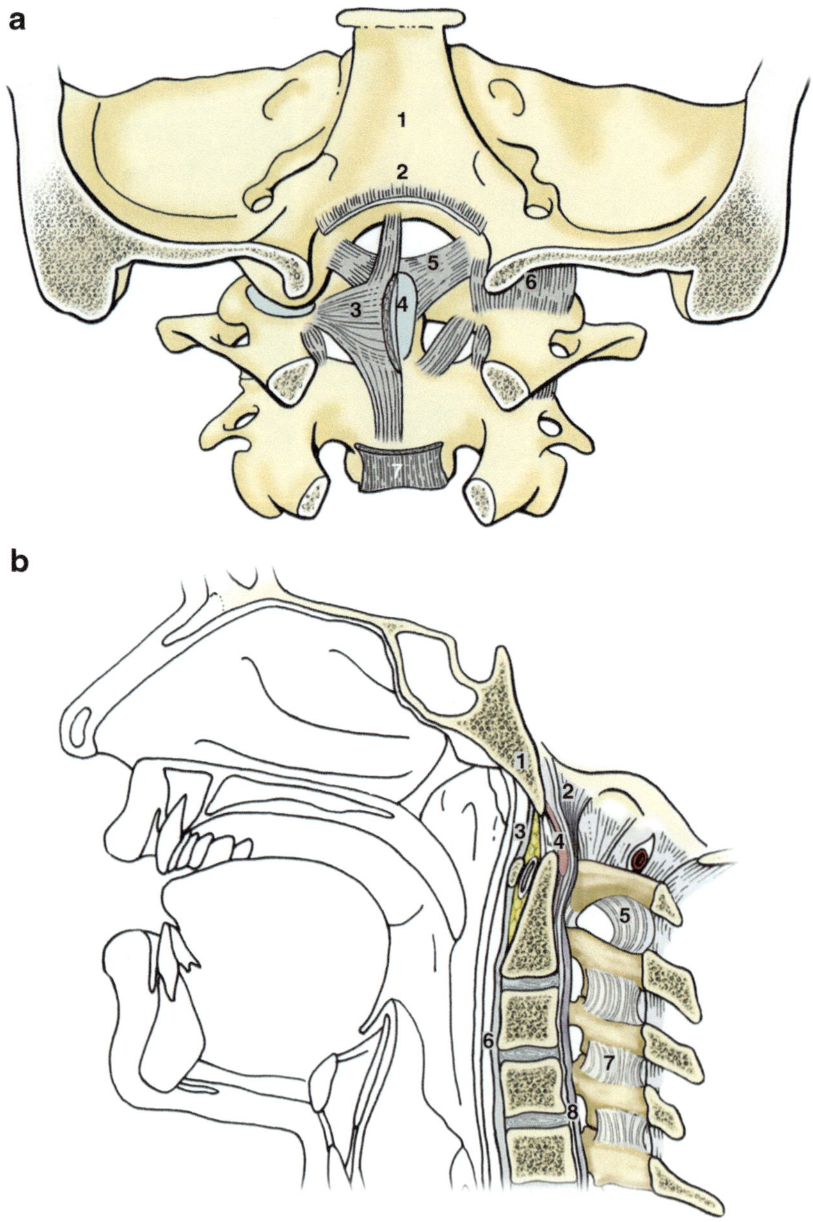

Fig. 2.9 Drawings of the occipito-cervical cardan. (**a**) Posterior view of a coronal section: 1. Clivus; 2. Transsected membrana tectoria; 3. Cruciform ligament with the superior longitudinal band fixing the transverse ligament to the occipital and to C2 by the inferior longitudinal band; 4. Articular surface of the dens of C2 and transverse ligament of C1; 5. Alar ligament; 6. Articular capsule of the atlanto-occipital joint; 7. Membrana tectoria. (**b**) Sagittal section: 1. Clivus; 2. Membrana tectoria in continuity with the posterior longitudinal ligament; 3. Anterior atlanto-occipital membrane in continuity with the anterior longitudinal ligament; 4. Cruciform ligament; 5. Posterior atlantoaxial membrane; 6. Anterior longitudinal ligament; 7. Yellow ligament; 8. Posterior longitudinal ligament

of freedom corresponding to the two axes (vertical and horizontal) of visual exploration [16, 18]. This system is a cardan joint, from the name of the philosopher and scientist Gerolamo Cardano who invented the method of fixing the compass on the boat. Indeed it has a vertical pivot on which is placed the needle indicating north. Therefore, the compass must always remain horizontal despite oscillations in roll and pitch of the boat. It is therefore necessary to increase the inertia of the compass by using oil and fix it to the boat by a system with two perpendicular axes of rotation. A combination of both axes may also result in a third DF with an instantaneous axis of rotation. This is exactly what you can see on the craniocervical cardan. It is needed in the future not to talk anymore for the craniocervical junction about a

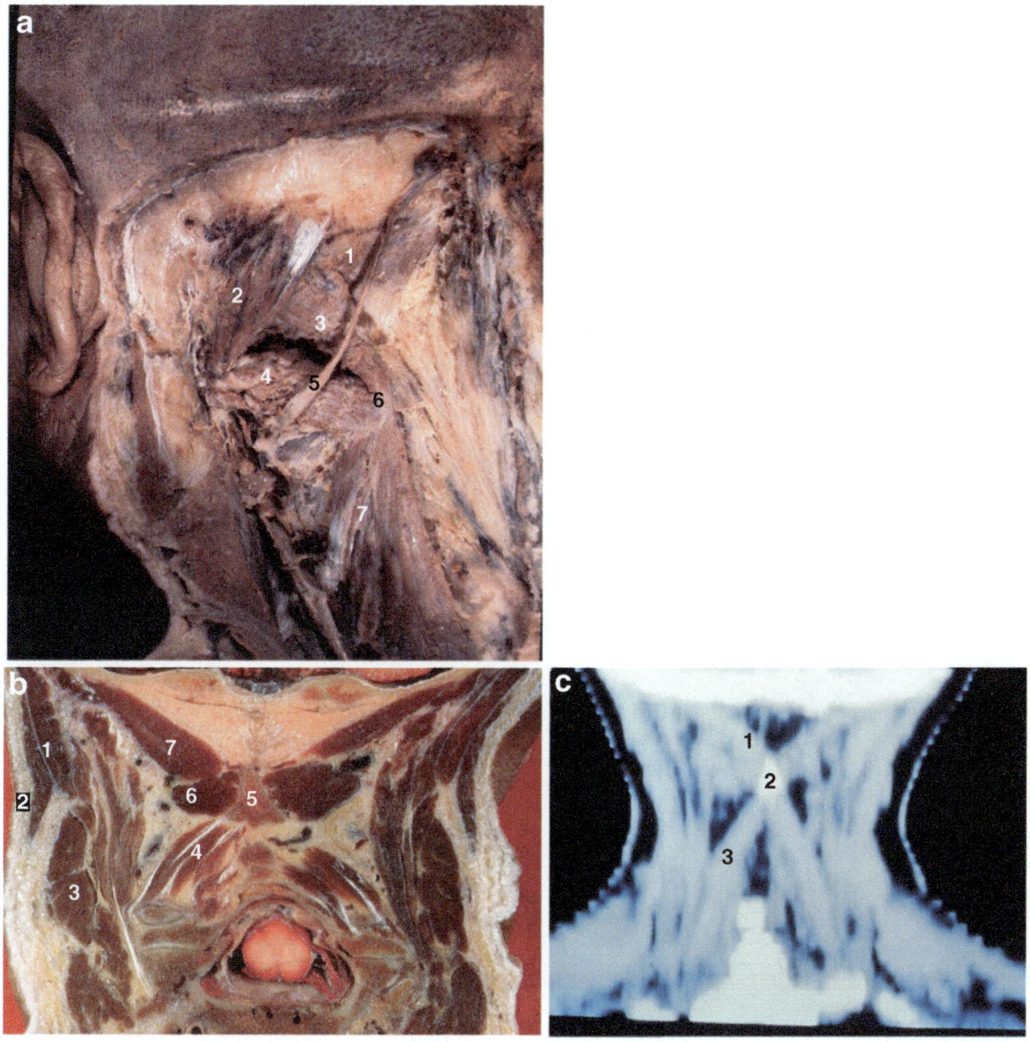

Fig. 2.10 Muscles of the cranio-cervical cardan. (**a**) Posterior view of the *left side*: 1. Small posterior rectus; 2. Superior oblique muscle; 3. Great posterior rectus; 4. Inferior oblique; 5. Great occipital nerve (Arnold); 6. Spinal process of C2; 7. Semispinalis cervicis muscle. (**b**) Coronal section: 1. Splenius capitis; 2. SCM; 3. Levator scapulae; 4. Semispinalis cervicis; 5. Spinal process of C2; 6. Inferior oblique muscle; 7. Great posterior rectus. (**c**) Scanner in coronal section (the "butterfly of the neck"): 1. Great posterior rectus; 2. Spinal process of C2; 3. Semispinalis cervicis

hinge which is a one degree of freedom joint (Figs. 2.8, 2.9 and 2.10).

2.3.2.1 Firstly, a Transverse Axis

Placed between the occipital condyles and the upper concave portion of the lateral masses of the atlas; it allows flexion/extension of the head in the order of 40° and corresponds to the vertical component of the visual exploration.

2.3.2.2 Secondly, a Vertical Axis

Placed between the odontoid process of the axis and the posterior part of the anterior arch of the atlas; it allows horizontal rotation of the head for about 30° corresponding to the horizontal component of the visual exploration. This micrometric system is operated by short posterior muscles: small and great recti muscles, superior and inferior oblique muscles.

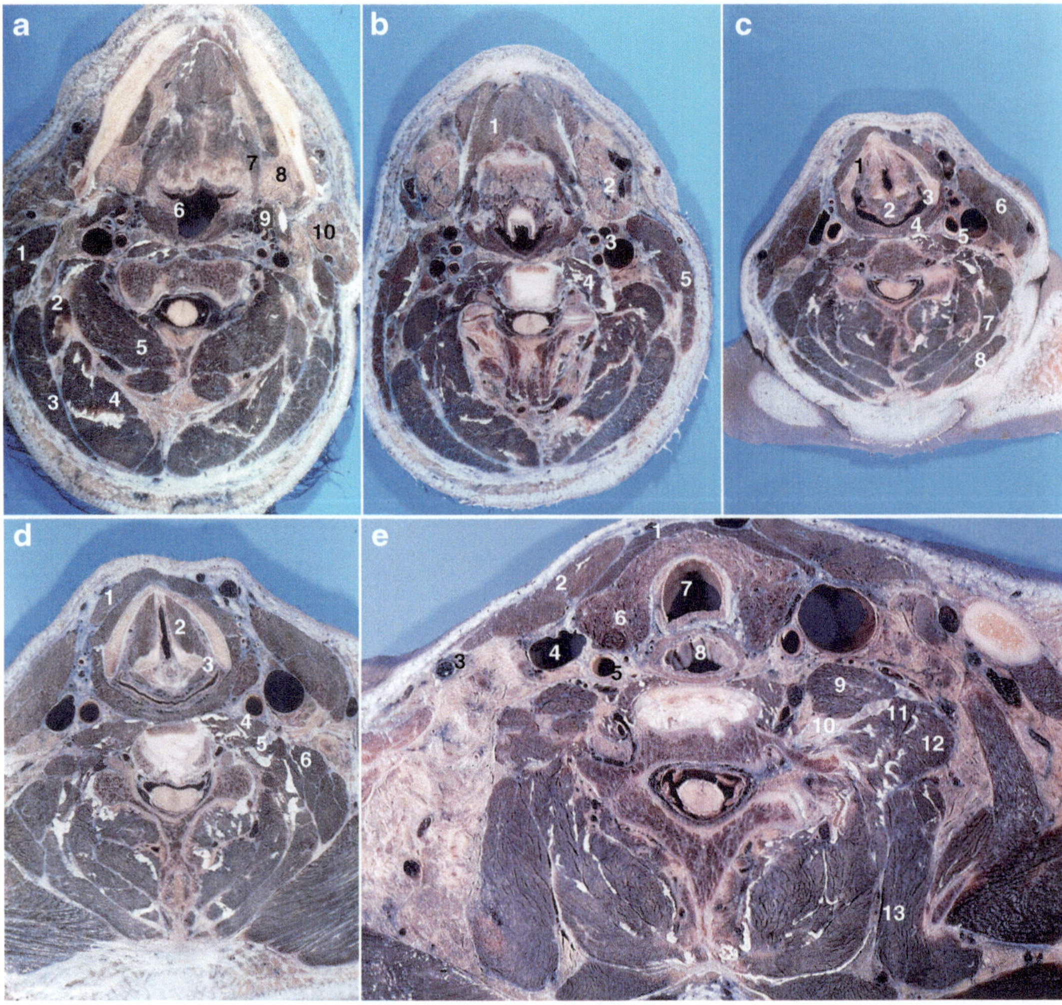

Fig. 2.11 Axial sections of cervical spine. (**a**) C2 axial section: 1. Sterno-cleido-mastoid muscle (SCM); 2. Splenius cervicis; 3. Splenius capitis; 4. Semispinalis muscle; 5. Oblique inferior muscle; 6. Rhinopharynx; 7. Hyoglossal muscle; 8. Submandibular gland; 9. Digastric muscle; 10. Parotid gland. (**b**) C3 axial section: 1. Anterior belly of digastric muscle; 2. Submandibular gland; 3. Carotid arteries and internal jugular vein; 4. Longus capitis muscle; 5. SCM muscle. (**c**) C4 axial section: 1. Thyroid cartilage; 2. Interarytenoid pad; 3. Piriformis sinus; 4. Constrictors of pharynx; 5. Carotid artery and internal jugular vein; 6. SCM; 7. Splenius capitis; 8. Trapezius muscle. (**d**) C4/C5 axial section: 1. Thyrohyoid and sternohyoid muscles; 2. Vocal cord and ligament; 3. Muscular process of arytenoid cartilage; 4. Anterior scalene muscle; 5. Median and posterior scalene muscles; 6. Levator scapulae. (**e**). C7 axial section: 1. Infrahyoid muscles; 2. SCM; 3. External jugular vein; 4. Internal jugular vein; 5. Left common carotid artery; 6. Thyroid gland; 7. Trachea; 8. Oesophagus; 9. Anterior scalene muscle; 10. Brachial plexus; 11. Middle scalene muscle; 12. Posterior scalene muscle; 13. Levator scapulae

For a larger amplitude of horizontal head rotation, a special muscle couple associates the sternocleidomastoid (SCM) on the one side and the splenius capitis on the other side (typical heterolateral synergy) (Fig. 2.11). These two powerful muscles cannot obviously be fixed horizontally in front of the head, which would give them a more high rotating efficiency. They therefore have an oblique direction and are fixed the more laterally possible on the skull and the more medially possible to their bottom: thoracic inlet (clavicle and sternum) to the SCM and

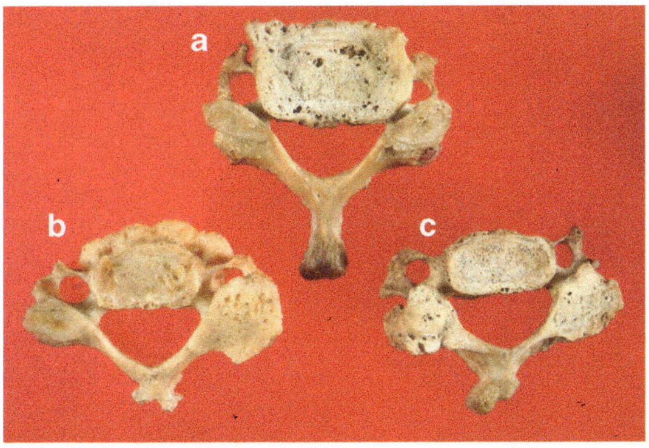

Fig. 2.12 Different types of cervical arthrosis. (**a**) Arthrosis of the vertebral body. (**b**) Arthrosis of the vertebral body and the right posterior articular process. (**c**) Arthrosis of both posterior articular processes

spinous processes for the splenius. This gives them a maximum torque, but imposes a spinal compression component that can explain even for small heads the frequency of cervical arthrosis (Fig. 2.12).

An interesting technical detail can be noted: the splenius cervicis, fixed on the transverse processes of C1 and C2, shrinks at the same time as the head rotation couple. It significantly reduces the difference of torque forces between the powerful head rotation couple and the weak horizontal rotation of the atlas by the inferior oblique muscle. Therefore, it is a protection for the cardan junction in case of sudden head movements.

To these muscles must be added the semispinalis capitis (formerly complexus) that maintains the head in extension, because when relaxed the head tends to fall in flexion (as it happens when one is sitting and falling asleep) considering the position of the centre of gravity of the head at the level of the clivus. This muscle is very visible especially in children and elderly people in the form of two vertical columns on the back of the head. Semispinalis cervicis attaches to the cervical spinous processes and laterally on the transverse processes from C2. Neck pain reported by those working hard in front of a computer is due to the contraction of these muscles required to maintain a distance eye/object consistent with the necessary visual accuracy. Typing on keyboard involves fixing the scapula by the trapezius and levator scapulae, which can be also the cause of a typical pain on the lateral side of the neck and upper chest. Similarly, drivers can charge for the same reasons posterior neck pain and high thoracic back pain, especially during night driving. Optical aberrations wrongly corrected can accentuate this phenomenon and should therefore be taken into account.

2.3.3 The Vertebral Artery

An important point of the programme of the cervical spine is the problem of the vertebral artery. It has a great functional value, because it vascularises the brainstem and in particular the respiratory centres that operate 24 h/24. If they stop functioning, that means death. It therefore seems necessary to protect its way up to avoid compressions and reductions in flow.

2.3.3.1 The Cervical Segment

It enters the foramen transverse process of the cervical spine at the level of C6 (with some possible variations). It undergoes mechanical stress in relation to the great mobility of the cervical spine, which are minimised by the existence of a hydraulic damper represented by the rich venous plexus surrounding the artery in his journey to the foramen magnum. This plexus communicates with the epidural veins and extra-spinal veins.

One can identify two channels at the intervertebral foramen of cervical vertebra two channels: one nervous for the passage of nerve roots and spinal nerve and the other venous for the passage of a large foraminal vein communicating with the epidural plexus (Fig. 2.13). This particular venous context explains that there is always a venous factor in cervicobrachial pain, which can increase in lying down position at night due to venous stasis. It is likely therefore that the specific venous richness of the cervical spine can be directly related to the presence of the vertebral artery (Fig. 2.14).

As we mentioned earlier, the uncus is an increase in the upper lateral edge of the cervical vertebra and should be considered as a protective wall of the vertebral artery. This avoids bulging of the disc and compressing artery in osteoarthritis pathology. Therefore, the normal fissuration

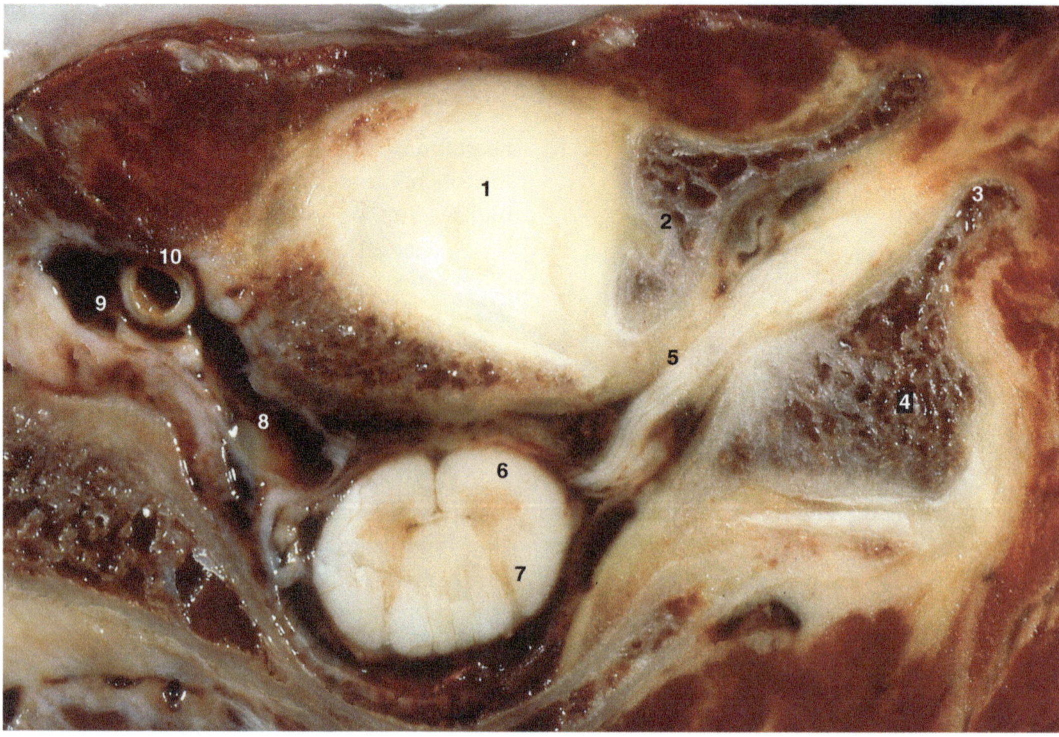

Fig. 2.13 Asymmetric axial section of the cervical spine. 1. Vertebral body and disc; 2. Uncus; 3. Posterior tubercle of transverse process; 4. Superior articular process; 5. Nervous part of the vertebral foramen (roots and ganglion); 6. Anterior motor horn of the spinal cord; 7. Posterior sensitive horn; 8. Venous part of the vertebral foramen; 9. Venous plexus of the vertebral artery; 10. Vertebral artery in its transverse foramen

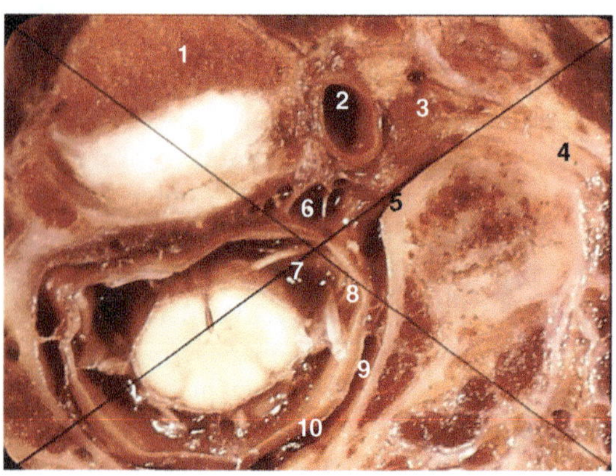

Fig. 2.14 Axial section of the cervical spine. 1. Vertebral body; 2. Vertebral artery; 3. Spinal ganglion; 4. Spinal nerve; 5. Nervous part of the intervertebral foramen; 6. Epidural space; 7. Anterior motor root; 8. Posterior sensitive root; 9. Dentate ligament; 10. Dura mater

of the disc under mechanical compression begins precisely very early at this level.

2.3.3.2 The Technical Problem of Going Through the Cardan

The most critical issue for the vertebral artery is the crossing of the craniocervical cardan (Fig. 2.15). Indeed, during rotation/extension of the head (e.g. watching the frescoes of the Sistine Chapel), one of the vertebral arteries is stretched and compressed with a substantial reduction of its flow [19]. That is in relation to the journey of the vertebral artery: with a first loop that goes to the vertebral foramen of the axis and then with a second loop passing horizontally behind the lateral mass of the atlas before entering the foramen magnum (Fig. 2.16). This manoeuvre particularly in elderly persons with arteriosclerosis can cause

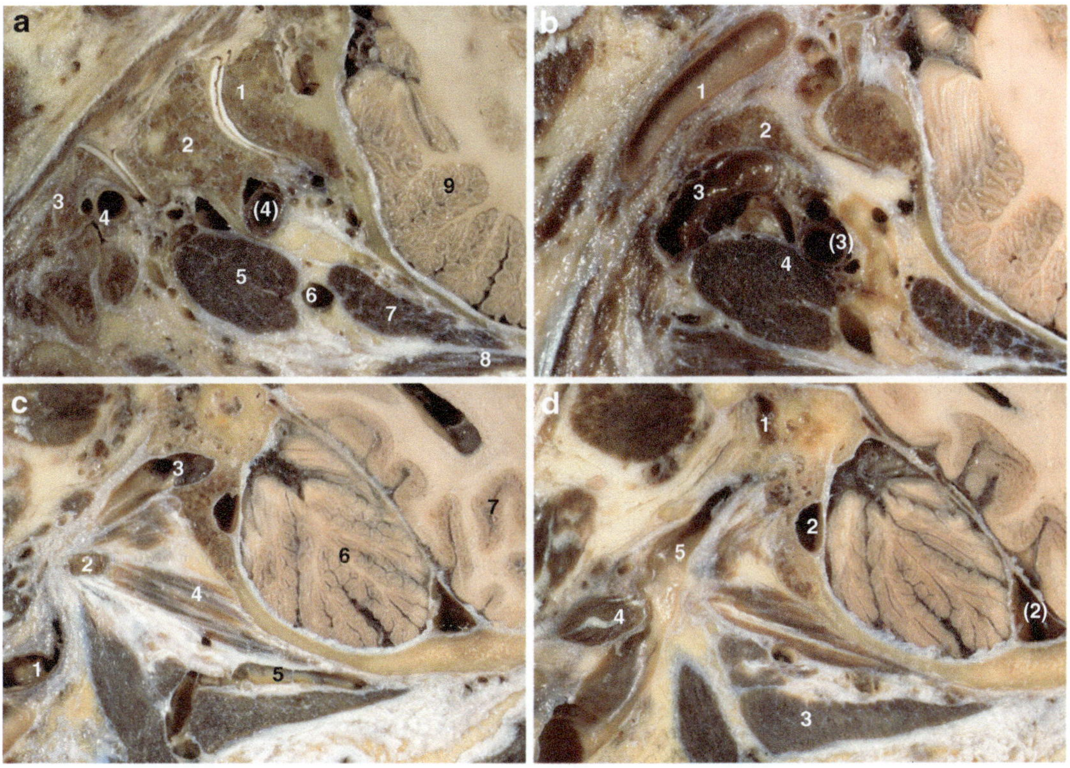

Fig. 2.15 Parasagittal sections of the cardan. (**a**) 1. Occipital condyle (C0); 2. Atlas; 3. Axis; 4. Vertebral artery; 5. Oblique inferior muscle; 6. Occipital vein; 7. Splenius capitis; 8. Trapezius; 9. Cerebellum. (**b**) 1. Internal carotid artery; 2. Atlas; 3. Vertebral artery and its venous plexus; 4. Inferior oblique muscle. (**c**) 1. Vertebral artery; 2. Lateral process of C1; 3. Internal jugular vein; 4. Superior oblique muscle; 5. Occipital vein; 6. Cerebellum below the tentorium; 7. Occipital lobe. (**d**) 1. Promontorium in the middle ear; 2. Lateral sinus; 3. Trapezius muscle; 4. Digastric muscle; 5. Internal jugular vein

vertigo (sign of Kleyn) due to lack of flow in the small internal auditory artery in relation to the vestibule (Fig. 2.17). Faced with this technical problem, the smart technical solution is to try to ensure a continuous blood flow to the brainstem by bringing the two vertebral arteries entering the foramen magnum in a single trunk: the basilar artery (Fig. 2.18). This anastomosis by convergence of two arteries is unique in the whole arterial system. Note that the same anastomosis is visible and for the same reasons on the anterior spinal artery issued by two branches from the vertebral artery. Some asymmetries of the vertebral arteries can be frequently observed, explaining some clinical features. The most important point to know and to check carefully for practitioners doing cervical manipulations is the crucial functional importance of the vertebral artery, which has imperatively to be respected.

2.4 Thoracic Spine and Respiratory Dynamics

The thoracic spine forms the rear portion of a **cage** closed forward with **costal bars** fixed to the sternal bone. It has a conical shape with an apex upward and a broad triangular opening on its front face. This mechanical structure limits the range of motion of the spine. It contains the heart and great vessels in the centre and lungs laterally (Fig. 2.19). Outside its **protective role** for these precious organs, the thorax may change its volume to ensure **respiratory vital function**.

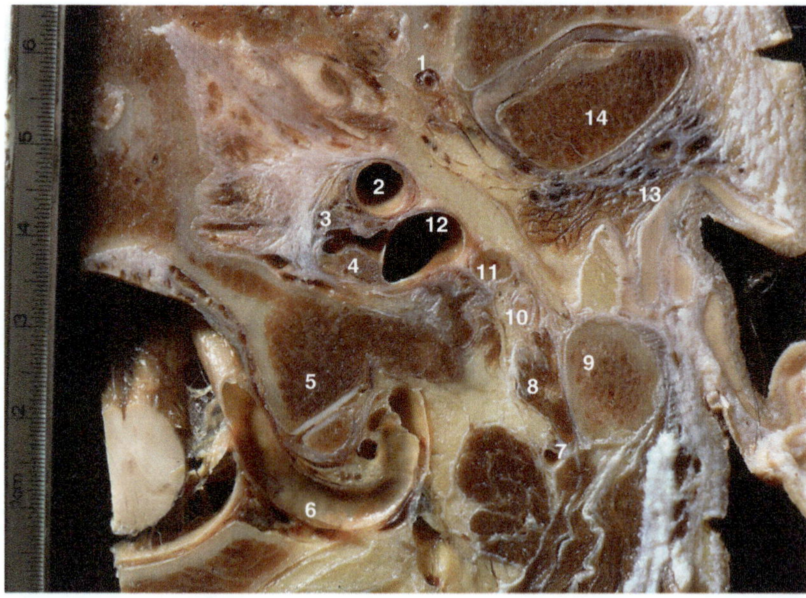

Fig. 2.16 Left axial section of occipital foramen seen from below. 1. Median meningeal artery; 2. Internal carotid artery; 3. Glossopharyngeal nerve (IX); 4. Vagal nerve (X); 5. Lateral part of occipital bone; 6. Horizontal loop of vertebral artery behind C1; 7. Occipital artery; 8. Posterior belley of digastric muscle; 9. Mastoid process; 10. Facial nerve (VII); 11. Styloid process; 12. Internal jugular vein; 13. Fibro-venous retrocondylar pad; 14. Mandibular condyle

The **ribs** typically have **three bends** and a **facet joint** with the vertebrae through two points: the head and the costal tuberosity. Any action muscle pulling the ribs upward increases the volume and down decreases, which is the basis of the definition of respiratory muscles and respiratory biomechanics (inspiration/expiration).

2.4.1 The Diaphragm

The centre piece of the thorax is the diaphragm which plays a dual role:

- firstly, it is the motor of the thoracic bellows bringing air into the lungs by its contraction (inspiration) and leaving out by its slackness (expiration);
- secondly, it separates two cavities: one on top containing the complex heart/lungs and one below containing the digestive tract and annexes until the pelvis. It allows vessels and nerves to pass from one cavity to another. It has the general shape of a dome rounded concave bottom modelled on the right of the liver (Fig. 2.20).

The muscular action of the diaphragm changes the abdominal volume and is strongly associated with the abdominal wall in a functional interaction important to remember. It works continuously but with alternating periods of contraction/relaxation avoiding fatigue. Its control in relation to the respiratory centres of the brainstem allows it to work either with a fully automated system since the first breathing out of the uterine cavity or in voluntary control of the pilot who can change pace and capacity (basics of breathing exercises).

It can be anatomically described in two parts: muscular in the periphery and tendinous in the centre (phrenic centre).

2.4.1.1 The Muscular Portion

It is fixed on the periphery of the thoracoabdominal cavity with different parts.

Lumbar Part

It is formed by the pillars of the diaphragm. The right is the longest, fixed on the front side of the lumbar spine (disc and bodies L2/L3) and only on L2 for the left. Their tendon fibres intersect to form the incompressible orifice of the aorta in the

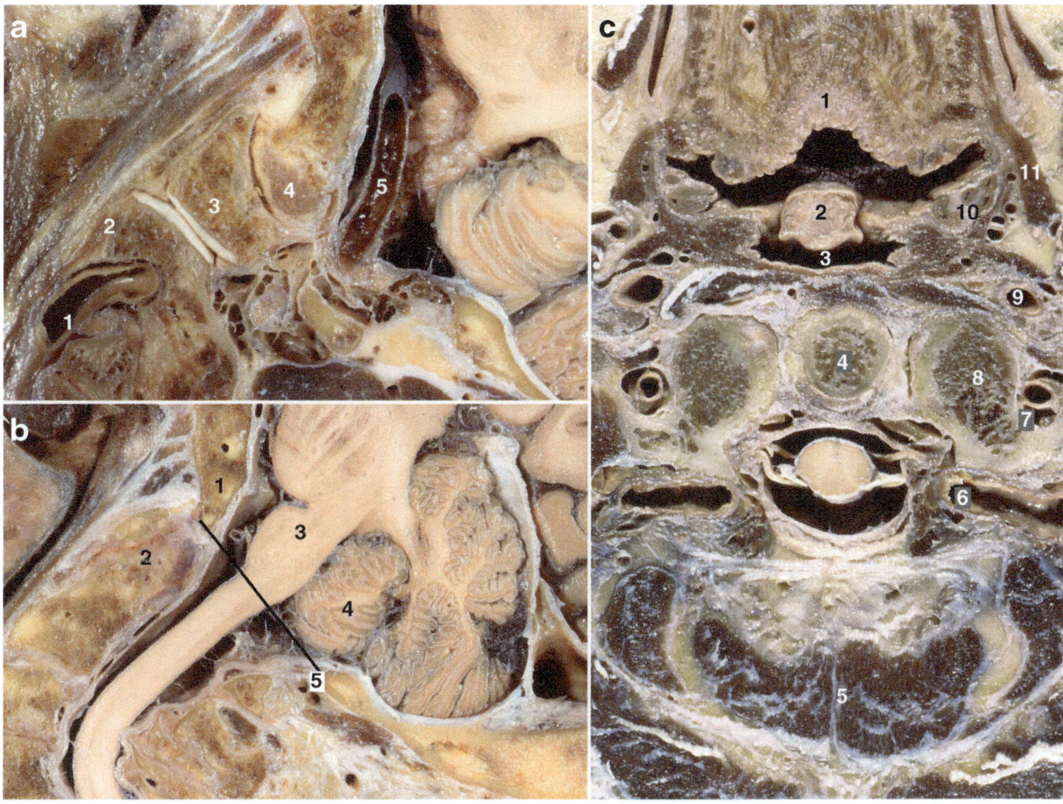

Fig. 2.17 Sagittal and axial sections of the occipito-cervical junction. (**a**) Left para-sagittal section: 1. Axial loop of vertebral artery; 2. C2; 3. C1; 4. Occipital condyle (CO); 5. Vertebral artery entering within the skull base. (**b**) Sagittal section: 1. Clivus; 2. Atlanto-odontoid junction; 3. Brain stem; 4. Cerebellar tonsils; 5. Occipital foramen. (**c**) Axial section of the cardan joint (superior view): 1. Tongue; 2. Uvula; 3. Rhinopharynx; 4. Dens of C2; 5. Neck muscles; 6. Transverse portion of the vertebral artery; 7. Vertebral artery in the transverse foramen of C1; 8. Lateral mass of atlas; 9. Internal carotid artery; 10. Pharyngeal tonsils; 11. Styloglossal muscle

front of T12. Muscle fibres intersect forwardly in an eight figure to determine the muscular orifice for oesophagus and vagus nerves (X) in the front of T10 (Figs. 2.21 and 2.22). There are accessory pillars placed laterally between which pass the greater splanchnic nerve and great azygos vein at the right and the hemiazygos at the left. They are two arched ligaments: medial on the relief of psoas and lateral forming a tendinous arch between the costiform process of L1 and the top of the 12th rib. At this level, there is a variable space: the costo-diaphragmatic hiatus.

Costal Part

It is inserted on the inside part of 7th to 12th rib between the fingering of the abdominal transverse muscle, forming a series of fibrous arches.

Sternal Part

It is made of two small muscle bundles separated by an avascular space which meet the posterior xiphoid process with the phrenic centre anterior leaflet. A slot of variable dimension described by Larrey is located laterally, which may sometimes contain an expansion of the gastric pouch generating specific pain and requiring eventually an endoscopic closure.

2.4.1.2 The Tendon Called Phrenic Centre

It has the form of a three-leaf clover with an inextensible fibrous orifice for the inferior vena cava.

2.4.1.3 The Innervation of the Diaphragm

It is provided by the phrenic nerve which is composed of a large majority of alpha and

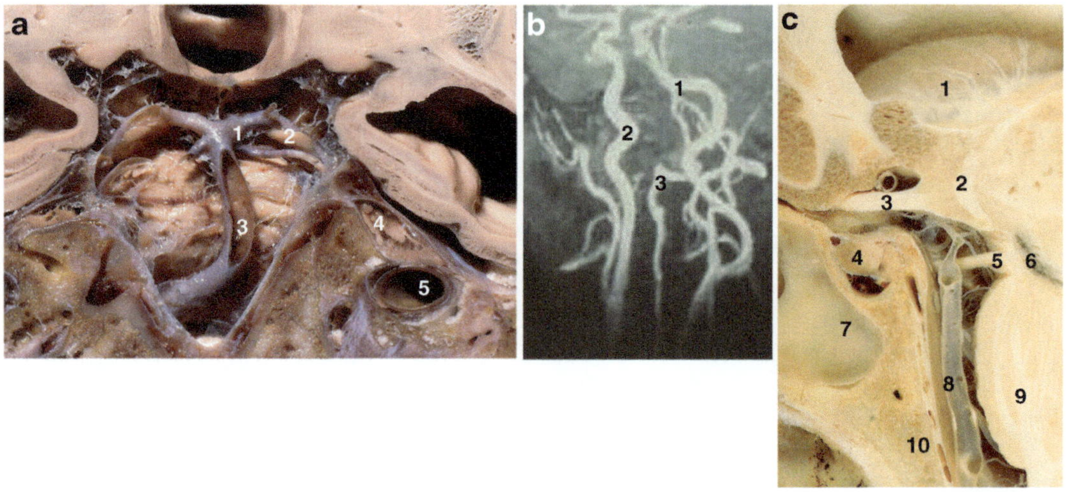

Fig. 2.18 Convergent anastomosis of the two vertebral arteries. (**a**) Coronal section seen from the front: 1. Vascular pincer: posterior cerebral artery above and superior cerebellar artery below; 2. Oculomotor nerve (III); 3. Basilar trunk; 4. Triangular plexus and trigeminal ganglion in its cavum; 5. Internal carotid artery. (**b**) Angiogram of internal carotid and vertebral arteries: 1. Basilar trunk and convergent anastomosis; 2. Internal carotid artery; 3. The three loops of the vertebral artery. (**c**) Sagittal section: 1. Nucleus caudatus; 2. Anterior pillar of the fornix; 3. Optic chiasm; 4. Hypophysis within the sella turcica; 5. Oculomotor nerve (III); 6. Substantia nigra (locus niger); 7. Sphenoid sinus; 8. Basilar trunk; 9. Pons; 10. Clivus

gamma motor nerve fibres (4/5) and also of sensory fibres for serous membranes (pericardium, peritoneum, pleura). It arises from a constant root C4 and some accessory roots, as C5 or sometimes C3. Its neck trajectory is inside the scalenus anterior aponeurosis where it may possibly be compressed by the hand. It follows a different mediastinal path on the right where it is shorter on the left. It can easily be isolated to put a stimulation electrode in order to create a diaphragmatic pacemaker.

The diaphragm muscle fibres are made of a majority of tonic type I (fatigue resistant and slow to contract) (55 %) and of phasic type IIA and IIB (sensitive to fatigue and fast to contract). Therefore, the diaphragm is a muscle designed to work for long. It is accompanied in its action by the rhythmic intercostal muscle action, mostly at the upper part.

In this regard, it is interesting to note the non-negligible thermoregulatory role of the chest wall, with an intercostal artery under each rib carrying arterial blood providing heat, a conveying effect to mitigate the variations of the outside temperature.

2.4.1.4 The Respiratory Control

It is particularly interesting because it must ensure rhythmic automation of breathing, essential for the survival of organs. There is a ventilatory oscillator containing automatically depolarised pacemaker neurons controlled by neural networks of the pre-Bötzinger complex, which constitutes the rhythm generator. These structures are placed in two groups of respiratory centres:

- the medullary dorsal group of the medulla oblongata, responsible for the inspiratory activity, with solitary tract nucleus, dorsal motor nucleus of the vagus (X) and area postrema of the lower part of the fourth ventricle;
- the medullary ventral group, with the nucleus ambiguus and periambiguus for expiratory activation;
- a coordinator is acting as an "orchestra conductor": the pneumotaxic centre controlled by the hypothalamus (emotions, fever) and the cortex (cough, voluntary apnoea) with reflexes at the starting point of the nasal mucosa (sneezing), larynx and bronchi (cough) and the skin (heat inhibitor shock to cold).

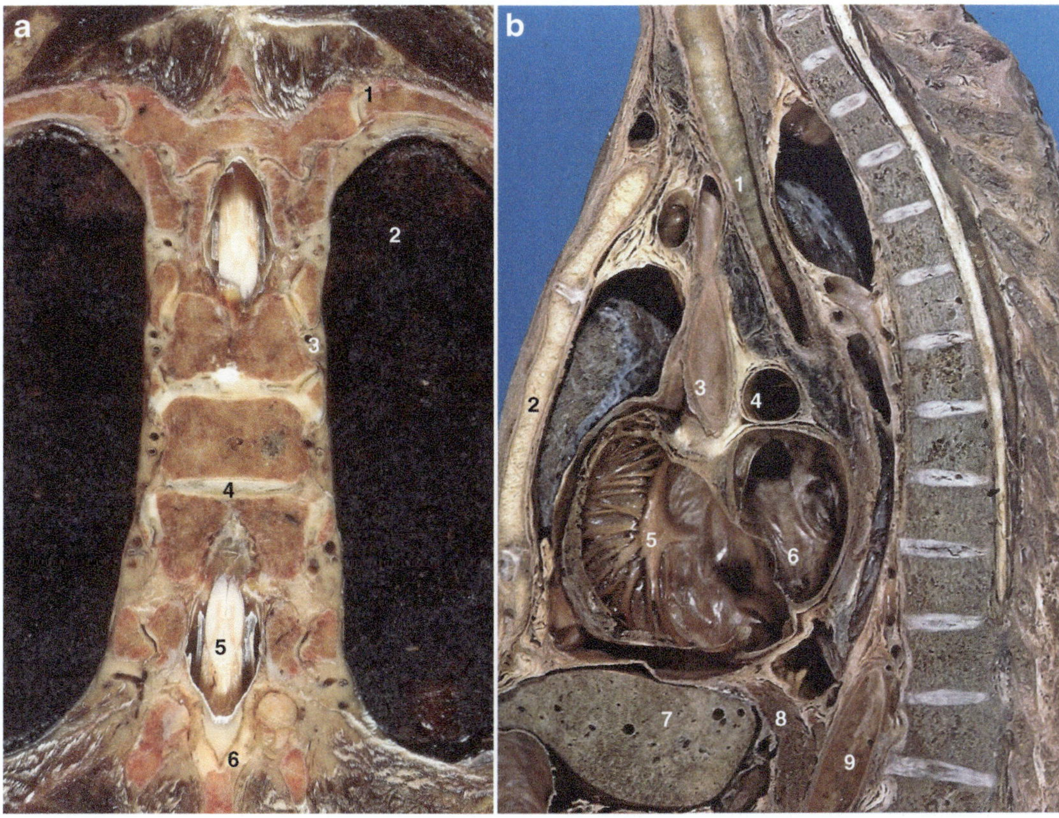

Fig. 2.19 Dorsal spine. (**a**) Coronal section seen from the front: 1. Costo-vertebral joint; 2. Left lung; 3. Intercostal artery; 4. Fissurated vertebral disc; 5. Spinal cord; 6. Yellow ligament. (**b**) Sagittal section of thorax: 1. Trachea; 2. Sternum; 3. Ascending aorta; 4. Right pulmonary artery; 5. Right auricle; 6. Left auricle; 7. Liver; 8. Diaphragm; 9. Descending aorta

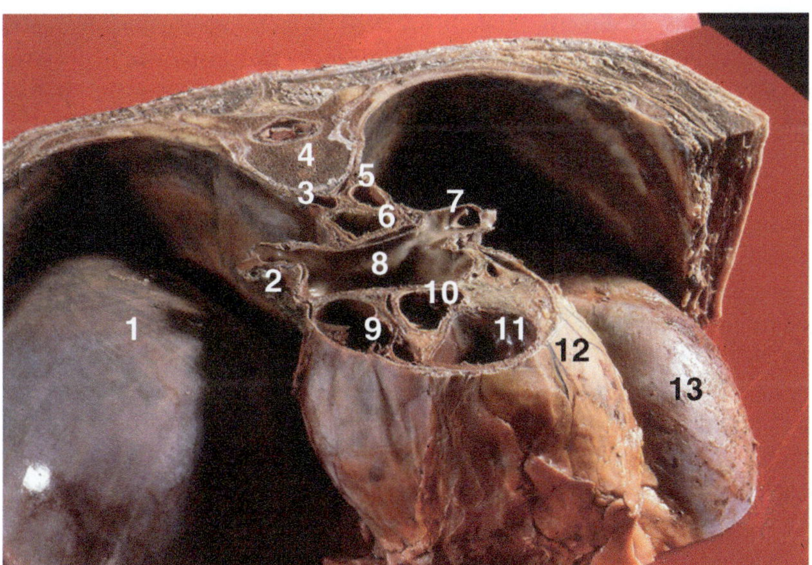

Fig. 2.20 Heart and diaphragm in three-quarter view. 1. Right diaphragmatic dome; 2. Right pulmonary veins; 3. Azygos vein; 4. Thoracic vertebra; 5. Descending aorta; 6. Oesophagus; 7. Left pulmonary veins; 8. Left auricle; 9. Superior vena cava; 10. Pulmonary artery; 11. Right ventricle; 12. Pericardium; 13. Left diaphragmatic dome

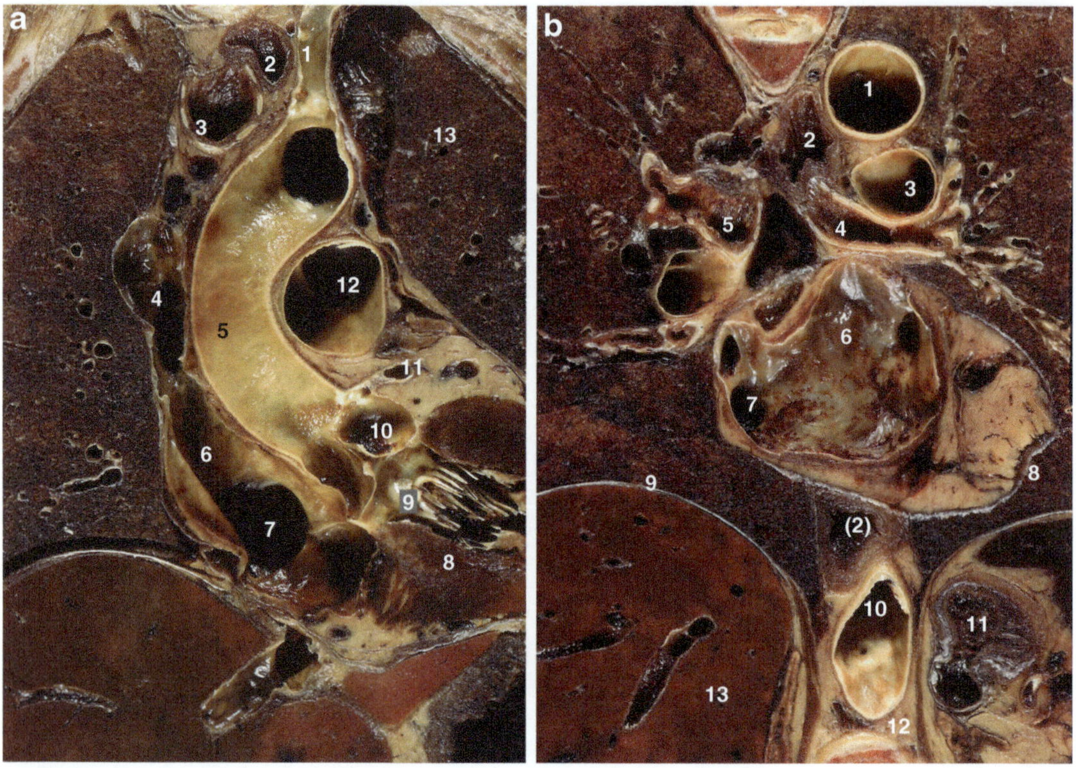

Fig. 2.21 Coronal sections of heart. (**a**) Anterior coronal section: 1. Common carotid artery; 2. Oesophagus; 3. Trachea; 4. Superior vena cava; 5. Arch of aorta; 6. Right auricle; 7. Inferior vena cava; 8. Left ventricle; 9. Mitral valve; 10. Aortic semilunar valves and sinus; 11. Left coronary artery; 12. Left pulmonary artery; 13. Superior lobe of left lung. (**b**) Posterior coronal section: 1. Aorta; 2. Oesophagus; 3. Left pulmonary artery; 4. Left main bronchus; 5. Right main bronchus; 6. Left auricle; 7. Right pulmonary veins; 8. Pericardium; 9. Diaphragm phrenic centre; 10. Abdominal aorta; 11. Stomach; 12. Pillar of diaphragm; 13. Liver

The respiratory control is directly related to two key parameters: the carbon dioxide partial pressure ($PpCO_2$) and oxygen partial pressure (PpO_2). Haematosis defines the diffusion of gases in the alveolar-capillary junction (area of about 170 m^2) based on the partial pressures of 40 mmHg for CO_2 and 100 for the O_2; O_2 combines for transportation to haemoglobin up to 97 % (oxyhemoglobin) and CO_2 up to 25 % (carbhemoglobin), the rest being provided by bicarbonates of Na and K.

Two types of regulation may be observed: firstly by fast humoral track by $PpCO_2$ action on brainstem central chemoreceptors in direct contact with the CSF (particularly in area postrema) and secondly by slow nerve way by the action of PpO_2 on peripheral chemoreceptors of the aortic arch and carotid sinus baroreceptors.

It is therefore a complex machine but cleverly effective in particular to ensure safe operation with the involvement of a defence system, real survival servomechanism, based on the only imperative alarm of the entire nervous system. It is possible to understand it in the case of voluntary apnoea for divers or involuntary in common sleep apnoea. The cessation of breathing causes a decrease in PpO_2 and an increase in $PpCO_2$ with pH shift. This determines a "red" alarm in the respiratory centres with specific action on central and peripheral chemoreceptors. That forces the resumption of breathing by direct diaphragmatic activation.

Sleep apnoea occurring primarily in obese snorers is due to the release of muscles maintaining the pharyngeal axis and the tongue,

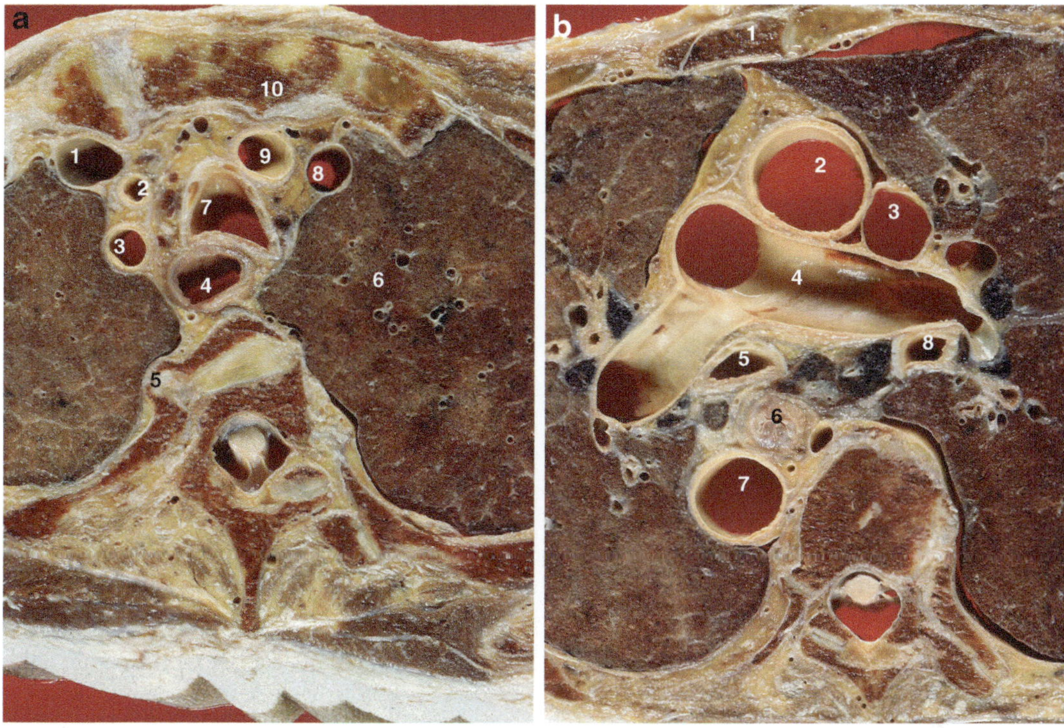

Fig. 2.22 Axial thoracic sections seen from above. (**a**) Superior part of the thorax: 1. Left brachiocephalic vein; 2. Left common carotid artery; 3. Left subclavian artery; 4. Oesophagus; 5. Costo-vertebral joint; 6. Superior lobe of right lung; 7. Trachea; 8. Right brachiocephalic vein; 9. Brachiocephalic trunk; 10. Sternum. (**b**) Middle part of the thorax: 1. Sternum; 2. Ascending aorta; 3. Superior vena cava; 4. Pulmonary artery fork; 5. Left bronchus; 6. Oesophagus; 7. Descending aorta; 8. Right bronchus

which creates a valve effect slowing and then stopping the respiratory flow. Triggering the alarm allows to restore breathing often with noise but maintaining tonic action, unfortunately for a transitional period before the resumption of sleep. Some patients make hundred breathing stops every night, which creates anoxic disorders particularly within the hippocampus with loss of memory, heartbeat disorders and drowsiness during the day which is responsible for many traffic accidents. The result is the death of 5,000 people every year in France. In general, the spouse indicates the phenomenon that requires an accurate diagnosis and treatment either by mask or, possibly, by surgical corrections.

Diver's apnoea is different, because it is a voluntary action that can even become a sport performance. The alarm mechanism is still present, but it is progressively modified by training. The static apnoea record in 2003 was 8 min 24 s, as the most common on average is around 5 min. Apnoea diving is not without risk, because besides stopping breathing, there are physical constraints due to the fluid pressure. This can give an idea, knowing that 1,000 l of air at surface level corresponds to 500 l at 10 m depth, 333 l at 20 m, 250 l at 50 m and 100 l at 100 m. Fortunately these depths are reached only by professionals, but there are still many accidents among amateurs, poorly trained or careless—accidents that mobilise all regional hyperbaric pressure chambers for saving purposes.

Finally, we should mention the importance of breath education for artists: lyric singers (without microphone) and wind instrument players using reeds or mouthpieces (trumpet, clarinet, oboe, tuba, horn, ...). Teachers often use personal empirical methods, the goal being always to increase breathing capacity and properly control the pressure and flow of air output by precise regulation of diaphragmatic and abdominal wall contraction. Another form of education is the use of these abdominal wall contractions with the diaphragm blocked in inspiration, in order to push on pelvic reservoirs for emptying them or, in a nobler and caring manner, to allow a newborn to discover the pleasure of breathing for the first time.

2.5 Lumbar Spine and Pelvic Ring

The lumbar spine is the segment that carries the highest load, and hence, the lumbar vertebrae have the most important dimensions. L3 is normally placed on top of the lumbar lordotic curvature and its two endplates are parallel. Above and below, the trays present an oblique inverted orientation, and L5 has a wedge aspect with a critical mechanical position.

It is important, in order to understand the frequent lumbar pathology, to know its exact organisation.

2.5.1 The Intermediate Vertebra T12

On top, in T12, is the thoracolumbar junction, which is not a hinge. T12 is called "intermediate" or "diaphragmatic" vertebra (Vallois) with, for the shape of facet joints, a superior thoracic type and inferior lumbar type. This is also the point of reversing direction of the spinous processes with cervicothoracic retrocliny and lumbar anticliny. In quadruped animals, the intermediate vertebra is located also in T12, which is the central part of the highly developed epispinatus muscle inserted into the spinous and transverse processes. This muscle is responsible for direction changing in locomotion by lateral bending of the spine. It also allows the quadruped to sniff its perineum, which normal human beings cannot do. The upright posture and the fact that direction changes in humans are made by hip rotation may explain why the epispinatus muscle centred at T10 in humans is clearly rudimentary, with no insertions on transverse processes.

2.5.2 The Four Muscular Columns

Another important aspect is that the lumbar spine is maintained in position by four muscular columns, according to the principle of the composite beam bone/muscle, which increases the mechanical strength [10].

2.5.2.1 On the Front

There are the two psoas polyarticular muscle columns starting from T12 up to join iliacus monoarticular muscle with an attachment onto the small femoral trochanter. This powerful set flexes the thigh in synergy with the polyarticular rectus femoris of the quadriceps to start walking. But feet being fixed on the ground, its action relies mainly on the lumbar spine.

2.5.2.2 On the Back

There is the lumbosacral part of the spinal muscles, which is not a common mass as described by ancient anatomists. In fact, as shown by Winckler, it is formed of three columns: medial with the spino-transverse group (multifidus), intermediate with the long dorsal and lateral with the iliocostal muscles. It is not possible to properly treat spine diseases without knowing well the lumbar muscles, not on diagrams or drawings as much artistic as they are, but only by doing an anatomical dissection.

One can therefore better understand the critical position of T12 located between the thoracic spine, mobilised by the powerful abdominal wall muscles and the lumbar column, flanked by four muscle columns. Maigne rightly stressed its specific pathology [20–22].

2.5.3 The Split Pelvic Ring

The lumbar spine is a part of an interactive mechanical assembly formed by lumbar spine applied on the pelvis (iliosacral). The concept of pelvic ring as proposed here should be clearly understood because it brings a smart technical solution to a specific mechanical problem. In general, we know that a mechanical rigid structure subject to mechanical stresses undergoes deformations which may extend to a rupture, whether ductile (after a large deformation like extending rubber) or fragile, as for glass, depending on the nature of the material.

One way to delay breaking and thus increase mechanical strength is to introduce within the ring a discontinuity allowing micro-controlled movements capable of absorbing energy into the rigid part. A good example is the inferior tibiofibular joint that is hinged with the talus in a 1 DF system that undergoes significant lateral stresses due to the calcaneus placed below with ground support. The deformation capacity of the bimalleolar clip, nevertheless controlled by powerful periarticular ligaments and tendons, absorbs energy and significantly increases strength. Regarding the pelvis placed between the lumbar spine and the lower limbs, it has a shape of a bony ring that can significantly increase the resistance by creating anatomical discontinuities.

2.5.3.1 The Pelvis

This is the concept of split ring with three discontinuities: one in the front by the pubic symphysis and two on the back by the sacroiliac joints. Finally, that allows defining two triangles: one for stress amortisation (symphysis/sacroiliac) and one for loading (L5/S1 junction and left and right cotyles). In fact, this represents a smart structure, but in order to avoid false interpretation, it is very important to remember that the three joints of the split ring are not made for movement but mainly to absorb stresses. The pubic symphysis is an amphiarthrosis with an interposed fibrous structure and powerful short ligaments on both sides.

2.5.3.2 The Sacroiliac Joint

It is [23, 24] a diarthrosis with an articular surface of cartilage shaped similar to an ear pavilion, thus making a broken articular junction line that severely limits the slip (Fig. 2.23). The top of the articular line shows often a change in articular cartilage due to mechanical stress. The joint is fixed by short and very resistant ligaments: anterior, interosseous and posterior sacroiliacs, sacrospinous and sacrotuberous. This strong synthesis tightly excludes any possibility of significant movement (sliding rotation) as some people still think, describing nutation/counternutation movements. It must be said, however, that at the end of pregnancy, these ligaments for endocrine reasons in relation with hormonal impregnation allow a little distension which facilitates the expulsion of the foetus.

2.5.3.3 The Lumbar Pathology

The anatochronesis, defined as a regressive evolution with time, of the anatomical structures must take into account the specific position of the pelvic ring interposed between the trunk supported on L5/S1 and the lower limbs that have inherently asymmetric activation in walking, running, climbing as well as in some sports like football, which requires a powerful foot impact on a ball. As with all amphiarthroses, the mechanical solicitations in compression and shearing forces create the mechanical cracking of the interposed fibrous material (Fig. 2.24), which is clearly seen in the intervertebral discs [14, 25] but also in the pubic symphysis, which can lead to a specific pubalgia.

The great frequency of lumbar pathology imposes to pay attention to the crucial role of the muscles. Without spinal muscles, the spine sag dramatically as in paralysed patients or in some neurodegenerative diseases, sometimes leading to death by progressive damage of respiratory function. In these cases, it requires orthopaedic treatment essentially by surgical correction and realignment of the spine.

The muscles acting on the spine are necessarily placed in an oblique direction, the origin being either medial or lateral and fixed on the posterior arch of the vertebra (spinous process, lamina, transverse process, dorsal angle of the ribs). As each vertebra is slightly mobile, we can define the action of muscles by a sort of muscular pliers, which requires a support to "open" the discal space. It is located on the posterior part of the disc, which support the maximum compressive stresses and represent the instantaneous axis of rotation during flexion/extension of the spine. This explains the preferential alteration and possible rupture of this portion of the annulus fibrosus, which can create a herniated disc, mainly in young patients with

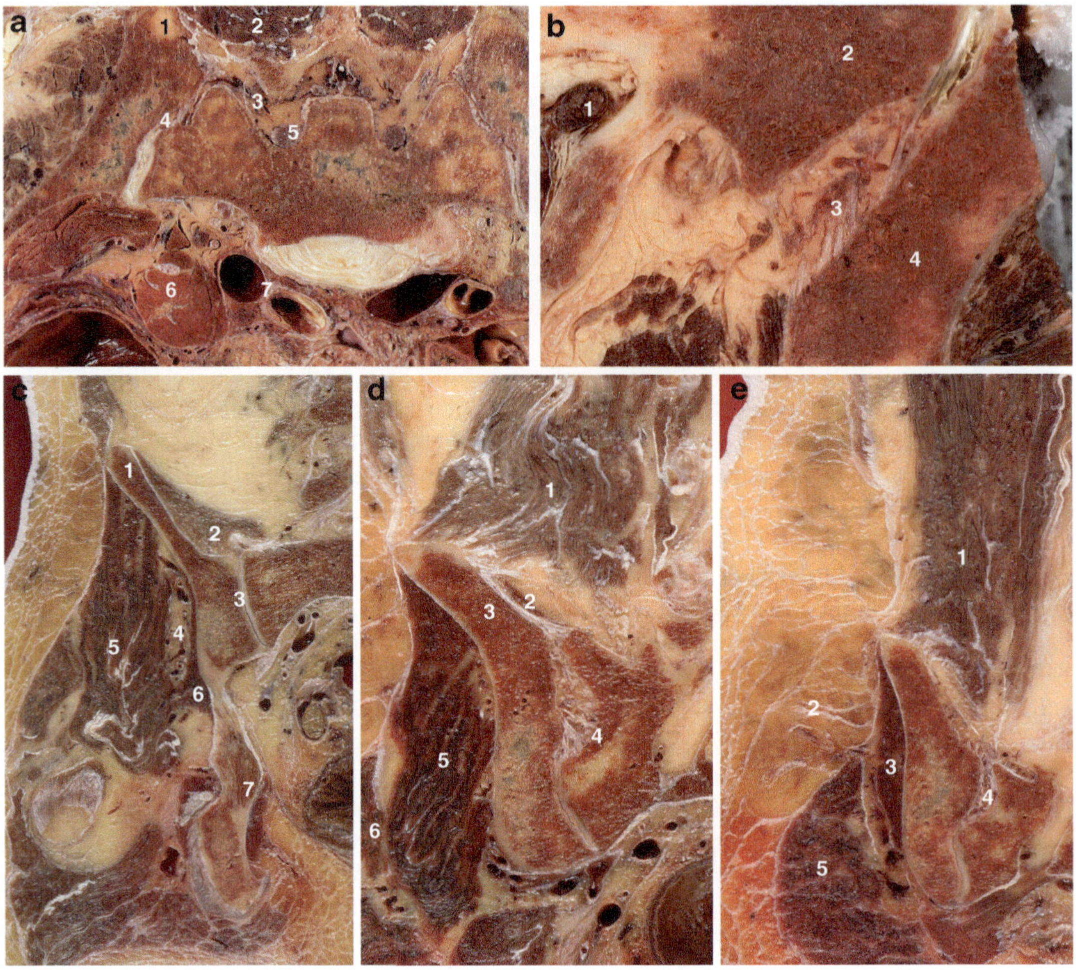

Fig. 2.23 Sacroiliac joint. (**a**) Axial section (S2): 1. Posterior part of iliac crest; 2. Spinal muscles within the sacral groove closed by the lumbo-sacral tendon; 3. Sacral foramen; 4. Sacroiliac joint; 5. Sacral nerve; 6. Psoas muscle with femoral nerve; 7. Vascular iliac pedicle. (**b**) Axial section (right half seen from above): 1. Sacral roots; 2. Sacrum; 3. Posterior ligaments; 4. Iliac bone. (**c**) Coronal section: 1. Iliac bone; 2. Iliac muscle; 3. Sacroiliac joint; 4. Branches of gluteal artery; 5. Gluteus medius muscle; 6. Gluteus minimus muscle; 7. Ischiatic branch. (**d**) Coronal section: 1. Spinal muscles; 2. Sacro-tuberal ligament; 3. Iliac bone; 4. Interosseous ligament; 5. Gluteus medius muscle; 6. Gluteus maximus muscle. (**e**). Posterior coronal section: 1. Spinal muscles; 2. Dermal fibrous mooring lines; 3. Gluteus medius muscle; 4. Sacroiliac joint; 5. Gluteus maximus muscle

sufficiently hydrated disc (Fig. 2.25). This "myopliers" has also an important support on the posterior articular processes, explaining their frequent degenerative osteoarthritis.

Load distribution on the front and the back of the spine has been variously calculated, and the figures are not scientifically valid, because in addition to the discal and joint support, it is important to consider the muscles which are viscoelastic bodies, able to change their stiffness and elastic modules and therefore the amount of charge they can hold and transmit [25–27]. The measurement of the load in living is very difficult to achieve and is usually done with more or less realistic dynamic models. Nevertheless, with age, the muscles tend to atrophy and sometimes after surgery, a muscular atrophy can be observed due to anoxia by intraoperatively prolonged compression

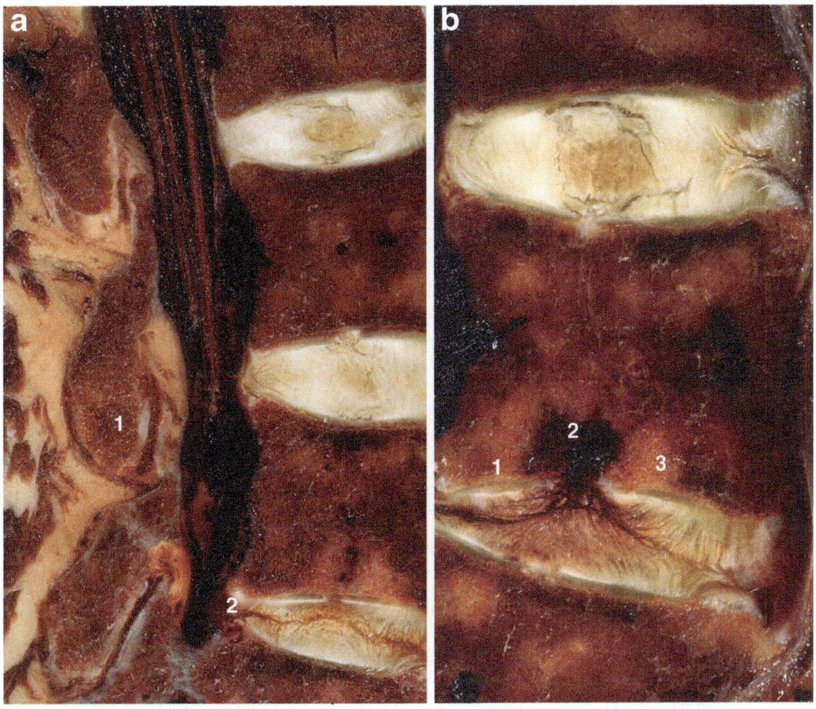

Fig. 2.24 Anatochronesis of the lumbar disc. (**a**) Sagittal section of lumbar spine: 1. Posterior articular process; 2. Complete mechanical fissuration of the disc without discal hernia (dehydration). (**b**) Sagittal section: 1. Degeneration of the cartilage plate with complete mechanical fissuration of the disc; 2. Schmorl hernia; 3. Fat involution of the bone marrow (Modic sign)

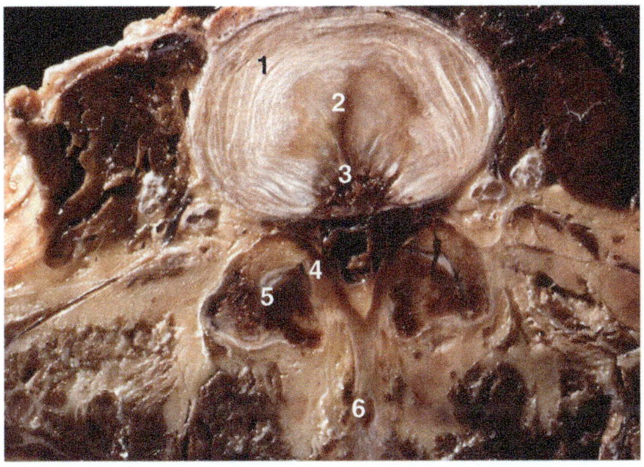

Fig. 2.25 Complete rupture of the posterior part of the disc (maximum compression zone in flexion/extension) in an axial section of lumbar spine. 1. Peripheral part of the disc (annulus fibrosus); 2. Central part of the disc with blood infiltration from the epidural venous plexus; 3. Complete rupture of the disc without hernia (elderly person); 4. Yellow ligament; 5. Facet joint; 6. Spinal process

of spinal muscles. Unfortunately for the surgeon, they are now perfectly visible on MRI. Anyway, the respect of spinal muscles is an absolute necessity as well as their maintenance by appropriate exercises without straining.

2.5.4 Functional Importance of the Abdominal Wall Muscles

These muscles are very much concerned with the spine mobility and also with breathing and emptying pelvic reservoirs in conjunction with the diaphragm. There are three large lateral muscles: external and internal oblique and transverse, plus a vertical median—the rectus. It is needed to well understand their functional organisation centred on three major aspects.

2.5.4.1 A Strong Muscle Corset for Abdominal Visceral Organs Contention

- These three large muscles have a fleshy part fixed on the external side of 5th to 12th ribs, and for the transverse, an additional fixation on costoid processes of lumbar vertebrae and a tendinous part represented by a flat tendon joining from both sides in the median part to create the raphe (white line). This tendinous line makes exchanges of fibres right/left which reinforce its mechanical resistance.

- The rectus muscle is fixed upward on the 5th to 12th rib cartilages and xiphoid process and downward on the pubis. It is a polygastric muscle with three to four tendinous intersections well visible under the abdominal skin of brawny athletes. It is included within a tendinous sheath made by the flat tendons of the large muscles that are separated on its lateral border in two anterior and posterior laminas and placed only on its anterior side in the inferior part reinforcing strongly this part of the anterior wall, more solicitated by gravity. On the back remains only the fascia transversalis that is the aponeurosis of the transverse muscle. This particular shape is clearly visible on the back, the Douglas arcade marking the transition zone. A little pyramidal muscle is placed in the front of the inferior part of the rectus muscle and plays the role of a tensor of tendinous sheath.

2.5.4.2 Rotation and Flexion Muscle Synergies of the Trunk

- The external and internal oblique muscles have an inverse oblique orientation of their fibres which generates a right/left complementary action to rotate the trunk and lumbar spine. In a spine rehab programme, it is recommended to pay attention to the critical position of T12 between the thorax—on which these muscles are fixed—and the lumbar spine. This heterolateral synergy of the two oblique muscles is improved by their tendinous fibres in continuity at the level of the median raphe. In addition a fibrous connective tissue is joining the two muscles, but without avoiding a separate action of both.
- The trunk flexion under resistance like done in lying position activates all the abdominal muscles (fitness basic principle). But the rectus muscle itself is integrated in a long superficial muscle chain going from the head by the SCM and from the mandible by supra- and infrahyoid muscles till the pubis, associated with a deep chain made by psoas muscles with shorter spine lever. Therefore, the combined action of the abdominal and spinal muscles determines the degree of lordosis of the lumbar spine.

Again, it is important to remember that to control this complex group of muscles, we consciously decide to activate movements and not muscles, the functional synergies being programmed within the unconscious part of the execution area of the nervous system.

2.5.4.3 A Wall Crossing for Spermatic Cord and Round Ligament: The Inguinal Canal

It is an anatomical problem because many descriptions exist in literature with some confusion, and it is requiring a precise clarification, which can easily be done, based on the Winckler research [28]. The classical description mentions that the oblique tendinous fibres of the external oblique muscles are winded round a specific ligament going from the anterior superior iliac spine (ASIS) to the pubic tubercle and called "inguinal ligament" or "Poupart's ligament". Then they are reflected on the pectineal crest to give the lacunar ligament described by Gimbernat. In fact, Winckler demonstrated that the inguinal ligament does not exist, nor the Gimbernat one. In reality, the oblique tendinous fibres are parallel to the inguinal furrow and are fixed, from up to down, first on the pectineal crest of the opposite side to create the posterior pillar of the inguinal canal, second on the pubic tubercle of the same side for the medial pillar and third on the pectineal crest of the same side to create the lateral pillar supporting the content of the inguinal canal. In order to maintain together the tendinous fibres, they are arciform fibres and, in addition, the aponeurosis of the external oblique muscle covering it continues with the fascia lata, the aponeurosis of thigh, and that closes the inguinal furrow (Fig. 2.26). Behind this anterior face of the canal are placed the internal oblique and the transverse muscles joining together to be fixed on the pubis by a strong tendon, which is a good anchoring zone to close a hernia.

Therefore, the inguinal canal has two openings: a superficial one made by the two medial and lateral pillars maintained by arciform fibres and a deep one placed at 3–4 cm laterally, which normally prevent an hernia by their opening in case of

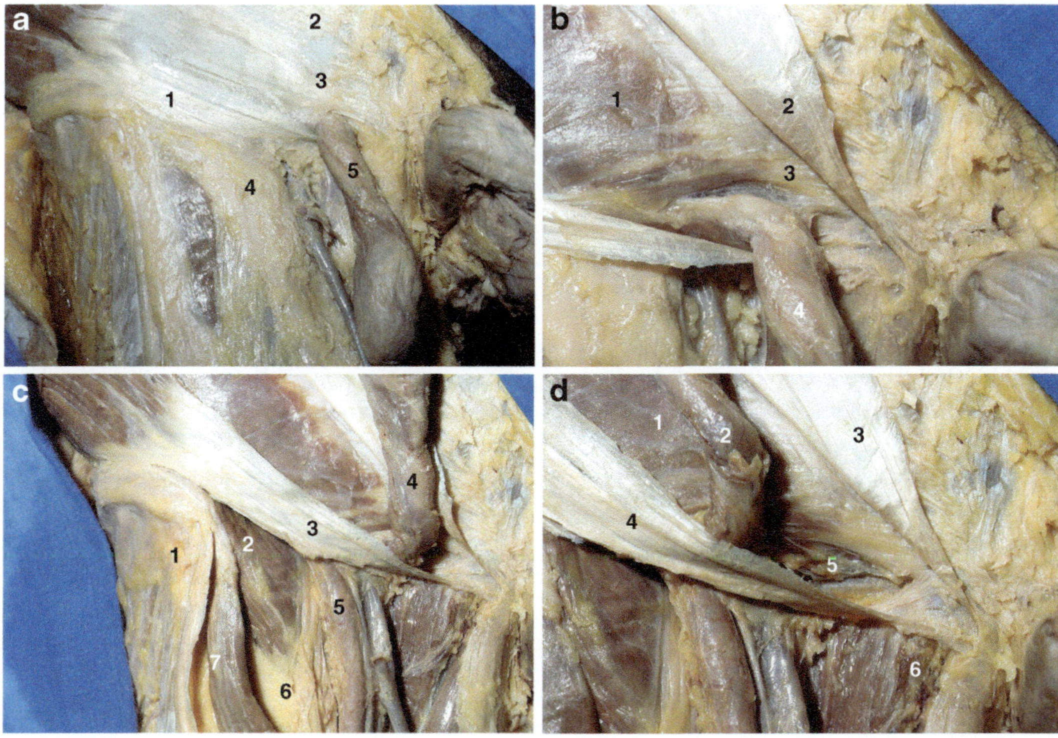

Fig. 2.26 Dissection of inguinal canal. (**a**) Superficial plan: 1. Lateral pillar; 2. Aponeurosis of external oblique muscle; 3. Medial pillar; 4. Aponeurosis covering the iliopsoas muscle; 5. Spermatic cord. (**b**) Section of medial pillar: 1. Oblique internal muscle; 2. Reclined medial pillar; 3. Common tendon of internal oblique and transverse muscles; 4. Spermatic cord. (**c**) Resection of the superior part of the tight aponeurosis: 1. Fascia lata; 2. Iliopsoas muscle; 3. Medial pillar; 4. Spermatic cord; 5. Femoral vein and artery; 6. Femoral nerve; 7. Sartorius muscle. (**d**) Medial part of the inguinal canal: 1. Internal oblique muscle; 2. Spermatic cord pushed up; 3. Reclined medial pillar; 4. Lateral pillar; 5. Iliopectineal band; 6. Pectineus muscle

hyperpressure in the abdominal cavity. This deep orifice is in relation with a fibrous band coming from the Douglas arcade (Hesselbach ligament) and with laterally the epigastric inferior vessels. The important point is also that the inguinal canal is really an intermuscular space compressed by muscles. Its origin is in man in relation with the migration during the growth of the testicle from the abdominal cavity to the scrotum outside, taking with it all the fibrous, serous (vaginalis tunica) and muscular sheaths (cremaster), as well as vessels, and in women with the round ligament fixing the uterus in its lateral angle to the skin of the labium majus. The abnormal lack of migration of the testicle (cryptorchidism) can generate sterility but can be treated surgically if detected early.

Below the inguinal canal there is a space for, laterally, the iliopsoas muscle and the femoral nerve surrounded by the iliopectineal fibrous band and, medially, for the femoral artery and vein and close to the pubis for the pectineus muscle covered by the posterior pillar from the other side. The crural hernia is in fact more frequent in women and the inguinal hernia in men. That can be treated surgically and more frequently now by laparoscopic mini-invasive approach with eventually the use of prosthetic material.

2.5.5 The Rachidian Pains

They are very frequent and represent one of the main reasons for consulting a physician. At the workplace, they are also the main cause for having people take a sick leave. Rachidian pains can take different forms (burn, electric discharge, skin hypersensibility, with or without irradiation, localised or diffuse, . . .), different intensities (high, low with pain scale used by patients), different durations (acute, chronic, day, night, . . .) and different origins. The careful and precise analysis of all these parameters is mandatory for the physician in order to get a precise diagnosis, even if the pain problems are mainly of a subjective nature. Pain has two important aspects to be noted:

- firstly it is an alarm system that is integrated within the protection system of biological structures. The pain signal has priority in the central nervous system and must go as fast as possible to the thalamus and the somatosensory parietal cortex, in order to be identified and localised in the affected body zone. There is no unconscious pain. At the same time, some withdrawal reflexes or escape reactions allow stopping the situation. This system exists in all animals;
- secondly, depending on the cortical organisation of the species and the volumetric importance of the brain, an emotional component is associated which is the suffering part of pain with particular neural pathways integrating amygdala, limbic and prefrontal cortex. There are many individual variations depending on the psychological profile and sometimes the emotional component is predominant, generating deep perturbations in social and professional life.

Therefore, regarding the therapies which are more and more diverse in the specialised pain centres, it is important to analyse the exact origin of rachidian pains in order to make the right choice, mainly between medical and surgical indications.

2.5.5.1 The Radicular Pain

The discal hernia represents the most frequent cause. It is possible to differentiate two different clinical profiles.

The first is a young fellow who, after a violent effort, feels in his back an acute pain blocking his spine. The clinical examination allows localising the level of lesion, which is validated by a scanner or MRI. We know now that the expulsed discal material is progressively absorbed by macrophages, but it can take at least six months to heal. The best solution seems to be to remove surgically the hernia by a mini-invasive approach, as commonly done nowadays.

The second profile is the case of an elderly person who suffers from the back for a long time, with a progressively increasing pain that perturbs seriously his social and professional life. Clinical examination identifies the level of an associated neurological deficit, and imaging investigation shows multiple discal alterations, going from back disc to discal hernia on different vertebral levels of spine. This is the typical aspect of what the anatochronesis is doing with age by shearing discs and articular capsule. The arthrosis also deforms the posterior joints as well as for the hip or knee. The dehydration of the disc by time avoids to observe massive hernia, but more protrusion or bulging of the discs.

In this particular clinical context, it is important to emphasize the exact role of the roots. Therefore, the compression of the roots can create an anoxia of nervous fibres or a lesion, which is expressed clinically by the pain and paresia of muscles if the motor anterior root is also concerned. But in case of chronic pain, they are not the roots which are responsible but the spinal ganglion. It is the first relay of the long sensitive pathways, containing the body neurons particularly those in relation with pain (C and Aδ fibres). The pain signal is transferred to the spinal cord where an inhibition can occur (Gate control of Melzack and Wall) by action on the transmission of the pain signal to the spinothalamic pathway and then to the thalamus and cortex.

The spinal ganglion from foetus to adult has the same position below the vertebral peduncle (Fig. 2.27). It is surrounded by a protective fat tissue, as it is the case for the eyeball within the orbit. The Lasegue manoeuvre by lifting up the lower limb is doing traction on the ganglion, and the flexion of the head makes traction on the roots. When the surgeon is operating on the disc, he tries to see the roots. But in reality he can only see and touch the short dural stump

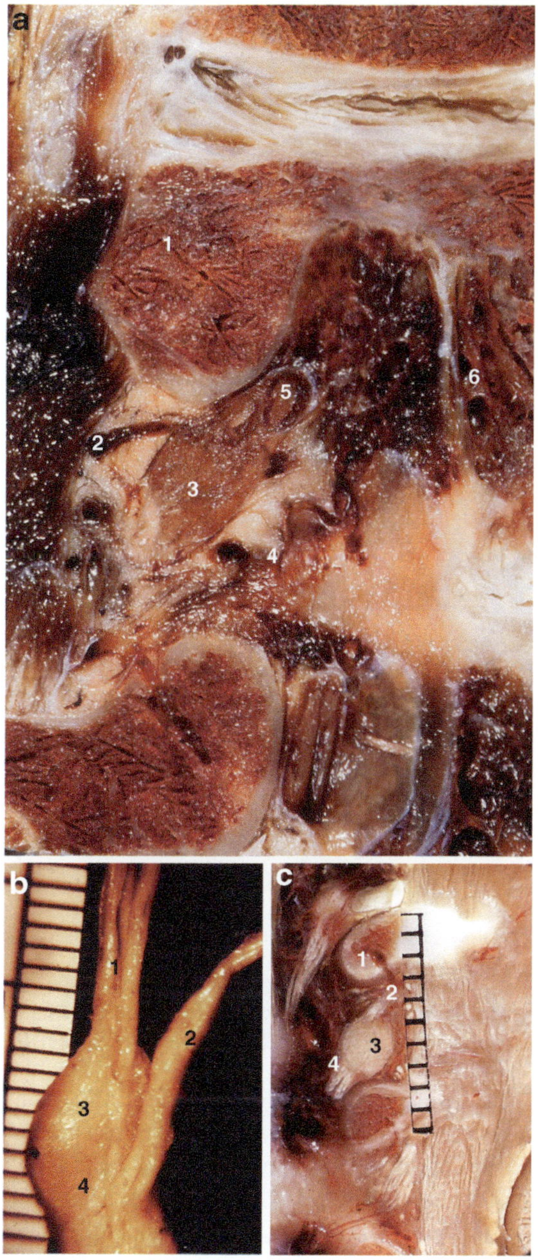

Fig. 2.27 The lumbar spinal ganglion in its infrapedicular space. (**a**) Coronal section of vertebral canal: 1. Vertebral pedicle; 2. Foraminal vein; 3. Spinal ganglion; 4. Foraminal fat; 5. Anterior motor root; 6. Epidural veins. (**b**) Dissection of lumbar spinal ganglion: 1. Posterior roots; 2. Anterior motor root; 3. Spinal ganglion; 4. Spinal nerve. (**c**) Coronal section of a foetus (J. Lang): 1. Vertebral pedicle; 2. Roots; 3. Spinal ganglion; 4. Spinal nerve

covering the roots and making its protection (Fig. 2.28).

In case of recurrence of pain after surgery and particularly in foraminal hernia, it is always important to think about the ganglion (Fig. 2.29). That is also true for cervical hernia with cervicobrachialgia in which we can talk about "millimetric" pathology. In this case, one can explain the benefit of soft traction on spine without the dangerous cracking rotation, yet still used by some practitioners.

2.5.5.2 The Discal Pain

Only the anterior part of the disc has some nervous fibres within the annulus. However they are no sympathetic fibres because there are no vessels inside the central part of the disc. Discitis exists with some particular pain and some possible modifications of the fibrillar organisation with penetration of vessels as reported recently.

2.5.5.3 The Facet Joints

The frequent arthrosis of facet joints generates a specific type of low back pain due to the distension and lesion of the articular capsule and ligaments. To validate their specific implication, the best approach is to infiltrate the joint with an anaesthetic drug.

2.5.5.4 The Muscle Pains

We previously saw the complexity of the spinal muscles which are normally working without pain. But in case of abnormal solicitations or conditions of contraction, lumbago can occur which is a typical myalgia in relation with the C fibres normally placed between the muscular fibres. It is associated with a painful movement locking which can be interpreted as automated protective system. These C fibres are sensitive to biochemical factors like the increasing rate of lactic acid during muscular exercise. They have to be called metaboreceptors. The interesting point is that we can feel our muscles only when they are painful.

Fig. 2.28 The infrapedicular space of the spinal ganglion at the lumbar intervertebral foramen. (**a**) Coronal anatomical section of the foramen: 1. Vertebral pedicle; 2. Roots; 3. Spinal ganglion; 4. Spinal nerve; 5. Foraminal fat pad; 6. Intervertebral disc. (**b**) Dissection of the lumbar dural sac seen from behind: 1. Dural sac; 2. Short dural stump containing the roots (seen by the surgeon during surgical approach); 3. Dural sac covering the spinal ganglion; 4. Intervertebral disc; 5. Spinal nerve within its epineural fibrous sheath; 6. Section of vertebral pedicle

2.5.5.5 The Cutaneous Pains

They are always present in spine diseases, and it is mandatory to try to identify them by precise palpation (pinch and roll) of the skin in different dermatomas as well demonstrated by Jarricot et Maigne [29, 30]. A particular treatment of these hyperalgic zones by different techniques can give excellent clinical results, at least for a short period of time without taking care of the vertebral lesion. An important clinical problem is to be able to make a clear distinction between a cutaneous pain due to spine lesion and one due to visceral pathology. Thinking about colon (left spasmodic colitis), caecum and appendix (grumbling appendix), ureter (nephritic pain), pancreas, gall bladder and some time coronary artery is the mandatory diagnostic reflex of a good practitioner.

At the end of this long chapter about the spine, it is necessary to give to manual medicine an important place within the spine therapies. The vertebral manipulations can be very useful for pain disorders, but have to be done by competent and knowledgeable practitioners in order to avoid prejudicial mistakes to patients. A very rich literature exists from different schools all over the world that is very useful to know and to carefully read before practising [19–22, 29–52].

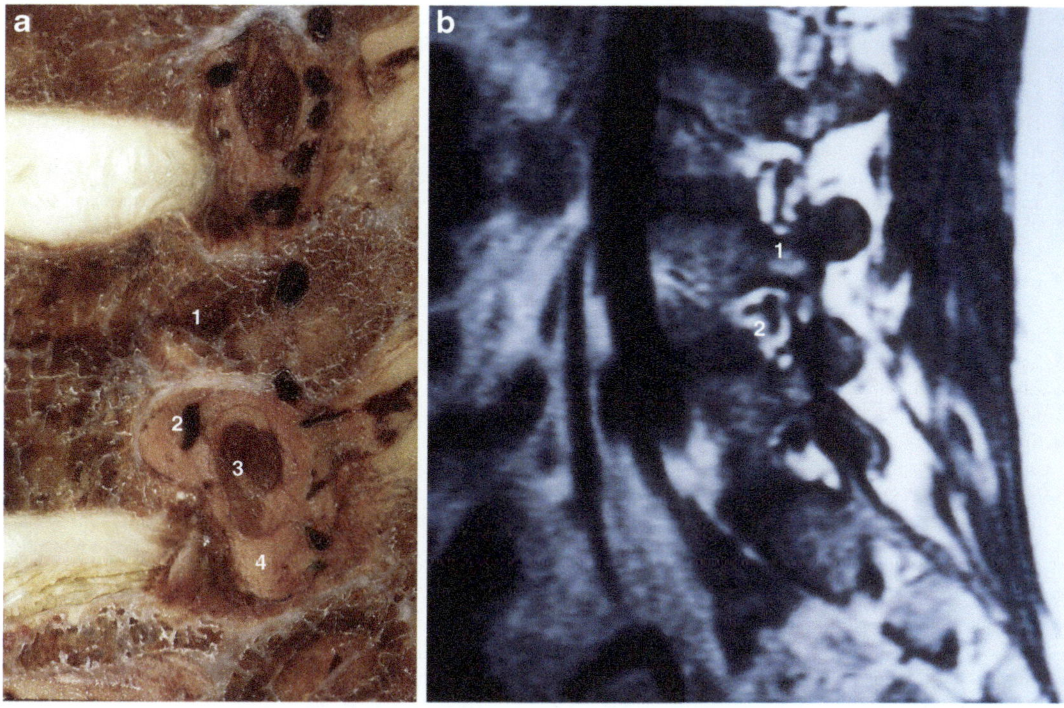

Fig. 2.29 Lumbar intervertebral foramen. (**a**) Sagittal section: 1. Vertebral pedicle; 2. Foraminal vein; 3. Spinal ganglion and roots; 4. Foraminal fat. (**b**) IRM sagittal section: 1. Vertebral pedicle; 2. Foramen with ganglion, fat and foraminal vein

References

1. Paillard J (1976) Tonus, posture et mouvement. Flammarion Paris, Tome III 6:521–7282
2. Massion J, Viallet F (1990) Posture, coordination, mouvement. Rev Neurol 146(10):536–542
3. Epstein BS (1976) The spine, 4th edn. Lea & Febiger, Philadelphia, PA
4. Karnath HO, Ferber S, Dichgans J (2000) The neural representation of postural control in humans. Proc Natl Acad Sci USA 97:13931–13936
5. Dolan P, Adams MA (1993) The relationship between EMG activity and extensor moment generation in the erector spinae muscles during bending and lifting activities. J Biomech 26:513–522
6. Bogduk N (1997) Clinical anatomy of the lumbar spine. Churchill Livingstone, London
7. Bogduk N, McIntosh JE (1984) The applied anatomy of thoracolumbar fascia. Spine 9:164–170
8. McIntosh JE, Bogduk N (1987) The morphology of the lumbar erector spinae. Spine 12:658–668
9. Bonneau D (2001) Les muscles du rachis : de l'anatomie à la modélisation. Thèse d'Université, Montpellier I
10. Rabischong P, Avril J (1965) Rôle biomécanique des poutres composites os-muscles. Rev Chir Orthop 51:437–458
11. Winckler G (1937) La structure du muscle spinalis dorsi chez l'homme. Arch Anat Histol Embryol 23:206–283
12. Bustami FM (1986) A new description of the lumbar erector spinae muscles in man. J Anat 144:81–91
13. Kricun ME (1988) Imaging modalities in spinal disorders. W.B. Saunders Company, Philadelphia, PA
14. Pritzker KPH (1977) Aging and degeneration in the lumbar intervertebral disc. Orthop Clin North Am 8:65
15. Argenson C, Dosch JC, Lemaire V, et al (2000) Imagerie du rachis cervical. Getroa XXVII Sauramps Médical, Montpellier
16. Bruniquel L, Hérisson C (2002) Pathologie mécanique de la jonction cervico-occipitale. Masson, Paris
17. Hérisson C, Vautravers P (2005) Rachis cervical et thérapies manuelles. Sauramps Médical
18. Rabischong P (2005) Anatomie compréhensive du rachis cervical. In: Hérisson C, Vautravers P (eds) Rachis cervical et thérapies manuelles. Sauramps Médical
19. Lederman E (1997) Fundamentals of manual therapy. Churchill Livingstone, New York

20. Beal MC (1982) Teaching of basic principles of osteopathic manipulative techniques. J Am Osteopath Assoc 81:602–609
21. Maigne R (1972) Orthopedic Medicine. Charles C. Thomas, Springfield
22. Maigne R (1989) Diagnostic et traitement des douleurs communes d'origine rachidienne, une nouvelle approche. Expansion scientifique française, Paris
23. Piedallu P (1952) Problèmes sacro-ilaques. L'Homme Sain 2:75–86
24. Lavignolle B, Vital JM, Senegas J et al (1983) An approach to the functional anatomy of the sacro-iliac joint in vivo. Anat Clin 5:169–176
25. Schultz AB, Anderson G, Ortengren R et al (1982) Loads on the lumbar spine. J Bone Joint Surg Am 64A:713–720
26. Styf J, Isyell E (1987) Chronic compartment syndrome in the erector spinae muscles. Spine 12:680–682
27. Nachemson A (1976) The lumbar spine: an orthopaedic challenge. Spine 1:59–71
28. Winckler G (1964) Manuel d'anatomie topographique et fonctionnelle. Masson, Paris
29. Maigne R (2006) Douleur d'origine vertébrale, Comprendre, diagnostiquer et traiter. Elsevier Masson, Paris
30. Soulier JM (2009) Ostéopathie orthopédique. Les techniques non forcées. Amphothérapie. Tome I Le rachis. Sauramps Médical
31. Berlinson G (1989) Précis de Médecine ostéopathique rachidienne, vol 1, 2, 3. Maloine, Paris
32. Le Corre F, Rageot E (1988) Manipulations vertébrales. Masson, Paris
33. Dvorak J, Dvorak V (1990) Manual medicine, diagnostics. Georg Thieme Verlag, Stuttgart
34. Gracovetsky S, Farfan HF (1986) The optimum spine. Spine 11:543–571
35. Greenman PE (1998) Principes de médecine manuelle. Pradel, Paris
36. Huguenin F (1991) Médecine orthopédique, médecine manuelle, diagnostic. Massson, Paris
37. Kaltenbron FM (1983) Thérapie manuelle pour les articulations des membres. Maloine, Paris
38. Still AT (1910) Osteopathy research and practice. ATT Still, Kirksville, MO
39. Lavezzari R (1954) L'Ostéopathie, 2éme ed. Doin, Paris.
40. Cyriax J, Schiötz EH (1975) Manipulation, past and present. Heinemann, London
41. Toole JF, Tucker SH (1960) Influence of head position upon cerebral circulation. Study on blood flow in cadavers. Arch Neurol 2:616–623
42. Niboyet JEH (1968) La pratique de la médecine manuelle. Maisonneuve, Paris
43. Soulier JM (2013) Ostéopathie orthopédique. Les techniques non forcées. Amphothérapie. Tome II Les articulations périphériques. Sauramps Médical
44. Vautravers P, Hérisson C (1994) Les manipulations vertébrales, collection de pathologie locomotrice. Masson, Paris
45. Bugnazet JM (2012) Ostéopathie ORL. DeBoeck Supérieur, Bruxelles
46. Schmorl G, Junghanns H (1971) The human spine in health and disease, 2nd edn. Grune & Stratton, New York
47. Hinshaw DB (1979) Epidural venography for lumbar disc disease. West J Med 130:444–445
48. Iaccarino V, Spaziante R, de Divitis E (1981) Dorsolumbosacral phlebography. Surg Neurol 15:198–203
49. Macnab I, St Louis EL, Grabias SL et al (1976) Selective ascending lumbosacral venography in the assessment of lumbar disc herniation. J Bone Joint Surg Am 58-A:1093–1098
50. Meijenhorst GCH (1977) Methods of transfemoral lumbar epidural venography in the diagnostic of lumbar disc herniation. Radiol Clin 46:439–457
51. Chazal J, Tanguy A, Bourges M et al (1985) Biomechanical properties of spinal ligaments and an histological study of the supra-spinal ligament in traction. J Biomech 18:167–176
52. Ross EC, Parnianpour M, Martin D (1993) The effects of resistance level on muscles coordination patterns and movement profile during trunk extension. Spine 18:1829–1838

3 Locomotion

Abstract

Human locomotion is based on a bipedal mode requiring two functions: the propulsion made by the scissor movement of hip joint increased by the additional force given by the feet and the dynamic automated stabilisation realised through four inputs: vision, skin, vestibule and muscles. Walking requires a high level of control for starting and stopping, speed and trajectory control and a low level of control for managing the gait sequences with an oscillator in the spinal cord for quadrupeds and in the brainstem for primates and humans. Posturology and gait analysis have made substantial progresses and represent at the moment very useful clinical tools to analyse normal or pathological locomotion.

3.1 Introduction

Locomotion is a complex function that must be clearly explained in order to get a good perception of the related technical problems. For human beings, moving means using a bipedal mode of locomotion, but he is not the only one.

The fowl stands also on two legs and although they almost all have the ability to fly, they move on the floor with an alternating sequence of two leg supports. As a special case, we can report the ostrich that walks and runs on two legs with a perfect dynamic balance, but cannot fly. Auk is another example of a bird with a special waddling walking as if it had a double dislocation of the hip. It cannot fly in the air, but perfectly in the water. This diversion is made to reaffirm that being a biped is not, as some still believe, the determinant factor in human emergence.

3.2 The Two Aspects of Locomotion

They are, firstly, the vertical posture preparatory to walking (stato-dynamic aspect) and, secondly, walking and its variants: running, jumping and climbing (dynamic and kinematic aspects).

3.2.1 Upright Posture

It can be done on two feet (bipodal) or on one foot (monopodal). In both cases, it is strongly automated, still owing to the same reason: the technical incompetence of the pilot.

Posture is defined as a mechanical condition of a stabilised body segment. Human being, unlike a lead soldier put on a table that really stands in a static domain, is characterised by an equilibrium

P. Rabischong, *Comprehensive Anatomy of Motor Functions*, DOI 10.1007/978-3-319-04169-8_3,

between forces and moments. Man is a polyarticulated viscoelastic structure that cannot remain in a vertical position, in a real static condition, because he continuously requires more or less important corrections of the oscillations, of the centre of gravity, which all can be measured and recorded. This represents the areas of posturology, posturography and stabilometry, which explore stato-dynamic field to comply with the rigour of mechanics [1–5].

As we have already seen about spinal stabilisation, it is really a postural servomechanism with an *input* and an *output* in the form of muscular contractions operating by impulses with anticipation.

3.2.1.1 Postural Servomechanism Inputs

We can define four inputs: visual, cutaneous, muscular and utricular. The last two have already been described in the spinal postural servomechanism, and it is necessary to focus on the two inputs able to ensure the balance: the vision and the skin through the plantar force plate.

Visual Input

We have already mentioned the power of neural foveal fixation of gaze which enables taking a fixation point in the space and to move around (turn head top or bottom, move the whole body in rotation, ...) by activating a number of muscles while keeping the same visual fixed point in space. However, this function is automatically used to ensure an unconscious equilibration of the body by taking a reference line in space (vertical or horizontal). To support this assertion, we can provide some evidence:

- firstly, if we put in a vertical position a person with a reduced skin sensibility of the foot, like in tabes, or even more easier to find, in diabetes with micro-angiopathies clinically detectable (retina or sole), we observe that the person is well balanced with eyes open, even while turning the head. However, if asked to close his eyes, the subject oscillates and loses his balance. This manoeuvre has been described by Romberg [1] who explains that the vision only cannot ensure balance;
- secondly, the subject feels a certain discomfort accompanied by dizziness when standing in front of a monument occupying the whole visual field and having a general inclination of several degrees (as is the case with the Guadalupe Church in Mexico City). This can be explained by the discordance between the visual input having a false reference line and the cutaneous plantar sole input getting the true reference from contact with the ground. We repeated this experiment in our laboratory by placing some subjects in a room without any indication of linear geometric angles or sides, and some others in a completely dark room with a vertical or horizontal light line inclined until the dizziness appears. Physiologically it is equivalent to a typical optical vertigo.

Similarly, an intersensory discrepancy including vertigo can be observed in a fisherman, focusing all his attention on trying to put a worm bait into his hook, while sitting in a boat moving at the low frequency of the oscillations from the surface of the water. The discrepancy between visual input and labyrinthine vestibular signals reflects a conflict within the brainstem, generating a vertigo with a possible activation of the central vomiting centre [6]. This approximates the seasickness of a tourist lying in his cabin and reading a book, while the boat oscillates in pitch and roll resulting in a dizziness coming with nausea and optionally vomiting. The solution is to go out and find a fixed reference line (horizon) to reset the vestibule.

Skin Input: The Plantar Force Plate

This entry also provides a full control of postural balance. The sole is equipped with a real dynamometric platform, which sensors are the strings of Pacinian corpuscles scattered throughout the plantar support area [7]. These corpuscles are similar to ant eggs with about the same dimensions (4 mm × 2 mm) that can then be dissected under a microscope. They present an ovoid shape formed by an external fibrous capsule with, inside, concentric lamellae in "onion bulb". The neural sole placed in the centre is formed by a nerve fibre spread having left its myelin sheath (Fig. 3.1).

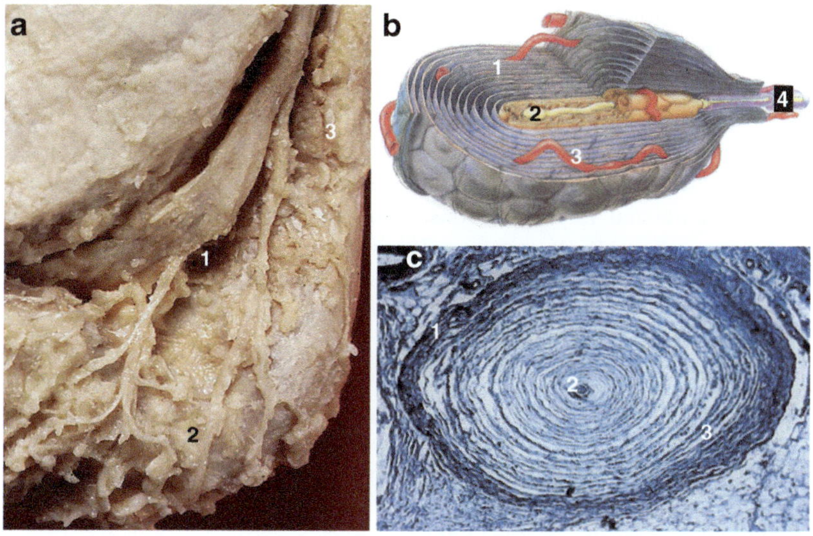

Fig. 3.1 The Pacini mechanoreceptors. (**a**) Microdissection of the Pacini corpuscles (2–4 mm): 1. Sensitive plantar nerve; 2. Corpuscles; 3. Triceps tendon. (**b**) Graphic reconstruction of a corpuscle (after Tood and Bowman): 1. Concentric lamellas; 2. Neural plate; 3. Arterial microvessels (responsible for renitency and sensibility of the transducer); 4. Myelinated sensitive fibre losing myelin after penetration within the corpuscle. (**c**) Histological section of a corpuscle: 1. Fibrous capsule; 2. Neural plate; 3. Concentric onion shape lamellas

The amazing fact is that it exists within the corpuscle fine capillaries with a liquid exudate filling more or less the sensor and creating thus a variable sensitivity. The more it is renitent, the more it is sensitive. It may therefore partly explain the impaired balance observed in diabetic patients with a microangiopathy affecting these fine capillaries and greatly reducing the sensitivity of the sensor [3]. These corpuscles are also present in the palm of the hand to ensure automatic control of grasping forces, as we shall see later.

It is clear that a proper posturologic investigation must take into account these facts altogether. It is not a problem to place a subject on a force plate for less than a minute, to record a posturogramme, analyse it and possibly make the Fourier transformed of collected data, but the conclusions are then very often extrapolated. It is rather necessary to study the subject eyes open and closed, in bipodal and monopodal stations, and what is even better, with a complete study of walking by measurement of centres of pressure, using if possible a treadmill in order to ensure exploring the automatic walking phase as well [8].

3.2.1.2 The Balance Corrections by Muscular Impulses

By placing a subject on a dynamometric platform using strain gauges and having EMG skin electrodes put on the anterior and posterior muscles of the leg, as was done in our Unit 103 of INSERM, it is possible to record the equilibration muscular reactions automatically triggered by imbalance measured by the foot platform. The muscle reaction appears with a certain threshold and reaches a maximum at the maximum of anterior and posterior unbalance. The acceptable limit of equilibration is possible up to about 7° of inclination. Over this limit, the gravity line no longer falls in the area of plantar lift and equilibration is not possible. It is mainly the leg muscles that are involved: tibialis anterior and toes extensor for posterior imbalance and tibialis posterior, soleus and flexor digitorum longus for anterior imbalance.

In the light of these experiences, it seems that the term "tonic-postural system" to describe these mechanisms of automatic balance corrections is not correct. The definition of tone has always been a subject of controversy, and it appears that it characterises the viscoelastic state of a muscle at rest, capable of exerting a minor elastic tension, very different from the active contraction of the muscle. Establishing a difference between tonic and phasic is not very clear with respect to the muscle reactions recorded on a platform, which are real contractions unrelated to muscle tone. This is why we prefer to speak about postural servomechanism.

3.2.2 Walking

Walking is the semi-automatic management of maintained imbalances. This definition deliberately focuses on the automatic sequence of operation, each phase being unstable and requiring a permanent stabilisation in a dynamic mode. This acquisition is pre-programmed in the sense that a normal child at a certain age, even abandoned, will stand up and walk after a few failed attempts resulting in a minor fall on the buttocks due to the mass distribution at this age, even if it is much more enjoyable for parents to believe that they had him make his first steps. After 1 year, the child walk better and it is clear that he has no knowledge in physiology which validates the automated character of walking. The early registration in the nervous system of this function explains also that it can be used up till the extreme stage of life, even with highly degraded systems (muscular atrophy, diverse arthrosis, use of crutches, . . .).

It is also interesting to observe that a natural ability exists to find the right functional compensation as may be seen in polio children, different lameness (hip, knee, foot) or amputation. However, it is clear that the movement-trained therapists (doctors and physiotherapists) are able to exploit the most of this natural ability.

There are two functions in walking: propulsion (put one foot before the other and again) and stabilisation, using control loops involving interacting complex systems.

3.2.2.1 Control Levels of Walking

This is one of the technical aspects we need to understand well. Specialists in automation control speak about high level and low level, which applies perfectly to walking.

High Level Control

It is what the "pilot" uses to manage consciously, regardless the automation of the sequence running, the progress including the numerous technical problems arising.

- Starting and Stopping
  It is a decision that must be made in the premotor area, mobilising also the subcortical structures containing the motor library. Thus, subjects with Parkinson have a clear hesitation; they usually start with a series of short and quick steps. Stopping is also a technical complex procedure since we slow down and have to end up in a stable position. We studied these processes in the INSERM Unit and observed that the brutal stop on an obstacle requires the involvement of many muscles and crossing the obstacle shows the important role of visual guidance. Therefore vision records unconsciously topographical parameters, allowing the movement of the foot, which is not seen by the subject, at 1 cm from the edge of the obstacle in the case of a box 60 cm high. This is the same problem when walking on a rough terrain, where the foot placement must avoid obstacles (emerging stone, depression, . . .), which is done without too much effort thanks to the anticipation power of visual guidance. The blind replaces it with a guide touch, using a stick more or less sophisticated. In this regard, advanced automation walking is limited in a sequence of obstacles like the steps of a staircase because, as we have seen, having in the sequence a step 2 cm higher (not visible by the subject) leads necessarily to stumbling or falling.
- Speed Control
  It is a semi-automatic control in the sense that one can voluntarily change its walking speed. Two parameters are changed: firstly, cadence (number of steps) and secondly, the length of the step.

  It should be remembered that a step is defined as being the same distance between two supports of the same foot (heel contact of the right and left foot). Some schools talk about a half-step to designate the distance between left and right support. This value can be asymmetric in some lameness. Trained walkers generally adopt a "comfort" rate level in flat ground, but it is interesting to note that there is an automatic regulation of the link cadence/step length, which is manifested in the rise of a slope usually with a decrease of both parameters or of one of them.

The military, normally prepared to obey without discussion, know how to walk in step, with or without music, but in this case, the perfect rhythmic answer demonstrates the power of the rhythm generation imposed by the hearing function. This is also the case for all forms of dance, from classical with strict traditional rules to modern with unbridled rhythmic fluttering and lonely on night clubs (like in a zoo).

- Trajectory Control

  Apparently simple, it is in fact quite complicated. We have seen that the quadruped performs its track changes by lateral bending of the spine (justifying the role of the epispinatus muscle). This also explains why their humerus and femur have no neck. Human beings in bipedal walking are changing direction with the hip, fully justifying the femoral neck lever operated by many muscle rotators. In reality, the changes are not always of large amplitude, but turning on the spot, which is apparently simple, requires the use of a differential length of left/right steps or vice versa, depending on the direction of rotation. We studied this problem when we produced, with the assistance of the Automation Division of Renault car factory, a "walking machine" operating in master-slave system for paraplegic patients. It was using a pneumatic splint with four hydraulic motors (hip/knee activation) driven by a physiotherapist coated with the same brace and with four potentiometers on hips and knees. In this model, the therapist allowed the patient to walk through appropriate control software. This allowed us to make two interesting observations:
  - firstly, the patient who had a complete paraplegia with no sensitivity in the lower limbs tended to look at his feet in order to see what was happening to the lower limbs. This represents a quite normal form of compensation for the lack of goniometric perception of movement. But it was perturbing the execution of the gait sequence. This was a valuable lesson for our research programme of restoration of locomotion in paralysed patients by implanted electrostimulation. In fact, the patient has to have full confidence in technology not to look at what is happening in their insensitive body area;
  - secondly, the artificial use of a right/left differential step length to change direction is too complicated and unreliable. So the best solution is to make a usable rotation on one foot by turning the trunk.

Low Level Control

It lies at an unconscious level located mainly in the spinal cord in which lies a complex functional architecture due to several factors:

- firstly, during embryogenesis, the segmentation of the neural tube in metamers (from Greek *meron*, the segment) gives rise to a series of spinal nerves. They are formed, because of the separation of sensitive inputs and motor outputs, with the junction of a sensory root with spinal ganglion (the first relay of the sensory pathway) and a motor root containing motor fibres coming from the anterior horn of the spinal cord (alpha and gamma motor neurons);
- secondly, the need to link the various segments by a functional interconnection system fully justifies the interneurons. This can be explained by the fact that most muscles have a multi-root innervation requiring to activate at the same time several metameric levels. The spinal cord has also an important reflex activity not only monosegmental but also plurisegmental like the flexor withdrawal

reflex of the lower limbs after painful stimulation of the skin which also exists in paraplegic patient or like alternated flexion/extension reflex;

- thirdly, this raises the issue of a possible gait pattern spinal generator, ensuring the automatic sequences of gait. This is still subject to controversy, and to understand, it is important to know the specific difference in the neurophysiology of an animal opposed to a man.

In animal, knowing nothing in neurophysiology, it is important to maintain a high level control as we defined before in order to enable them to voluntarily choose speed and trajectory, and still without being capable of complex voluntary conscious actions with limbs. That may explain why the corticospinal tract connecting the cortical motor keyboard to spinal cord stops at the cervical level in ungulates and herbivores, and at the level of lamina V (in the terminology of Rexed) of the sensory dorsal horn in carnivores and rats.

It happens only in humans and primates that the corticospinal tract is connected to each metameric level of the motor anterior horn of the spinal cord. This indisputable biological fact explains the predominance in the animal of a spinal cord automation, able to generate the sequence of forelimbs and hindlimbs in several possible modes used in quadrupedal locomotion, very useful to catch mobile provender. But for a quadruped walking on two hind legs, it is not a natural phenomenon and it requires a careful and very prolonged training by a human trainer.

In humans, the direct connection to each metameric level of motor control centre explains the wide variety of actions of upper limbs and lower limbs, although the relatively small size of its cortical representation implies less variability of cortical control.

Therefore, it is clear that in the human central nervous system it necessarily exists a walking oscillator delivering rhythmic sequences. Several experimental arguments allow to find out its location:

- firstly, a beheaded duck, frequently experienced when living in the countryside, can do a few steps before collapsing completely;
- secondly, in the hands of a good physiologist, a decerebrate cat can still walk on a treadmill;
- thirdly, thanks to the French doctor Guillotin, inventor during the French Revolution of a cleaner and more efficient instrument than the axe, we have in France a great experience for the beheading of humans. A headless man standing up and making a few steps has never been observed (except in some exceptional miracle). Nowadays, public beheading is still practised in some countries. A man carrying a sword standing behind a condemned man executes the task, but nobody reported the victim running afterward;
- fourthly, the newborn automatic gait described by André Thomas (1952) can be observed by putting a newborn immediately after delivery on his feet. The alternating movements of flexion of the lower limbs observed are in reality not comparable to real walking, but are, as the phenomenon disappears at about the age of 2 months, spinal flexor withdrawal reflexes in connection with an incomplete myelination of the major pathways of the nervous system.

This allows us to understand that animals have their locomotion automation centre within the spinal cord, but man has his walking oscillator into a supraspinal level in the brainstem. It is therefore unrealistic to expect to stimulate the spinal cord of a paraplegic patient to restore locomotion. In addition, the vision, so often shown, of a spinal transected rat, and even with the inclusion of various neuronal components, walking on a treadmill is not particularly convincing except for researchers who continue to work in these quadruped animals without showing positive results in humans.

The newborn automatic gait described by André Thomas (1952) can be observed by putting a newborn immediately after delivery on his feet. The alternating movements of flexion of the lower limbs observed are in reality not comparable to real walking, but are, as the phenomenon disappears at the age of 2 months, spinal flexor withdrawal reflexes in connection with an incomplete myelination of the major pathways of the nervous system.

3.2.2.2 Biomechanics of Walking

Without going into tedious technical details, it is important to know the various phases of walking and the role of each motor unit: hip (cyclic motor), knee (length control) and foot (stabiliser and additional propellant).

Gait Phases

There is a certain consensus in all gait analysis laboratories to define the different phases of a step as follows:

- starting out: bipodal support;
- pendular (or swing) phase on the starting side (R or L) with opposite monopodal support;
- bipodal stance phase;
- pendular (or swing) phase of opposite side with leg stance;
- bipodal stance phase.

Action of the Three Motor Assemblies

The Hip

This is a 3-DF joint and it is the main motor of gait, producing alternative active scissor leg movements. This movement is due to the alternate action of the hip flexors (iliopsoas and especially iliac and rectus femoris) and extensors (hamstrings and gluteus maximus on uneven ground) (Figs. 3.2 and 3.3).

A bilateral thigh amputee can still walk with an orthosis but in a very tiring manner due to a significant reduction in muscle mass.

In single leg stance, the hip can guide the path through the lever of the femoral neck. If fixed on the ground, the hip is responsible for the important function of the lateral stabilisation of the pelvis, avoiding the lateral bending of the trunk. The group responsible for that is formed by the gluteus medius, gluteus minimus and tensor of fascia lata. The paralysis of these muscles either requires the use of crutches to compensate or produces a walk with the lateral sway of the trunk during the swing phase, as can be seen in the gait of congenital hip dislocations. Analytical testing (foot without ground contact) of these three muscles of the lower limb shows internal rotation and abduction movements, which create, during the monopodal stance phase, a forward rotation of the pelvis, which is called pelvic step (Figs. 3.4 and 3.5). This movement may be increased by some women for reasons that are not related to biomechanics.

The launch of the lower limb in the forward swing phase should be done if possible in the axis of the path. That is the problem of the iliopsoas (the pitcher's leg) being still discussed regarding its internal or external rotator action. In reality, it has a bistable position and a low strength given by the piriformis muscle and possibly the internal obturator in external rotation directs, despite the rotation done by the pelvic step, the swing foot in path axis or with a slight external rotation (Fig. 3.6). An internal rotation of the foot touching the ground is frequently seen in Japanese population in relation with a particular abnormal femoral neck.

The Knee

It is a joint that allows controlling the length of the lower limb [9–11]:

- from flexion giving the maximum shortening (active hamstrings and twins from 140° to 160° in relation to the extension of the hip, and passive ones up to 160°);
- to extension by the quadriceps.

This allows all kinds of sitting postures, sometimes with extreme postures that our Japanese friends readily use. Moreover, it is the knee which gives the elegance of walking as it is well known by puppeteers.

The anatomical organisation of the knee responds to mechanical requirements that must be understood by comparing the elbow and the knee. These are intermediate articulations between the distal limb end (hand/foot) and the proximal joint (shoulder/hip). This situation imposes a special mechanical torque for the hand with a fixed grasping or tool handling and for the foot to be fixed to the ground with trunk rotation.

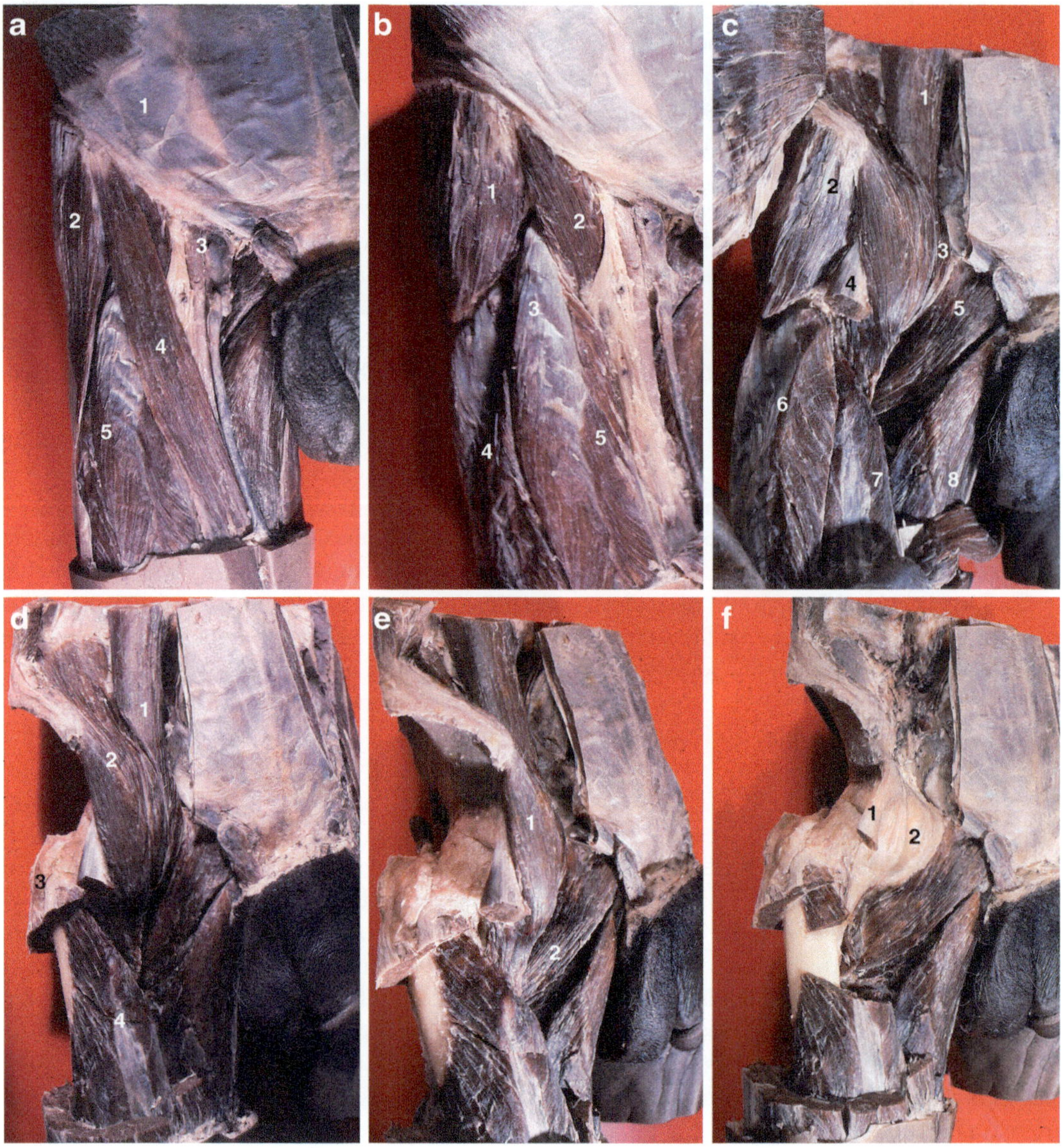

Fig. 3.2 Dissection of the anterior muscles of the hip joint. (**a**) Superficial aspect: 1. Abdominal muscular wall; 2. Tensor fascia lata muscle; 3. Vessels and nerve (femoral artery and vein, femoral nerve, internal saphenous vein); 4. Sartorius muscle; 5. Rectus femoris muscle (polyarticular part of the quadriceps). (**b**) Resection of tensor and sartorius: 1. Gluteus medius muscle; 2. Iliac muscle; 3. Rectus femoris muscle; 4. Vastus lateralis muscle; 5. Vastus medialis muscle. (**c**) Section of the right abdominal wall: 1. Psoas muscle; 2. Gluteus medius muscle; 3. Iliac muscle (monoarticular part of the iliopsoas muscle); 4. Tendon of the rectus femoris; 5. Pectineus muscle; 6. Vastus lateralis muscle; 7. Vastus medialis muscle; 8. Adductor longus muscle. (**d**) Resection of gluteus medius and vastus lateralis: 1. Psoas muscle; 2. Iliac muscle; 3. Superior part of the vastus lateralis muscle; 4. Vastus medialis muscle. (**e**) Resection of iliac muscle: 1. Psoas muscle; 2. Pectineus muscle. (**f**) Resection of the psoas: 1. Tendon of the rectus femoris muscle fixed on the anterior and inferior iliac spine; 2. Articular capsule reinforced by the anterior ilio-femoral ligament (limitation of tight extension)

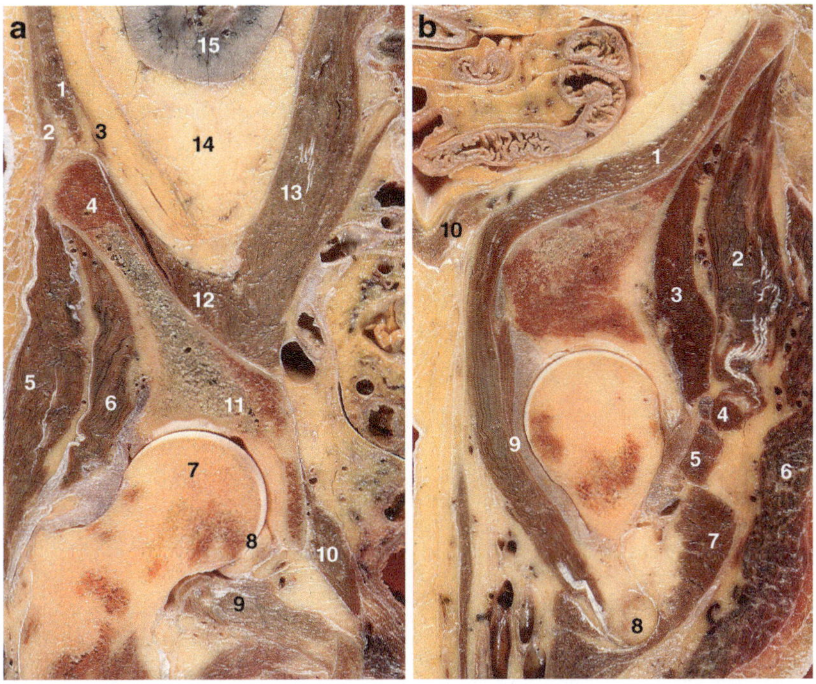

Fig. 3.3 Anatomical sections of the hip joint. (**a**) Coronal section: 1. Obliquus internus abdominis muscle; 2. Gluteus medius muscle; 3. Gluteus minimus muscle; 4. Piriformis muscle; 5. Gemelli muscles; 6. Gluteus maximus; 7. Quadratus femoris muscle; 8. Transversum acetabuli ligament; 9. Obturatorius externus muscle; 10. Levator ani muscle; 11. Acetabulum; 12. Iliac muscle; 13. Psoas muscle; 14. Corpus adiposum pararenalis; 15. Right kidney. (**b**) Sagittal section: 1. Iliac muscle; 2. Gluteus minimus muscle; 3. Gluteus medius muscle; 4. Gluteus maximus muscle; 5. Piriformis muscle; 6. Gemelli muscles; 7. Quadratus femoris muscle; 8. Small trochanter; 9. Articular capsule reinforced by the ilio-femoral ligament; 10. Obliquus externus abdominis muscle

It is interesting to analyse the smart technical response to this common problem.

For the Upper Limb

At this level, where it is unquestionably the mobility of the hand in space that is predominant, it exists the special joint system described before of the two forearm bones: a double inverted trochoid joint allowing axial rotation of the hand in pronation to grasp an object and in supination to support it. This brings the solution to absorb torsional stresses and allows the elbow joint to use a trochlea-type joint, a single DF type for the flexion/extension of the elbow.

For the Lower Limb

The loss of prehensile capacity of the foot does not justify an axial rotation of the foot, but rather a search for mechanical stability in support and movements in a two-separated DF system placed within a cardan joint. The torsion problem mentioned above must therefore find its solution in the knee itself. This will be done by a few technical tricks.

- The joint is large.
- Flexion/extension is given by two condylar associated wheels operating by two long articular tread covered with a thick cartilage. The reason for having two wheels is the difference between a car and a bicycle. The latter has 2 DF allowing it to roll but also to tilt sideways, which a four-wheel system has not. This solves efficiently the problem of the lateral stability (Fig. 3.7).
- Given the torsion constraints mentioned above, an articular synthesis is made by the original anti-torsion system formed by oblique ligaments roughly oriented at 45° that correspond to the maximum torsion stresses at 45°. These oblique ligaments are:

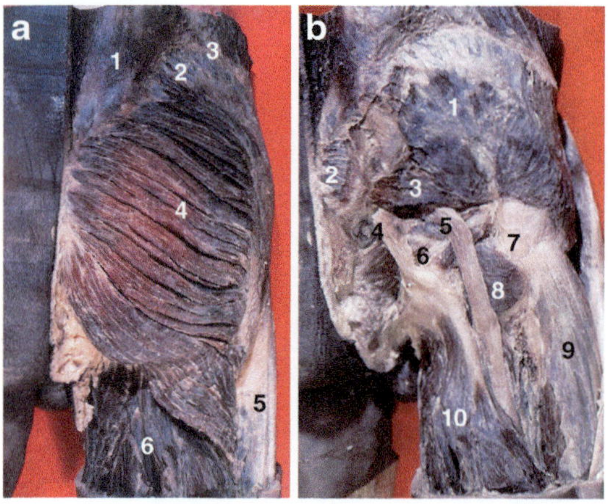

Fig. 3.4 Posterior aspect of the hip. (**a**) Superficial plan: 1. Spinal muscles; 2. Tendinous sheath of the gluteus medius muscle; 3. Iliac crest; 4. Gluteus maximus muscle; 5. Fascia lata; 6. Hamstrings muscles. (**b**) Posterior aspect after resection of the gluteus maximus muscle: 1. Gluteus medius muscle; 2. Section of the gluteus maximus; 3. Piriformis muscle; 4. Sacro-tuberal ligament; 5. Ischiatic nerve; 6. Tendon of the obturatorius internus muscle with the two superior and inferior gemelli; 7. Posterior aspect of great trochanter; 8. Quadratus femoris muscle; 9. Vastus lateralis of the quadriceps muscle; 10. Hamstrings muscles

  - the medial collateral ligament, the most resistant one due to the tendency of medial articular space to open due to the elastic deformation in compression of the curved femur;
  - the lateral extra-capsular ligament;
  - the anterior and posterior cruciate ligaments, placed in an extra-synovial compartment and securing forward and backward the movements.
- In the posture and gait stance phases, it is absolutely required to have a stable locked knee in extension. But in a knee semi-flexion, it exists the possibility of an axial rotation controlled by the muscles. To achieve this, there are two mechanical adjustments: a flattening of superior tibial articular surfaces and a cam effect. A cam is an eccentric mechanical system with two radii of curvature commonly used in the engine camshaft and transforming a rotation into a translation movement. This is, for example, the case with condylar wheels: their curvature has a spiral shape increasing from the back to front, from 17 to 40 mm for the medial condyle and from 12 to 60 mm for the lateral one. This also explains a lateral automatic rotation at the end of a knee extension (Fig. 3.8).
- In extension, the cam opposes to the tibial tray its large radius of curvature that generates a tightening of ligaments and, consequently, a locking of the joint. But in a flexion, it is the small radius that is opposed to the tibial tray and only that way can the ligaments be released. A foot placed on a stool and tight in flexion allows to measure the amplitude of this axial rotation which is comprised between 30° and 40°. Therefore, the semi-flexion position is stabilised, and mainly by muscles. It may become dangerous in particular in people practising ski wherein rotation with a high torque given by the skis fixed to the end of the foot can create ligament injuries. This is due to insufficient attention and often maladjustment of the safety bindings of the boots with the ski.
- Due to this axial rotation, the receiving tibial glenoid cavities cannot be congruent and are slightly concave or convex on lateral side. But a viscoelastic congruence is given by two menisci: O-shaped for the medial and C-shaped for the lateral (Fig. 3.9). They are partially attached at their periphery to the joint capsule. Their anterior and posterior horns are fixed to the two bone spurs located in an extra-synovial space. A median jugal or transverse ligament is linking them forward. In this way, these menisci accompany condylar displacements. However, undergoing significant compression and shearing forces can explain their frequent avulsion, sometimes even putting them like a wedge under a wheel by blocking the condylar movement. Fortunately, knee surgical experts can repair through endoscopy much of the damage [12].
- Finally, the powerful strength of the quadriceps and the need to guarantee its perfect traction in the axis of the leg regardless of the position of the thigh draw a clear advantage of the sesamoid patella (Fig. 3.10). It slides into the groove of the femoral trochlea

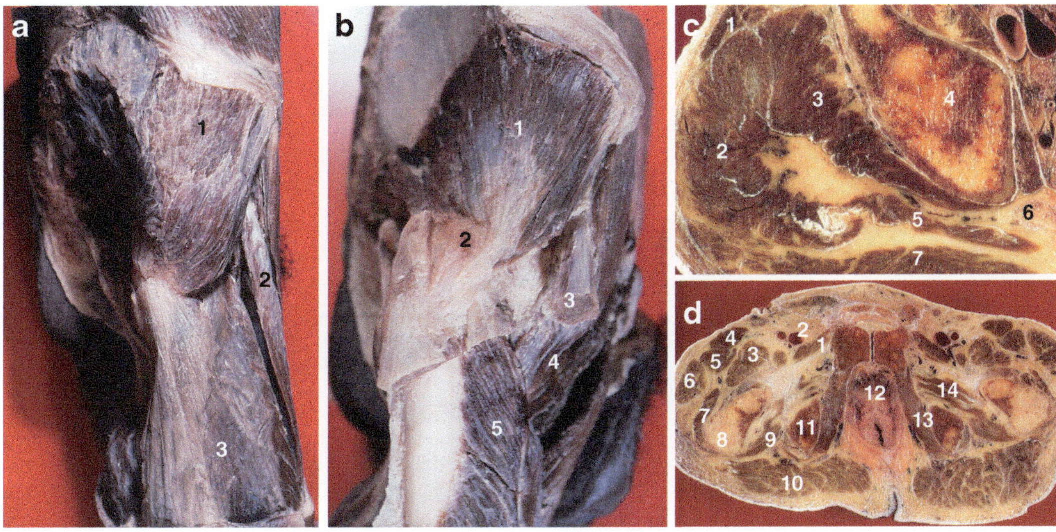

Fig. 3.5 Gluteus medius and minimus. (**a**) Lateral view of the hip after resection of gluteus maximus: 1. Anterior sagittal part of the gluteus medius; 2. Rectus femoris muscle; 3. Vastus lateralis muscle. (**b**) Lateral view after resection of gluteus medius: 1. Gluteus minimus with its tendon; 2. Great trochanter; 3. Resected rectus femoris; 4. Adductor brevis muscle; 5. Vastus intermedius muscle. (**c**) Axial section above left hip joint: 1. Tensor fascia lata muscle; 2. Gluteus medius muscle with its anterior sagittal part and posterior transversal part; 3. Gluteus minimus muscle; 4. Acetabulum; 5. Piriformis muscle; 6. Ischiatic nerve; 7. Gluteus maximus muscle. (**d**) Axial section below hip joint at the level of inferior part of pubic symphysis: 1. Pectineus muscle; 2. Femoral artery and vein; 3. Ilio-psoas muscle; 4. Sartorius muscle; 5. Rectus femoris muscle; 6. Tensor fascia lata muscle; 7. Gluteus minimus muscle; 8. Great trochanter; 9. Quadratus femoris muscle; 10. Gluteus maximus muscle; 11. Ischion; 12. Prostate gland; 13. Obturatorius internus muscle; 14. Obturatorius externus muscle

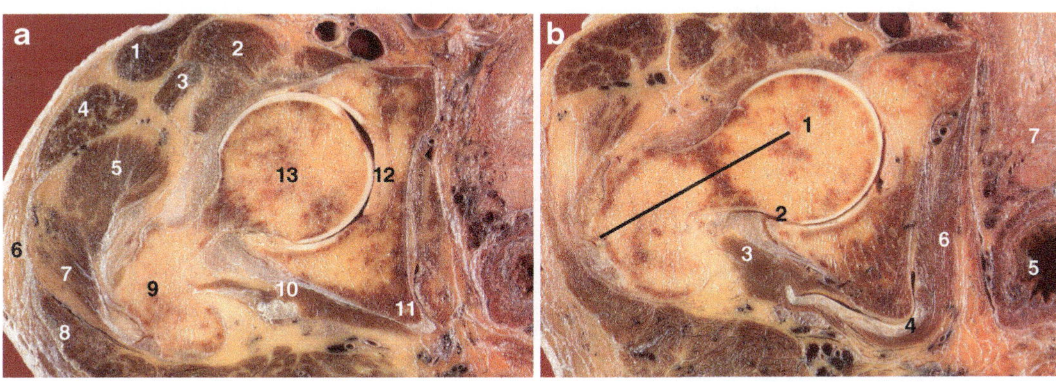

Fig. 3.6 Axial sections of left hip joint showing neck angulation and uncovered anterior part of the femoral head. (**a**) Superior section: 1. Sartorius muscle; 2. Iliopsoas muscle protecting femoral head; 3. Rectus femoris tendon and muscle; 4. Tensor fascia lata muscle; 5. Gluteus minimus muscle; 6. Fascia lata; 7. Gluteus medius muscle; 8. Gluteus maximus muscle; 9. Great trochanter; 10. Tendon of the obturatorius internus muscle and superior gemellus; 11. Sciatic spine; 12. Acetabulum; 13. Femoral head. (**b**) Inferior section: 1. Oblique axis of the femoral neck with large uncovered anterior part of the femoral head; 2. Acetabular labrum; 3. Superior gemellus muscle; 4. Incisura ischiatica minor; 5. Inferior part of rectum; 6. Obturatorius internus muscle with gliding connective bursa; 7. Bladder

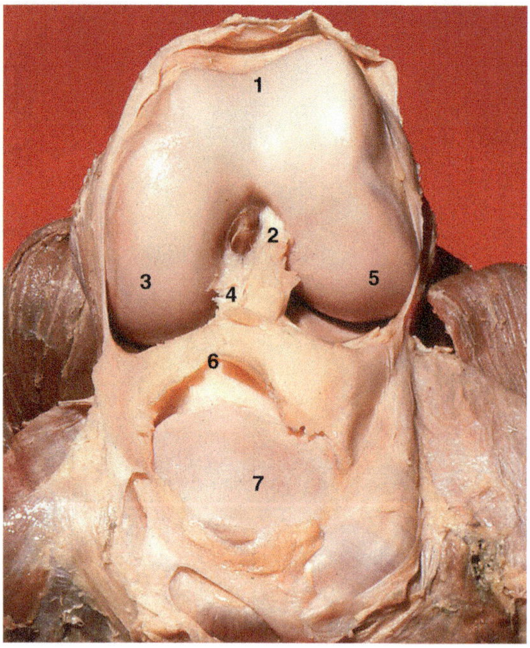

Fig. 3.7 Opening of the left knee joint in flexion.1. Trochlea with asymmetric patellar surfaces; 2. Anterior cruciate ligament; 3. Medial condyle; 4. Posterior cruciate ligament; 5. Lateral condyle; 6. Infrapatellar synovial fold; 7. Patella (sesamoid of quadriceps)

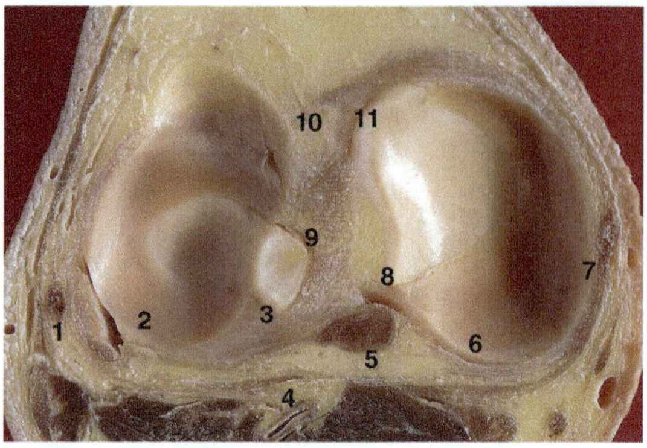

Fig. 3.9 Superior view of meniscus of left knee. 1. Fibular collateral ligament; 2. Lateral meniscus; 3. Lateral intercondylar tubercle; 4. Articular capsule; 5. Posterior cruciate ligament; 6. Medial meniscus; 7. Tibial collateral ligament; 8. Medial intercondylar tubercle; 9. Anterior cruciate ligament; 9. Infrapatellar corpus adiposum; 11. Transverse ligament of the knee

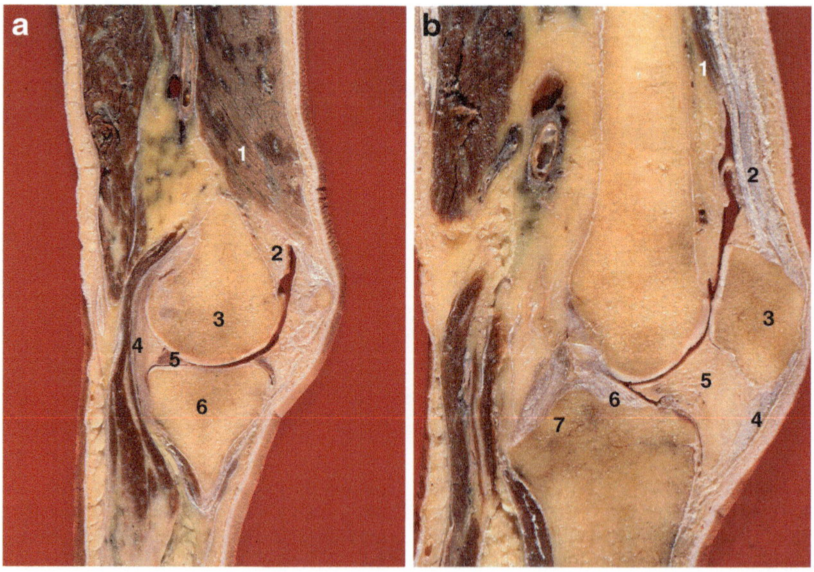

Fig. 3.8 Sagittal sections of the knee joint. (**a**) Section at the level of medial condyle: 1. Vastus medialis of the quadriceps muscle; 2. Superior recessus of articular capsule; 3. Medial condyle; 4. Reinforced posterior capsule (passive extension limitation) with medial gastrocnemius muscle behind; 5. Medial meniscus; 6. Superior part of tibia. (**b**) Sagittal section through the intercondylar space: 1. Knee articular muscle (tensor of the anterior articular capsule); 2. Quadriceps muscle tendon; 3. Patella; 4. Patellar ligament; 5. Infra-patellar corpus adiposum; 6. Anterior cruciate ligament; 7. Intercondylar medial spine

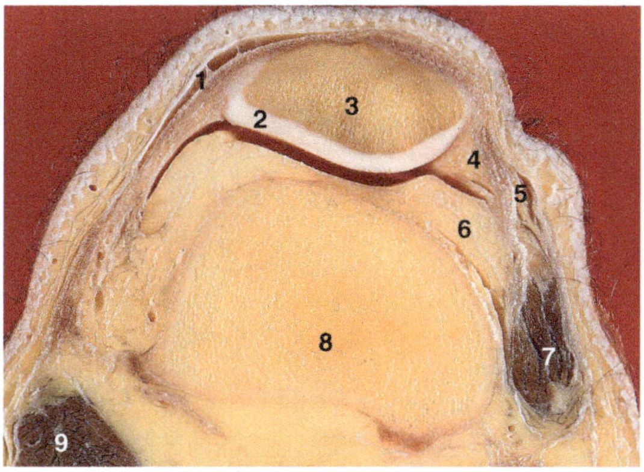

Fig. 3.10 Transverse section of the patella of the left knee joint. 1. Lateral patellar retinaculum; 2. Lateral articular surface of the patella; 3. Patella; 4. Alar medial plica; 5. Medial patellar retinaculum; 6. Corpus adiposum; 7. Vastus medialis of the quadriceps; 8. Inferior part of femur; 9. Vastus lateralis of the quadriceps

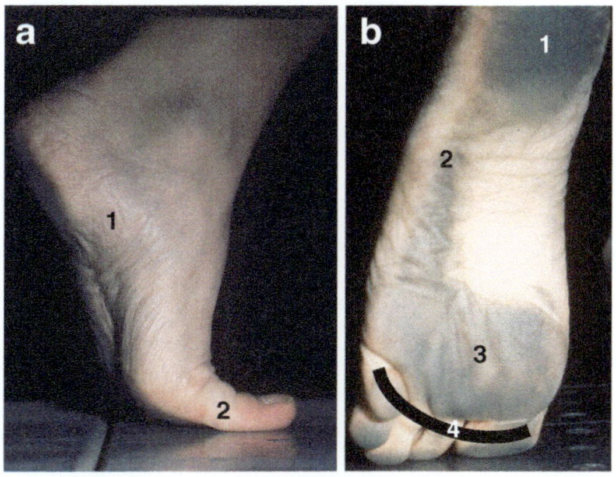

Fig. 3.12 Gait propulsion phase. (**a**) Medial view of the foot: 1. Medial raised longitudinal arch; 2. Great toe ground support. (**b**) Plantar foot seen from behind: 1. Heel ground contact zone; 2. Lateral plantar roll; 3. Anterior metatarsal ground support; 4. Metatarso-phalangeal curved hinge

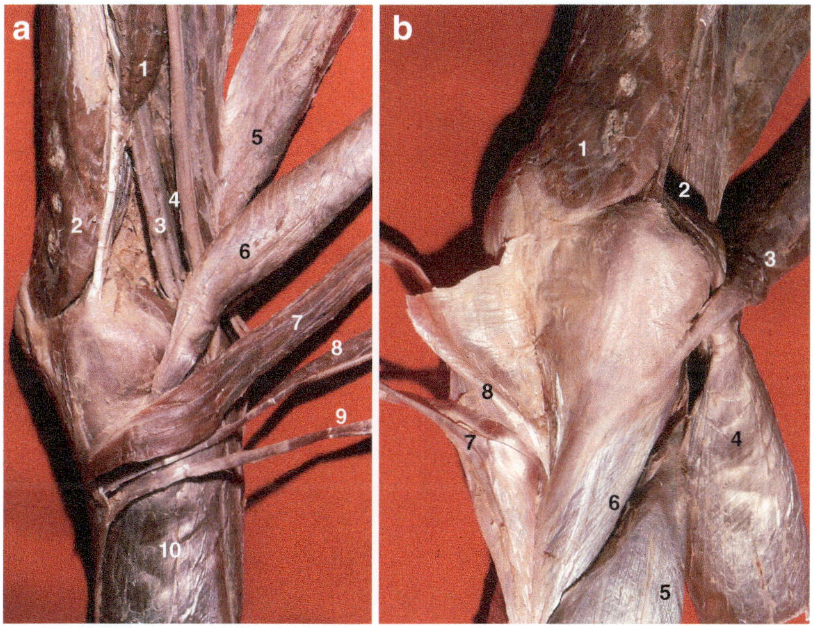

Fig. 3.11 Dissection of the medial part of the knee joint. (**a**) Some medial muscles are cut off and reclined outside: 1. Great adductor; 2. Vastus medialis of quadriceps; 3. Femoral artery and vein; 4. Ischiatic nerve; 5. Biceps femoris muscle; 6. Semi-membranous muscle; 7. Sartorius muscle (normally in lateral position); 8. Semi-tendinous muscle; 9. Gracilis muscle; 10. Medial gastrocnemius muscle. (**b**) Three muscles are reclined to show their particular tendinous feature: 1. Vastus medialis of the quadriceps; 2. Great adductor; 3. Semi-membranous muscle (with two tendons); 4. Medial gastrocnemius muscle; 5. Solear muscle (monoarticular part of the triceps muscle); 6. Popliteus muscle; 7. Gracilis muscle; 8. Semi-tendinous muscle

with a lateral portion more prominent than the medial and is held in position by lateral fins associated with some fibrous beams from the menisci. A large synovial sac under the quadriceps and two latero-patellar expansions facilitate travel jerks.

We can therefore consider that the knee is one of the most complicated joints of the motor

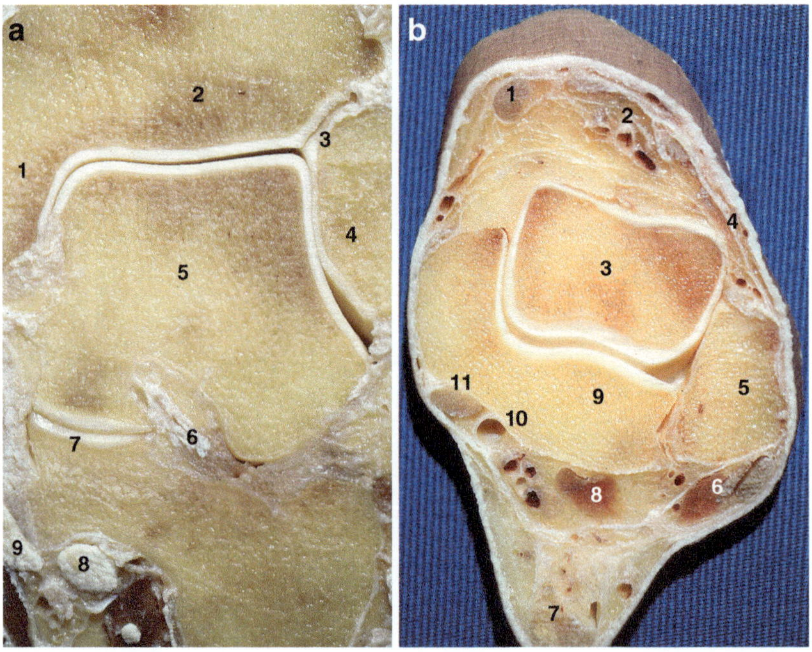

Fig. 3.13 Coronal and transversal anatomical sections of the right suro-podal junction (**a**) Coronal section of the suro-podal cardan: 1. Medial malleolus; 2. Inferior part of the tibia; 3. Inferior tibio-fibular joint; 4. Lateral malleolus articulated with the lateral articular surface of the talus; 5. Talus; 6. Interosseous ligament within the tarsal sinus; 7. Sustentaculum tali of the calcaneus articulated with inferior surface of talus; 8. Tendon of the flexor hallucis longus; 9. Tendon of the flexor digitorum longus. (**b**) Transversal section of the ankle joint: 1. Tendon of the tibialis anterior muscle; 2. Tendons of the extensor muscles; 3. Talus larger in the front than in the back (physiological diastasis); 4. Inferior retinaculum of extensor muscles; 5. Lateral malleolus; 6. Tendons of the short and long fibular muscles; 7. Tendon of the triceps muscle (Achille tendon); 8. Tendon of the flexor hallucis longus muscle; 9. Inferior part of the tibia; 10. Tendon of the flexor digitorum longus muscle; 11. Tendon of the tibialis posterior muscle

system, showing a very clever design making its pathological deviations often difficult to treat well (Fig. 3.11). It is often painful and can cause a "somatisation" (mental somatic fixation) important to recognise in order to properly diagnose and treat with the same kind of intelligence.

The Foot

It is an organ that usually lives at the surface of the ground. Its functional programme of construction is perfectly consistent. It intervenes actively in both stabilisation (it is in the first line) and propulsion of the whole body. Several technical details are peculiarities to remember [13–19].

- It must ensure a stable support from which it has to stabilise the entire body along two anteroposterior and lateral axes necessarily existing in its connection with leg. For safety reasons, it is safer to separate the 2 DF and use a cardan joint.
- The plantar support is done by two triangles: posterior for stabilisation and anterior for propulsion (Fig. 3.12). The foot is not a tripod, which represents mechanically an orthostatic or isostatic support by three points, but a hyperstatic support in more than three points, not geometrically measurable. It is given by the skin that is firmly attached to the fibro-skeletal basement especially on the plantar tendinous sheath in continuity with the triceps tendon. This solid footbed set is commonly used by many people walking barefoot. The skin can then thicken by surface keratinisation.
- The junction with the leg is made by the leg/foot cardan with a separation of the two axes of stabilisation:
  - the first transverse axis of flexion/extension (dorsiflexion/plantar flexion) is given

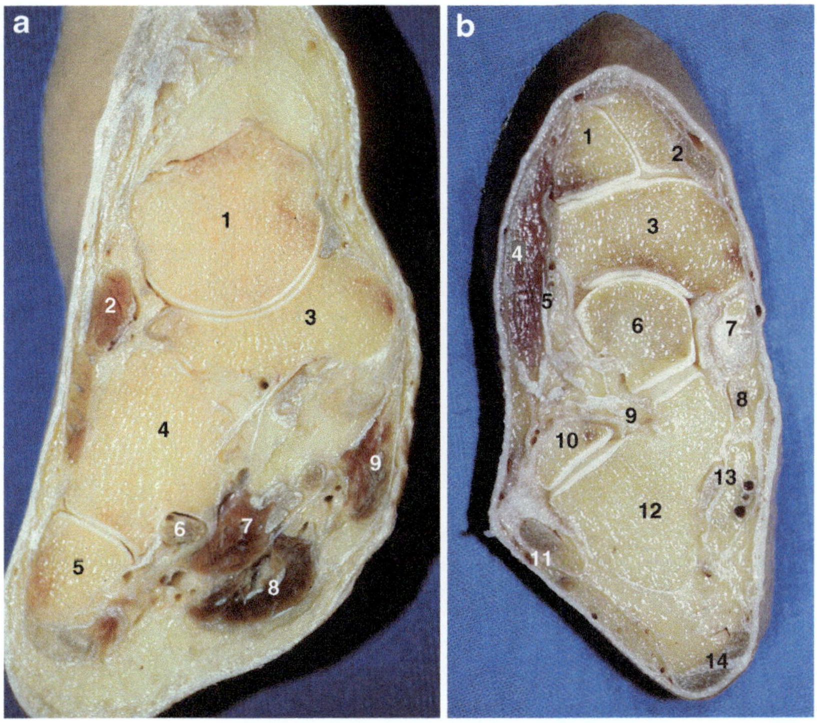

Fig. 3.14 Coronal and transversal sections of the tarsus. (**a**) Oblique coronal section seen from the front: 1. Head of the talus (second DF of the suro-podal cardan); 2. Extensor hallucis and digitorum brevis muscles; 3. Navicular bone with its medial tuberosity; 4. Cuboid bone; 5. Fifth metatarsal bone; 6. Tendon of the long fibular muscle within the groove of cuboid; 7. Quadratus plantar muscle; 8. Flexor digitorum brevis muscle; 9. Abductor muscle of hallux. (**b**) Transverse section of left tarsus seen from above: 1. Medial cuneiform bone; 2. Tendons of the tibialis anterior and long extensor of hallux muscles; 3. Navicular bone; 4. Short extensor digitorum; 5. Short extensor of hallux; 6. Head of talus; 7. Tendon of the tibialis posterior muscle; 8. Tendon of the long flexor digitorum; 9. Tarsal sinus; 10. Lateral part of talus; 11. Tendons of fibular muscles (short and long); 12. Calcaneus; 13. Tendon of the long flexor of hallux; 14. Tendon of the triceps muscle

by an articulation between tibia/fibula and talus. It is the intelligent answer to a mechanical problem that we have already discussed (Fig. 3.13). A rigid structure is fragile and the bi-malleolar clamp [14] (mortise) enclosing the talus (pin) is subject to significant lateral forces from the ground support of the calcaneus. Consequently, the mechanical solution to increase the resilience of the system is to separate the two jaws of the clamp by an articulation accepting small transversal movements (diastasis) and allowing the absorption of energy and the delay of the rupture.

Comparative anatomy can help to understand this rationale. The animals that walk on the fingers, such as horse, chicken or lamb, do not exert transverse stresses on talus and have therefore a totally rigid mortise with a deep trochlea and a complete synostosis of tibia with fibula; fibula itself might be incomplete by lack of lower link and, as happens in chicken, this creates a pointed thin bone that can be dangerous for the pharynx during its consumption. In human beings, the superior tibiofibular joint allows very low movements since its role is to secure the fixation of the top fibular shaft. The upper articular surface

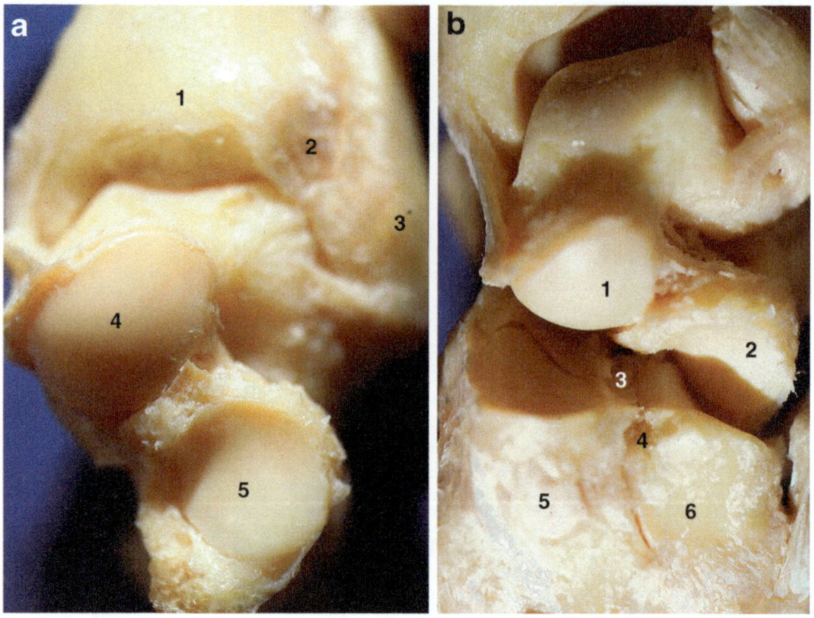

Fig. 3.15 Posterior articular view of the foot: the superposed talo-calcaneus junction (suro-podal cardan joint). (**a**) Seen from the front: 1. Tibia; 2. Tibio-fibular inferior joint; 3. Lateral fibular malleolus; 4. Convex and oblique head of talus; 5. Concave anterior articular surface of calcaneus. (**b**) Superior view of the opened medio-tarsal joint (Chopart): 1. Head of talus; 2. Anterior surface of calcaneus; 3. Articular line in S shape; 4. Bifurcated ligament; 5. Navicular bone; 6. Cuboid bone

of the talus is wider in the front than behind, which creates a slight diastasis during physiological movements of dorsiflexion and helps to maintain the small mobility of the inferior tibiofibular junction;

– the second oblique axis (subtalar joint) corresponds to the complex movements of inversion/eversion of the foot (or pronation/supination, which in reality are not possible as it is the case for the hand) [18]. The calcaneus can operate a transverse rotation along the oblique axis of a double reversed trochoid formed by two posterior articular surfaces: a concave on talus and a convex on calcaneus and four anterior articular surfaces (two convexes on talus and two concaves on calcaneus, the more medial being on an calcaneal overhang called sustentaculum tali) (Fig. 3.14). During the movement, the oblique axis described by Henke is moving laterally, which requires to speak about instantaneous axis of rotation.

• One of the originality of the foot construction is in fact the double offset (superimposed model) on its posterior part with in superior and medial positions the talus and in lateral and inferior positions the calcaneus (Fig. 3.15). This allows to articulate the two bony columns of the foot: the first medial formed by the navicular/three cuneiforms/first three metatarsal rays and the second by the lateral one cuboid/two last metatarsal rays.

Therefore the medial column is raised in the shape of a longitudinal arch. This implies, firstly, a strong attachment of the calcaneal "abutment" by the inferior calcaneocuboid ligament (long and short plantar ligaments) and of the talar head by the plantar calcaneonavicular ligament (spring ligament) and secondly an active suspension for holding it, which is provided primarily by the tibialis anterior muscle fixed on the navicular bone representing the top of the arch (Fig. 3.16); its paralysis leads to the collapse of the longitudinal arch as seen in polio patients. In addition, the head of the talus is convex and the anterior end of the calcaneus concave, resulting in an S shape at their junction (Chopart joint) which is locked by the bifurcated ligament (Y) with a navicular and cuboid fixation from the calcaneus.

The talocalcaneal synthesis is carried by a strong interosseous ligament made by a double fibrous row fixed in the obliquely oriented sinus tarsi. Finally, we must remember the particular form of the tarsometatarsal joint (Lisfranc) with a mortise between the medial and lateral cuneiforms.

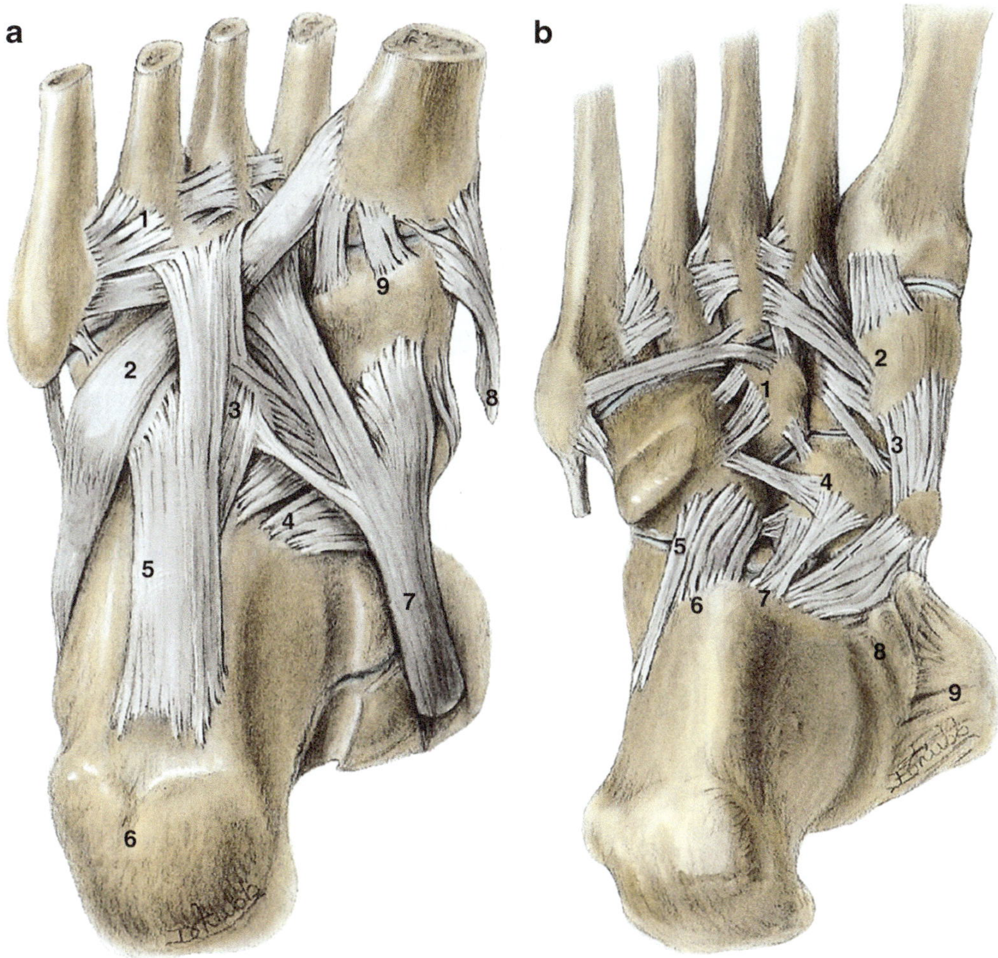

Fig. 3.16 Drawing of the inferior aspect of tarsus. (**a**) Superficial plan: 1. Plantar intermetatarsal ligaments; 2. Tendon of the long fibular muscle; 3. Calcaneo-cuboidal ligament (short plantar); 4. Plantar calcaneo-navicular ligament (spring ligament); 5. Long plantar ligament; 6. Calcaneus; 7. Tendon of the tibialis posterior muscle; 8. Tendon of the tibialis anterior muscle; 9. Medial cuneiform bone. (**b**) Deep plan: 1. Lateral cuneiform bone; 2. Medial cuneiform bone; 3. Plantar cuneo-navicular ligament; 4. Navicular bone; 5. Short plantar calcaneo-cuboidal ligament; 6. Anterior tubercle of calcaneus; 7. Spring ligament; 8. Sustentaculum tali; 9. Medial malleolus

At the front, the five metatarsal rays are "landing" forming a curved hinge curve as described by de Donker [19]. This has two consequences: a false image of the anterior arch in a cross section due to the unequal length of the metatarsals and an interesting functional benefit by allowing forward propulsion in all directions from the anterior foot triangle. Ground support of the hallux is made on two little bony balls, the sesamoid bones attached to the joint capsule (Fig. 3.17). In the middle, we find the tendon of the hallux flexor that facilitates the correct orientation of traction on the hallux. A frequent deformation called hallux valgus is mainly due to the wearing of "deforming" shoes (specifically for women) with a tapered front portion or triangular support as well as a calcaneus strongly raised by raised heels. Surgery is often required but the surgical correction is not so simple due to the particular configuration of the hallux metatarsophalangeal joint.

Adaptive flexibility of the foot is due to its variable mechanical compliance in relation with the many muscle tendons passing front and back

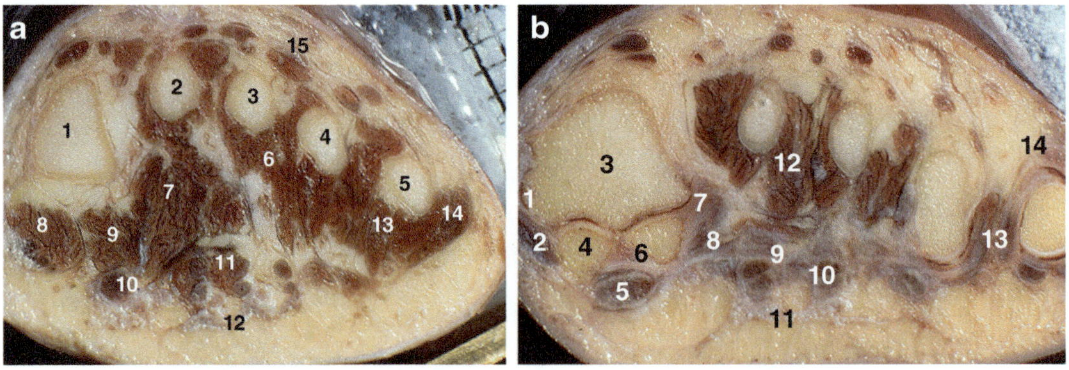

Fig. 3.17 Anatomical transverse sections of left forefoot seen from the front. (**a**) Middle part of the forefoot:1.2.3.4.5. Metatarsal bones; 2. Physiological axis of the foot (two dorsal interosseous muscles); 6. Interosseous muscles; 7. Adductor of hallux; 8. Abductor of hallux; 9. Short flexor of hallux; 10. Long flexor of hallux; 11. Lumbrical muscles attached to flexor tendons; 12. Plantar tendinous sheath; 13. Short flexor of fifth toe; 14. Abductor of fifth toe; 15. Extensor digitorum muscles. (**b**) Anterior transverse section with false image of anterior arch: 1. Articular capsule of the first metatarso-phalangeal joint; 2. Tendon of the abductor of hallux; 3. Head of first metacarpal bone; 4. Medial sesamoid bone; 5. Tendon of the long flexor of hallux fixed between the two sesamoids; 6. Lateral sesamoid with the intersesamoid ligament; 7. Tendon of the short flexor of hallux; 8. Adductor of hallux with two parts: oblique and transverse; 9. Deep transverse intermetacarpal ligament; 10. Tendons of superficial and deep flexor muscles; 11. Anterior extremity of the plantar tendinous sheath; 12. Interosseous muscles; 13. Tendon of the short flexor of the fifth toe; 14. Extensor apparatus of the fifth toe

and ensuring a variable clamping joint and stiffness of joints. This also creates a viscoelastic damping of multidirectional mechanical solicitations (Fig. 3.18).

The functional classification of these muscles is in relation with the two axes of stabilisation:

- in front of the transverse axis are the dorsiflexors (tibialis anterior, extensor digitorum longus and hallucis and third fibular);
- behind the transverse axis are the plantar flexors (triceps sural, long and short peroneal, posterior tibial, long flexor of toes and hallux);
- on the lateral side of the anteroposterior oblique axis are the fibular pronators and on the medial side the tibial supinators.

The efficiency of this tendinous damping is significantly improved by fibrous bands maintaining tendons in contact with osteoarticular structures: superior and inferior retinaculum of extensor muscles, superior and inferior peroneal retinaculum and flexor retinaculum.

Special attention should be given to the tricipito-plantar tendinous sheath (plantar fascia) that marks the direct continuity with the triceps tendon. It prolongs the action to the middle phalanges of the toes by longitudinal strips with beams of transverse fibres. It doubles the flexor digitorum brevis muscle and serves a solid mooring for the plantar skin. Its break is possible in sports, especially in jumping when starting or running. Note the similarity of the palmar fascia that has the same role for the attachment of the palmar skin and is tensioned by the palmaris longus muscle.

Finally, we can see on a cross section of the foot the three muscle columns. One might think that the muscles attached to the hallux and particularly to the fifth toe have a disproportionate volume regarding the amount of abduction they can achieve in relation to the functional axis of the foot made by the second ray and not the third, as it is the case for the hand. They must then be considered as stiffeners [17] of medial and lateral edges of the foot playing an important role in the adaptation of the foot to the ground.

3.2.2.3 Kinematics of Gait

The kinesiological EMG of gait gave us accurate data on temporospatial coordinates of muscle actions playing a role in the course of walking [20–26].

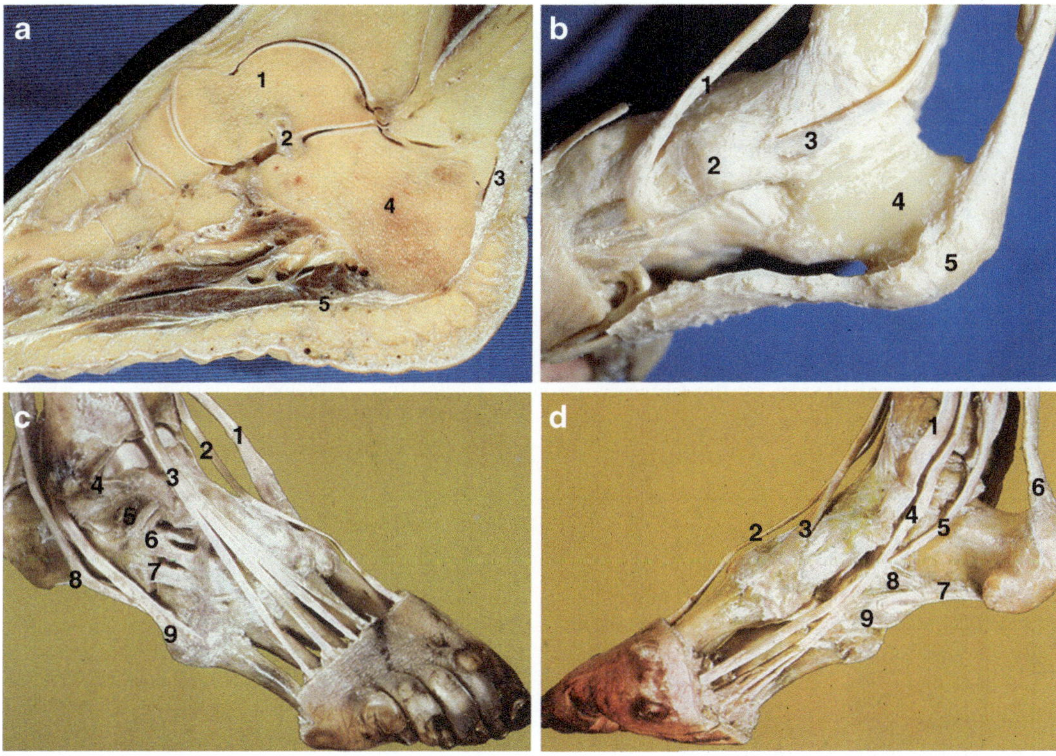

Fig. 3.18 Tendons and ligaments of the right foot. (**a**) Median sagittal section of tarsus: 1. Talus; 2. Interosseous ligament; 3. Triceps tendon (Achille); 4. Calcaneus; 5. Plantar tendinous sheath applied on the short plantar flexor of toes muscle and on which is strongly fixed the plantar skin. (**b**) Medial side of tarsus with tendons of supinator muscles: 1. Tibialis anterior muscle; 2. Medial tuberosity of navicular bone; 3. Tibialis posterior muscle; 4. Medial side of calcaneus; 5. Tricipito-plantar tendinous sheath. (**c**) Dorsal view of foot tendons and ligaments: 1. Tibialis anterior muscle; 2. Long extensor of hallux muscle; 3. Long extensor of toes muscle; 4. Talo-fibular anterior ligament; 5. Talo-calcaneus interosseous ligament; 6. Bifurcated ligament (Chopart); 7. Dorsal calacaneo-cuboid ligament; 8. Long fibular muscle; 9. Short fibular muscle. (**d**) Medio-plantar view: 1. Tibialis posterior muscle; 2. Long extensor of hallux; 3. Tibialis anterior muscle; 4. Long flexor of toes muscle; 5. Long flexor of hallux muscle; 6. Triceps tendon; 7. Great plantar ligament; 8. Plantar calcaneo-navicular ligament; 9. Plantar calcaneo-cuboid ligament (fixation of the longitudinal arch)

Start Up

It is made by a hip flexion which determines the step length by the action of the iliopsoas (mainly iliacus) and rectus femoris (polyarticular portion of the quadriceps).

Swing Phase

It is given by the hip flexors with quadriceps contraction initiating the extension of the leg and tibialis anterior and extensor of the toes raising the foot upward to prevent the friction of the soil (which can be observed in hemiplegic patients).

Monopodal Stance Phase with Two Important Actions

- Firstly, locking of the knee which is done by cocontraction quadriceps/hamstrings. This explains that the vertical axis thigh/leg during this phase is bending for about 10°, which protects the knee which has no bony blocking of extension and prevents hyper-extension generating a recurvatum.
- Secondly, the lateral stabilisation of the pelvis by contraction of the group gluteus medius, gluteus minimus and tensor fascia lata, all innervated by the same superior gluteal nerve and of the trunk by contraction

of the ipsilateral spinal muscles. In this phase, the trunk makes its forward motion (with swing phase of the opposite side) with pelvic rotation (pelvic step) and extension of the thigh (gluteus maximus if slope and hamstrings).

Starting a New Swing Phase

The foot has made progress to end up on the anterior propulsion triangle with passive hallux followed by propulsive thrust by the long flexor hallucis, foot dorsal flexion by the tibialis anterior and flexion of the leg by the hamstrings and in case of slope combined action of twins.

Conclusions

All these actions are automated during the gait sequence with variations on demand due to some obstacles on the ground. Classic or modern dances require infinitely varied supports and movements, but they need a long training often putting a big strain on the joints (foot peak, splits, . . .).

The shoes to go in rough terrain or mountains must be strengthened, but for walking or running on flat ground, they must not hamper the anterior triangle dorsiflexion during the propulsive phase and especially avoid hard spots or friction, generators of bulbs or localised hyperkeratosis (corns). However, a fairly rich foot pathology is directly related to often unsuitable shoes (too high or too low) or biomechanically inappropriate (too high heels, fore foot in snake head, . . .).

References

1. Romberg MH (1853) Manual of nervous diseases of man. Sydenham Society, London
2. Massion PBC, Viallet F (1990) Posture, coordination, movement. Rev Neurol 146(10):536–5421
3. Wolfson L (2001) Gait and balance dysfunction: a model of interaction of age and disease. Neuroscientist 7(2):178–183
4. Baron JB (1982) Statokinésimétrie. Etude de la posture verticale humaine. Ann Kinésith 9: 377–388 V1
5. Gagey PM (1996) La stabilométrie en podologie. In: Villeneuve P (ed) Pied, equilibre et posture. Frison Roche, Paris, pp 69–75
6. Igarashi M, Black FO (1985) Vestibular and visual control on posture and locomotor equilibrium. Karger 7th international symposium of the international society of posturography, Houston
7. Hunt CC (1974) The Pacinian corpuscle. In: Hubbard SJ (ed) The Peripheral Nervous System, vol 13. Pergamon Press, London, pp 405–422
8. Enjalbert M, Rabischong P, Micaleff JP, et al (1983) Sensibilité plantaire et équilibration. XIX Entretiens De Rééducation Et Réadaptation Fonctionnelles, Masson, pp 9–29
9. Bonnel F, Jaaeger JH, Mansat CH (1984) Les laxités chroniques du genou. Masson, Paris
10. Liorzou G (1990) Le genou ligamentaire: examen clinique. Springer, Berlin
11. Segal P, Jacob M (1983) Le genou: anatomie, cinématique, sémiologie, pathologie, diagnostic et traumatologie sportive. Maloine, Paris
12. Chassaing V, Parier J (1985) Arthroscopie du genou. Masson, Paris
13. Inman VT (1976) The joints of the ankle. Williams and Wilkins, Baltimore
14. Le Coeur P (1938) La pince malléolaire. Physiologie normale et pathologique du péroné. Librairie Louis Arnette, Paris
15. Rabischong P (1991) Anatomie compréhensive du pied. In: Chiron PH, Utheza G (eds) Le Pied. Actualités en Médecine, Chirurgie et Réeducation. Masson, Paris
16. Rabischong P (2004) Evolution des idées sur la biomécanique du pied. In: Hérisson C, Aboukrat P, Claustre JE (eds) Progrès en médecine et chirurgie du pied. Sauramps Médical, Montpellier
17. Rabischong P (2011) Anatomie compréhensive du pied dans la locomotion. In: Morvan G, Bianchi S, Bouysset M et al (eds) Le Pied. SIMS. Sauramps Médical, Montpellier
18. Viladot A, Lorenzo JC, Salazar J, Rodriguez A (1984) The subtalar joint: embryology and morphology. Foot Ankle 5(2):54–66
19. De Doncker E, Kowalski C (1979) Cinésiologie et rééducation du pied. Masson, Paris
20. Rabischong P, Avril J, Aubriot H et al (1967) L'électropodographie: une méthode nouvelle de mesure des forces plantaires. Revue de pédicurie 44:9–24
21. Péruchon E, Jullian JM, Rabischong P (1989) Wearable unrestraining footprint analysis system. Applications to human gait study. Med Biol Eng Comput 27:557–565
22. Arcan M, Brull MA (1976) A fundamental characteristic of the human body and foot, the foot-ground pressure pattern. J Biomech 9:453–457
23. Elftman H (1984) A cinematic study of the distribution of pressure under the foot. Anat Rec 59:481–492
24. Clot J, Rabischong P, Péruchon E, et al (1975) Principle and applications of the artificial sensitive skin. Proceedings of the 5th international symposium on external control of human extremities ETAN, pp 211–220
25. Dimegliio A (2002) La marche de l'enfant. Sauramps Médical, Montpellie
26. Ducroquet RJ, Ducroquet P (1965) La Marche et les Boiteries. Masson, Paris

Prehension

4

Abstract

Prehension is a vital function according to its important role in serving the oral pole—for feeding—and the genital pole—for grooming and sexual activities. Everything is designed to give a maximum mobility to the terminal organ, the hand, which is in humans, unlike in monkeys, a multi-grasp unspecialised organ. Three different functional systems have to be described. The wrist/hand complex shows a very intelligent organisation of polyarticulated finger chains with an opposable thumb and all corresponding muscles allowing a great quantity of grasp and manipulation procedures. Then, the humero-radioulnar complex can regulate the length of the limb and put the hand in a supine or prone position thanks to the special trochoid radioulnar joint. Finally is the cleido-scapulohumeral complex that allows the upper limb to get 5 degrees of freedom (DF) for its spatial orientation resulting from only one junction to the axial skeleton: the sternoclavicular joint. Full responsibility lies then in the very powerful shoulder muscles for stabilising and moving the upper limb.

4.1 Introduction

This specific function of the upper limb will impose a number of technical requirements [1].

(a) **A high mobility,** allowing a wide spatial excursion of the grasping terminal organ. This is particularly evident when looking at the shoulder in comparison to the hip. The clavicle has only one articular junction with the axial skeleton (sternoclavicular joint) which gives 5 DF for the spatial orientation of the upper limb: 2DF for clavicle/sternum and 3 DF for scapulohumeral.

(b) **An unspecialised multi-grasp end-organ** able to use any tools, allowing the development of all forms of technology and with some specific features: opposable thumb, mesaxonic symmetry with two dorsal interossei attached to the IIIrd finger, defining the functional axis according to Vialleton and Anthony. This is not the case in monkeys who have two types of specialised autopods: the hook of brachiators monkeys (hylobatidae, pongids, new world monkeys, semnopithecids) with reduced thumb and paraxonic symmetry (between IIIrd and

P. Rabischong, *Comprehensive Anatomy of Motor Functions*, DOI 10.1007/978-3-319-04169-8_4,

IVth finger) and the hand of arboreal grasping monkeys like lemurians, for example, with their opposable thumb, palmar tactile balls, reduced index and ectaxonic symmetry (IVth finger).

In general, each animal must have a gripper, essential for nutritional survival, which can take four different shapes:

- **the clamp,** like the multiform beak of some birds (sharp, curved, in spatula) or like the clamp jaw for plant grinder of ruminants or like carnivore jaw with a 1 DF temporomandibular joint and protruding canines to catch mobile provender or like pliers of shellfish;
- **the winding** that can take the form of pseudopods in amoeba, spine in snakes or trunk in elephants;
- **the suspension by gravity,** by a hook formed by the fingers or a prehensile tail or a plane cutaneous surface;
- **the adhesion,** using a mucus with adhesive properties as in the chameleon when it throws its tongue onto the prey.

In total, the human hand has the ability to operate on demand all these forms of grasping and thereby is not specialised [2–5] (Fig. 4.1).

(c) Grasping and manipulating objects

The hand has an easy-to-use segment length (arm/forearm) to serve the oral pole (feeding) and the genital pole (grooming, sexual activity). A rich innervation with an area of wide central control projection sensorimotor cortex to give to the hand a major palpatory role (haptic sense) (Fig. 4.2) and a clear strategy for taking and handling objects in three major phases:

- approach: visual, tactile or on order;
- grasping in various ways;
- grasp releasing (Fig. 4.3).

Therefore, the upper limb is tailored to its non-specialised terminal organ, capable of taking and manipulating (the etymological meaning of the term) any form of objects in a three-dimensional space and also bear or cling to external structures by mobilising the entire body. This programme requires a basic quality: mobility with a major functional requirement to serve the holes of biological survival (mouth, genitals).

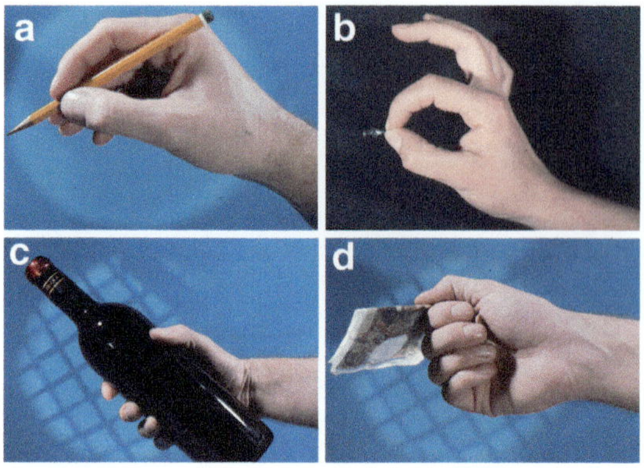

Fig. 4.1 The different types of grip. (**a**) Tridigital grip (very stable due to the fixed three first metacarpal bones). (**b**) Bidigital grip (very variable from nail grip to pulp grip). (**c**) Grip by hand seizing (typically French). (**d**) Lateral grip between pulp of thumb and lateral side of index

(d) Robotics inputs

The robotics that built more or less complicated multi-DF manipulators (2–6 DF and more) have taught us a lot about command and control problems of these polyarticulated tools technically identical to the human upper limb [6]. Controlling the motion of the terminal organ can be done either by an open loop without possible regulation or by a closed loop with sensors (touch, force, temperature, . . .) in order to get the feedback from the effector to the control system. We can create automatic regulation and servomechanisms allowing some form of adaptive intelligence to scheduled tasks and environmental changes. Some scientists reserve the term of "robot" for these intelligent machines (Fig. 4.4).

However, to write algorithms and control equations for these robots, two information are essential: the state of the motors measured by the level of electrical power or hydraulic pressure and the angles of the segments measured by potentiometers, either linear or angular.

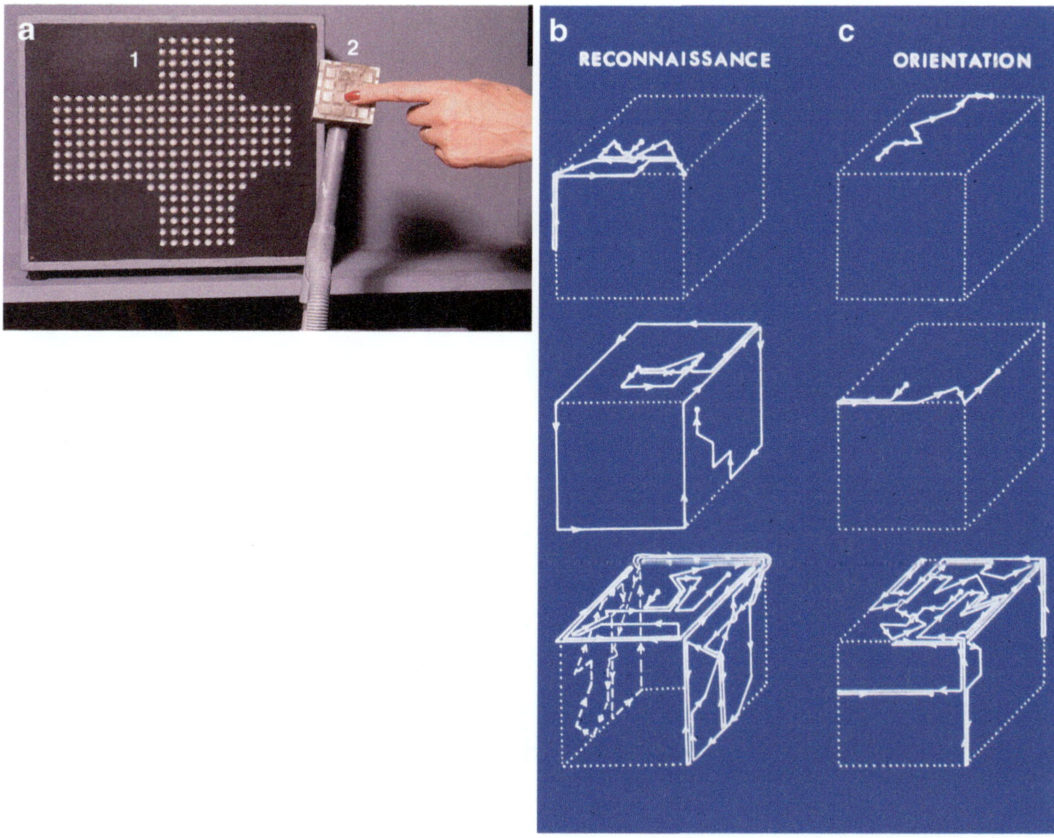

Fig. 4.2 Recognition test of nature and spatial orientation of a test-object (cube) with no vision. (**a**) Instrumentation: 1. Recording of the contact transducers for the four sides and the twelve borders of the cube; 2. Cube-test with contact transducers to analyse the palpation strategy. (**b**) Recognition test with extreme profiles of subjects, from up to bottom (subjects no. 5, 9, 14 out of 40 subjects): 1. Palpation only of one face and one border with potential risk of mistake; 2. Optimal recognition by palpation of two faces and seven borders; 3. Redundant palpation of faces and borders. (**c**) Orientation test with extreme profiles of subjects, from up to bottom (subjects no. 8, 16, 4 out of 40 subjects): 1. Palpation of one face; 2. Palpation of one face and two borders; 3. Redundant palpation of faces and borders

It is the same technical problem with the human upper limb. The motor muscles are viscoelastic, nonreversible and non-linear. Here is their originality which makes them work without any noise, which is not the case of industrial motors. The central nervous system (CNS), in order to coordinate them in a 3D-space, needs also two relevant information: the state of the muscular actuators and the angles of the joint segments.

Striated muscles are able to change their module of elasticity by the shortcut of their dark disc of contractile proteins without exceeding one-third of the length of the striated fibres, which is an important parameter fixing the construction of muscle plans. Regarding their three possible states: relaxed, contracted or stretched—data which have to be known by the CNS for their control—they can be measured by a mechanical stiffness transducer.

Regarding the measurement of the joint angles, muscles or ligaments cannot do it but the skin itself can; it is the goniometer for all the body segments as we saw previously and as Moberg indicated [7–9].

The upper limb innervation is provided by the brachial plexus located in the neck and

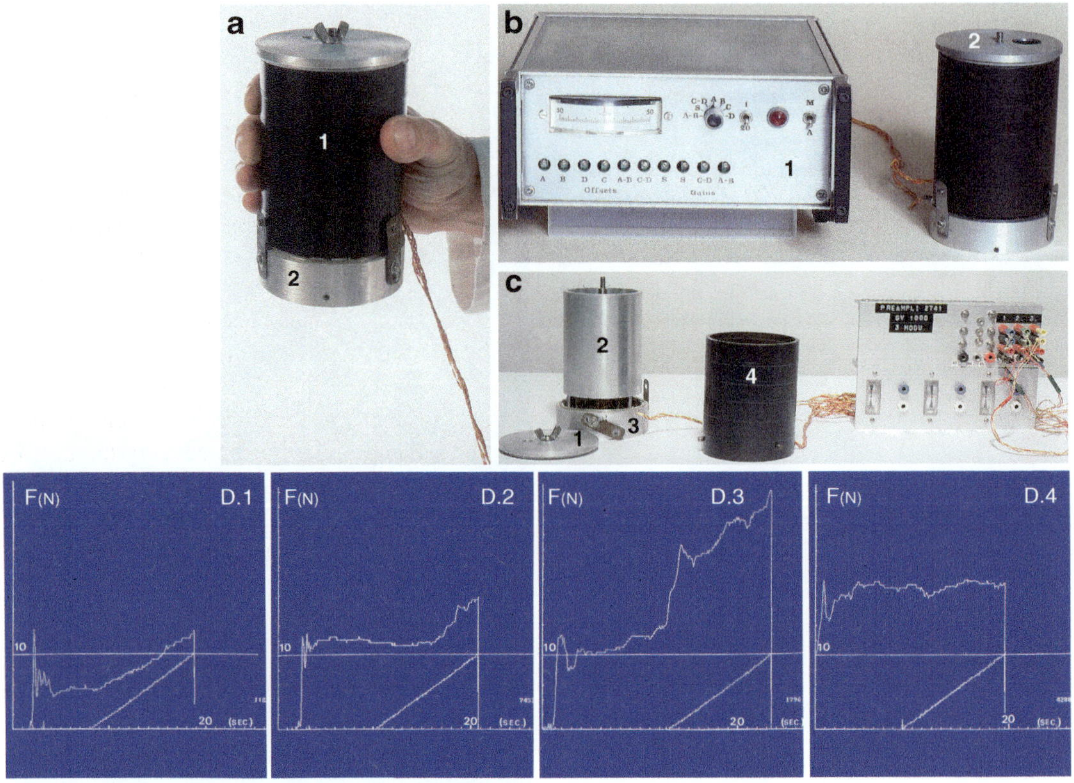

Fig. 4.3 Measure of grasping force during prehension tests (INSERM, Unit 103). (**a**) Palmar grip of the test-object: 1. Force transducer equipped with strain gauges bridges; 2. Base with strain gauges accelerometers. (**b**) Instrumentation: 1. Electronic measuring and recording system; 2. Cover of the test-object with a hole to change the weight by filling it with lead powder. (**c**) The test object: 1. Removable cover; 2. Metallic internal cylinder for filling; 3. Accelerometers; 4. Force transducer. (**d**) The different profiles of grip force regulation in relation with the growing weight of the test-object: *superior line*: force level and *inferior line*: weight of the object. 1. First type: Proportional regulation of grasping force; 2. Second type: regulation with threshold; 3. Third type: regulation in stages; 4. Fourth type: complete lack of regulation of grasping force in relation with the weight of the test-object

Fig. 4.4 Anthropomorphic robotics in Japan. (**a**) Professor Kato lab: Professor Rabischong visiting the laboratory of Professor Ichiro Kato from Waseda University in Tokyo. (**b**) Professor Kato's humanoid robot playing on organ: (two hands, one foot) and reading the score of a Concerto of Haendel during the opening ceremony of the International Fair of Tsukuba in 1985

formed by the roots C5/T1. It passes between the scalene muscles to enter into the limb and is divided into primary and secondary trunks providing all nerves. The plexus is not done by anastomoses of fibres—which is only possible for vessels—but instead by nervous fibres juxtapositions as in a marshalling yard, according to the multi-root innervations of most of the muscles [10–15].

We counted the number of nerve fibres in the roots and different trunks of the brachial plexus in 20 subjects [16]. There are an average of about 110,000 nerve fibres to innervate the entire upper limb, but we were amazed by the large individual variations of these fibres ranging from one to three. This is largely in favour of inequality of individuals with regard to the size of the nervous potential and, consequently, the level of skills.

The upper limb is in fact a *high-tech* manipulator [17] with 9 DF up to the hand, which has 20 DF and the carpal joint 10. Its construction is organised starting from the hand in three interactive sets: the carpo-metacarpal-phalangeal complex, then, the humero-radioulnar complex and finally, the cleido-scapulohumeral complex, hanging all the upper limb at the axial skeleton.

4.2 Wrist/Hand Complex

4.2.1 The Human Hand

It is a non-specialised multi-grasps tool, which is not the case in brachiates or grasping arboreal monkeys as we saw before. It comprises five digital rays and one of them, the thumb, lives in the opposition of the others. This allows a series of pollici-digital clips and a secure grip by hand seizing. Each finger is a sensitive probe richly innervated (about 350 mechanoreceptors by sq. mm in the pulp of the index).

4.2.1.1 The Four Middle Fingers

They are formed by a polyarticulated chain with three phalanges: proximal (P1), intermediate (P2) and distal (P3) with three different joints (two 1-DF trochlear interphalangeal type—distal (DIP) and proximal (PIP)—and a 2-DF condylar metacarpophalangeal joint (MP)).

The movements of the chain comprise flexion/extension and lateral bending. The biomechanical functioning principle which justifies the complexity of muscle organisation is the mechanical coupling of the interphalangeal joints and the decoupling of the metacarpophalangeal. In other words, it is not normally possible to flex one interphalangeal joint and expand the other, thanks to the blocking of the linear chain extension by palmar ligaments, powerful for PIP (volar plate) and extensible for DIP, that can be extended in a pulp passive support (which is a fair protection of the article). On the contrary, it is possible to put the MP in flexion with an extension of IP or in extension with flexion of the IP (metacarpophalangeal decoupling).

The Muscular Organisation

This important principle explains the organisation of muscle that we can understand by a few schematic points to remember:

- the flexor muscles that close the fingers on an object by wrapping the digital chains are more powerful than extensor which open by releasing the hand grip;
- there is one flexor muscle by phalanx: deep flexor for P3, superficial for P2 and interosseous for P1. For maximum efficiency, the long flexor tendons should stay as close as possible to the digital chain. This explains why the deep flexor tendon attached to P3 has to pass through the superficial flexor tendon split into two tongues before its fixation on P2. In addition, it is needed to fix the tendons by highly resistant digital fibrous sheaths having a synovial sliding system in order to allow tendons to slide. The first fibrous sheath is attached to P1 and overflows on the palmar aspect of MP and is lined by distal fibres crossed in X on PIP (Fig. 4.5);
- the antibrachial extensor muscle tendons are necessarily fixed in the axis of the digital chain by fibrous bands on each side of the MP. This indeed limits their excursion

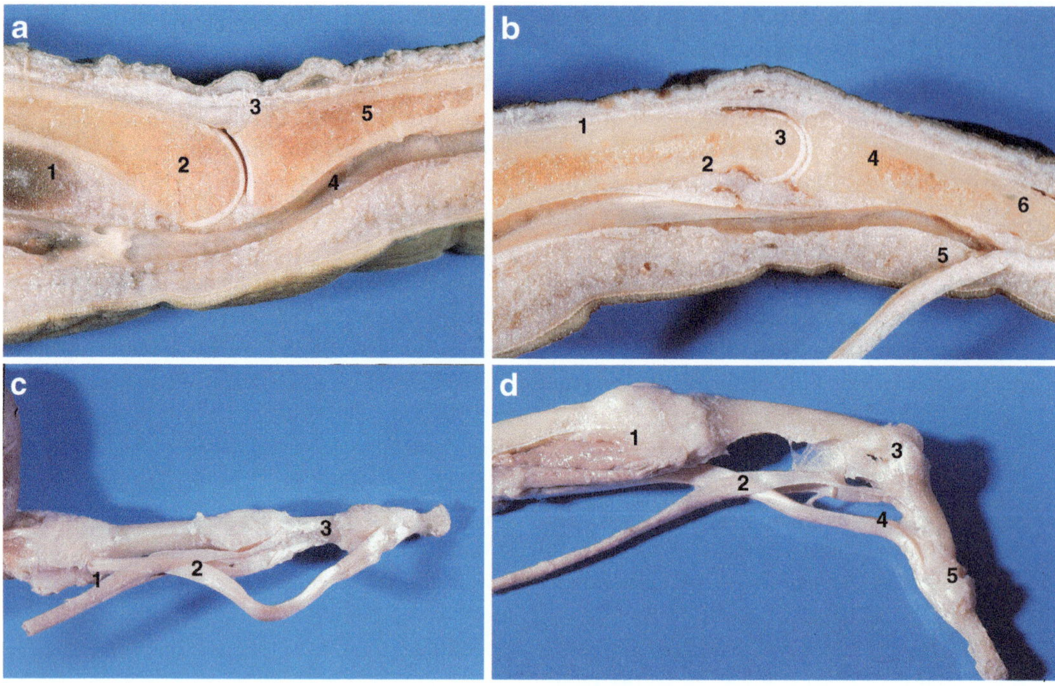

Fig. 4.5 The long flexor tendons of the digital chain. (**a**) Sagittal section of the metacarpophalangeal joint (MP): 1. Interosseous muscle; 2. Metacarpal bone; 3. MP joint with the extensor apparatus on its dorsal part; 4. Annular fibrous pulley for flexor tendons; 5. P1. (**b**) Sagittal section of the IP joints: 1. P1; 2. Palmar plate (strong ligament limiting the extension); 3. PIP joint; 4. P2; 5. Profundus flexor tendon; 6. DIP joint. (**c**) Semi-flexed finger: 1. Superficial flexor tendon inserted on P2 with its vincula (with blood vessels); 2. Profundus flexor tendon inserted on P3 and attached to P2 by its annular pulley; 3. The two bands of the superficial flexor tendon attached to P2. (**d**) Semi-flexed digital chain with two flexor tendons: 1. MP joint; 2. Fixation of the superficial tendon on the middle of P2; 3. PIP joint; 4. Profundus tendon; 5. DIP joint

(Fig. 4.6). The extensor tendon attaches to the proximal portion of P2 without bony attachment on P1 except a fibrous adhesion to the articular capsule of MP. Its contraction opens hand incompletely, but sufficiently to release a grip.

A very clever mechanical system will ensure the full chain extension as well as the flexion of P1:

- firstly, the tendinous dorsal lozenge formed by the two lateral bands of the extensor communis placed on either side of the median tongue inserted on P2, on which are attached fibrous expansions on the dorsal interosseous muscles. These lateral bands are maintained in lateral position by the retinacular ligament described by our friend Landsmeer [18–20]. They are also closed distally by transverse fibres (triangular ligament);
- secondly, fibres in sling shape come from the interosseous muscles on each side, creating a true flex belt of P1 called the interosseous back plate. They can, by the contraction of the two interosseous, flex P1 and put P2 and P3 in full extension. Contracting at the same time the long extensor tendon allows to get an extension of the entire chain, including the MP, and to play with the flexion/extension system of MP.

There is also a special extensor tendon doubling the common extensor for the index finger, to show, and to the fifth finger. Finally the total extensor digitorum unit is a large tendinous strip located directly under the dorsal skin and applied on curved back of phalanges like a saddle on back horse (Fig. 4.7).

The Physiological Axis of the Hand

It is the third finger on which two dorsal interossei generating abduction of this finger are fixed. There are three palmar interossei, which

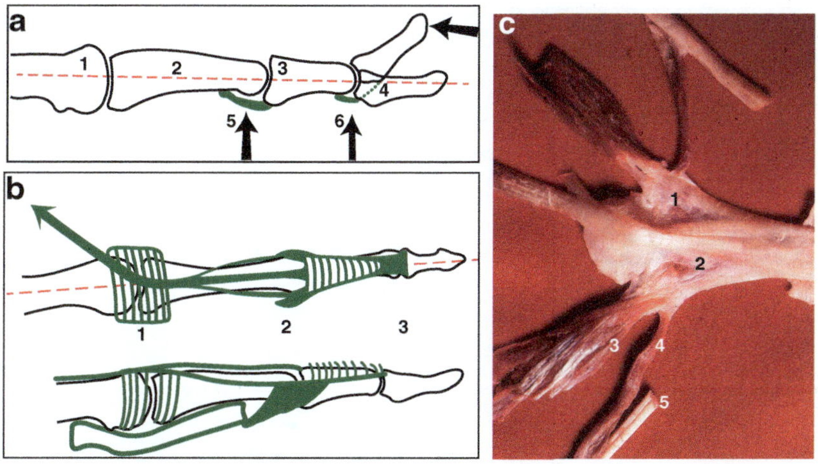

Fig. 4.6 Extension of digital chain. (**a**) Drawing of lateral aspect of the digital chain: 1. First metacarpal; 2. Proximal phalanx (P1); 3. Intermediate phalanx (P2); 4. Distal phalanx (P3); 5. Palmar plate limiting the hyperextension of PIP except in hyperlax subjects; 6. Palmar ligament less strong allowing the hyperextension of P3 in compression. (**b**) Drawing of the fixation of the extensor tendon on dorsal side of finger (dorsal and lateral view): 1. Collateral ligaments of MP joint centering the tendon traction, limiting its excursion and giving incomplete extension of finger but enough for grasp release; 2. The retinacular ligament of Landsmeer fixing laterally the lateral bands of extensor tendon; 3. Triangular ligament fixing before their fixation on P3 the lateral bands of extensor tendon. (**c**) Dissection of the third finger extensor system at the level of MP joint: 1. Collateral ligament of MP joint; 2. Median band of the long extensor tendon; 3. Interosseous muscle; 4. Lumbrical muscle; 5. Long flexor tendon

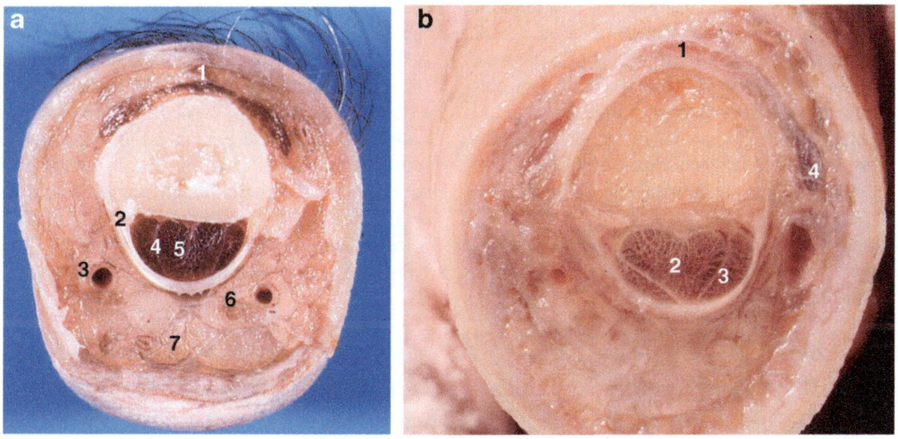

Fig. 4.7 Transverse sections of P1 (dorsal extensor apparatus and ventral flexor tendons). (**a**) Proximal section: 1. Extensor apparatus on the dorsal convex shape of phalanx; 2. Annular fibrous pulley of flexor tendons; 3. Non compressible collateral artery during grips; 4. Lateral band of superficial flexor tendon; 5. Profundus flexor tendon; 6. Collateral nerve; 7. Palmar skin thicker than dorsal for contact grip. (**b**) Distal section: 1. Medial band of extensor tendon; 2. Tendon of the profundus flexor between the two lateral bands of the superficial flexor; 3. Lateral band of the superficial flexor within the annular fibrous pulley; 4. Lateral band of extensor tendon (tendinous fibres from interosseous and lumbrical muscles)

are attached only on the corresponding metacarpal and not on both for the dorsals. They are making adduction of the IInd, IVth and Vth fingers which have an extreme medial position requiring a specific abductor and P1 flexor associated with the fourth dorsal interosseous.

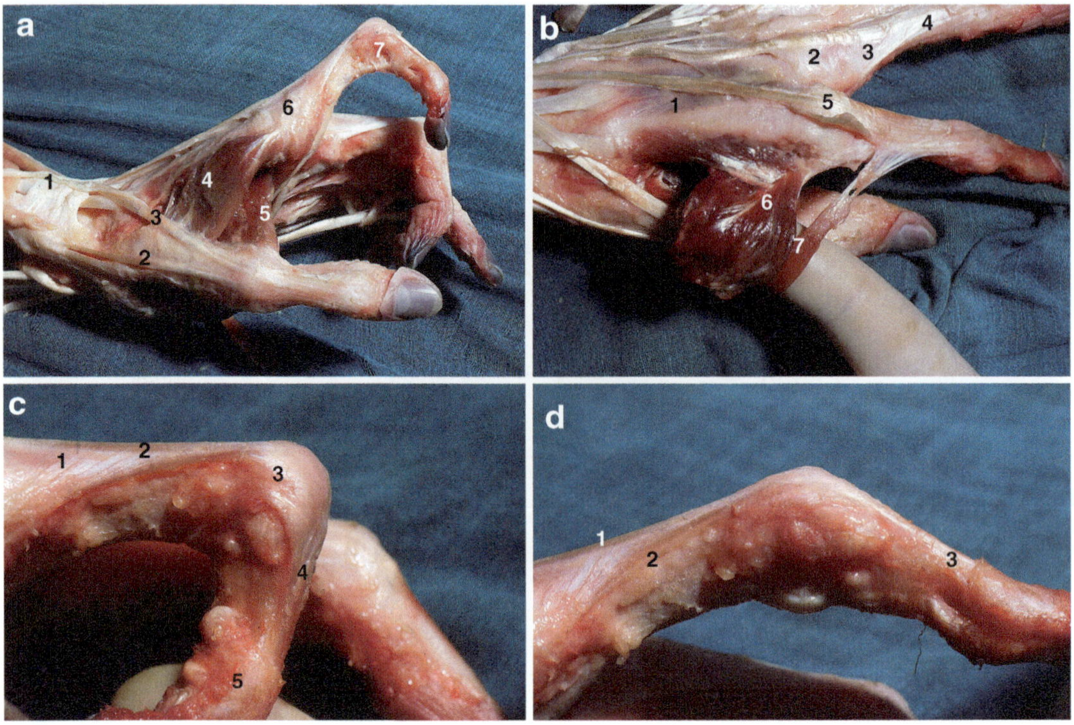

Fig. 4.8 Dissection of the extensor apparatus of fingers on a left hand (fresh cadaver). (**a**) Extension of MP joint and flexion of IP joints (decoupled movement): 1. Retinaculum extensorum; 2. Short extensor of thumb; 3. Long extensor of thumb; 4. First dorsal interosseous muscle (DIM); 5. Adductor pollicis; 6. Dorsal tendinous fibres of interosseous muscle; 7. Contracted flexor tendons. (**b**) Resection of the first DIM: 1. Tendons of the two extensors of index finger; 2. Fibrous lateral metacarpophalangeal strap fixing laterally the extensor tendon (centering traction along the finger axis); 3. Dorsal tendinous fibres of interosseous muscle (active flexion of P1); 4. Tendinous extensor lozenge; 5. Transsected lateral MP strap; 6. First DIM; 7. First lumbrical muscle separated from the flexor tendon with its tendinous fibres attached to the lateral extensor band. (**c**) Index finger in flexion: 1. Dorsal tendinous fibres of interosseous and lumbrical muscles (flexion of P1 and full extension of IP joints); 2. Median band of extensor fixed on base of P2; 3. Lateral fixation of tendinous lozenge by retinacular ligament (Landsmeer); 4. Lateral bands of extensor both joined dorsally by the triangular ligament and fixed on the base of P3 (distal phalanx); 5. P3. (**d**) Index finger in semi-flexion: 1. Medial band fixed on P2; 2. Lateral band fixed on P3; 3. Distal fixation on P3

The Specific Role of Lumbrical Muscles

One can only admire the fantastic possibilities of the digital play so useful for pianists and other musicians. But we miss in our description the need for a viscoelastic tension regulator between flexor and extensor muscles, which is represented by the lumbrical muscles, so named because of their resemblance with earthworms. These are the only muscles of the body which are intertendinous, fixed on the side of the four flexor tendons and on the extensor tendinous lozenge without any bone attachment. They are the richest of all the body in proprioceptive receptors like muscle spindles, simple and complex Golgi organs, Golgi–Mazzoni corpuscles. They are genuine proprioceptive tensiometers between two opposite functional systems and also capable of ensuring a complete extension of the digital chain regardless of the position of the MP (Fig. 4.8).

It should be noted that the fourth and fifth metacarpals are mobile in flexion/extension of their carpal base, while the second and third are fixed, which guarantees the mechanical stability of the tridigital grip. In addition, the deep transverse metacarpal ligament maintains metacarpal heads together securely.

4.2.1.2 The Thumb

A special place must be made for this lateral finger able to oppose its pulp to the four middle fingers pulp because of its special lateral position. It has only two phalanges but the first metacarpal on which they are articulated is moving through its articulation with the trapezium bone of the wrist. The joint surfaces are of reversed condylar type with two concave shaped saddles [21]. The question is why such a special joint?

The answer is that the thumb has to move along two oblique axes for antepulsion/retropulsion and abduction/adduction. This can only be done by a 2-DF condylar joint, but when a pollicis grip is decided, the saddle shape prevents the axial rotation of metacarpal and guarantees a great stability of grip. In this locked position, a low mobility of the MP and IP allows viscoelastic adaptation of the pollici-digital grip.

Eight muscle motors ensure the mobility of the thumb (6 for middle fingers), with four antibrachials:

- flexor pollicis longus whose tendon passes, as for the foot great toe, between two sesamoids in MP ensuring the alignment of the tension with respect to the phalangeal column;
- adductor longus;
- long and short extensor;
- four intrinsic muscles placed in the thenar mass: short abductor taking off the thumb from palmar plan/short flexor/opponent responsible for an axial rotation of the metacarpal—all those three innervated by the median nerve and adductor innervated by the ulnar nerve like all the interosseous (Fig. 4.9)

Note that the first dorsal interosseous, easily palpable in the first interosseous space, is mainly active on the index finger and the privileged witness, in case of its atrophy, of a clinical impairment of ulnar nerve.

4.2.2 The Carpus

It has eight bones in two rows.

- The proximal row is a condylar surface with two radii of curvature corresponding to 2 DF: flexion/extension (or dorsiflexion) of the hand and abduction/adduction (or radial and ulnar deviation) (Fig. 4.10). It is articulated with the inferior extremity of the forearm and mainly with the radius which is connected with the ulnar styloid process by the very strong triangular ligament covered with articular cartilage.
- The second row is articulated by the mediocarpal joint to the first row and participates in part to the amplitude of flexion/extension of the wrist. A mortise made by the first row encloses os capitatum and blocks the transverse mobility.

The carpus has a curvature which is generally concave anteriorly and closed by a powerful carpal ligament, the flexor retinaculum, fixed on the lateral side on scaphoid tubercle and trapezoid and on medial side to pisiform and hamate. This creates an osteofibrous canal allowing the fingers flexor tendons and the median nerve to go through. This is often compressed by inflammatory synovial reactions that require surgical decompression which can be done nowadays by means of endoscopy (Fig. 4.11).

Mobilisation is provided by muscles attached on both palmar and dorsal sides:

- palmar side: flexor carpi radialis fixed on the base of second and third metacarpals/palmaris longus on the palmar fascia [22] which in reality is, as in the plantar sole, a tendinous sheath protecting the palm of the hand/flexor carpi ulnaris, on the ulnar metacarpals and the pisiform bone (a true sesamoid);
- dorsal side: long and short extensor carpi radialis on the dorsal base of the second and third metacarpals/extensor carpi ulnaris on the basis of the fifth metacarpal.

  Their action can easily be remembered considering two perpendicular planes of action: those activated on palmar side are flexors and backside extensors (or dorsiflexors) and those on the ulnar side make an ulnar deviation and on the radial side make a radial deviation;
- fragmentation of the carp (as tarsus) into a mosaic of small bones allowing small movements limited by many short inextensible ligaments greatly increases its strength, because

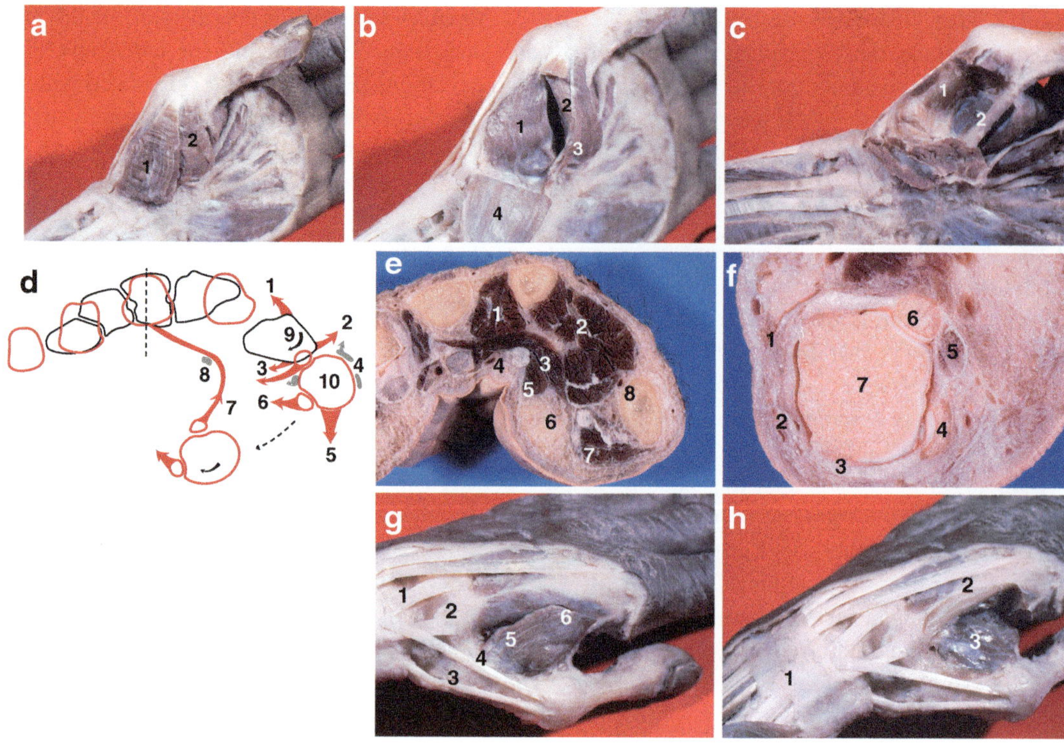

Fig. 4.9 Muscles of the thumb. (**a**) Left hand, superficial plan of the thenar muscles: 1. Abductor pollicis brevis responsible for thumb anteposition; 2. Flexor pollicis brevis. (**b**) Middle plan: 1. Opponens (only metacarpal fixation); 2. Deep part fixed on lateral sesamoid (ulnar nerve); 3. Superficial part of flexor pollicis brevis fixed on proximal phalanx of thumb (median nerve); 4. Reclined short abductor. (**c**) Deep plan: 1. First metacarpus; 2. Tendon of long flexor of thumb. (**d**) Diagram of thumb muscular actions: 1. First interosseous muscle; 2. Long abductor pollicis; 3. Opponens muscle; 4. Long and short extensors; 5. Short abductor; 6. Short flexor; 7. Adductor with its reflexion pulley; 8. Index flexor tendons; 9. First metacarpal bone; 10. Head of first metacarpal. (**e**) Transverse section of right hand seen from the front: 1. Second dorsal and first palmar interosseous muscles; 2. First dorsal interosseous muscle; 3. Adductor pollicis making reflexion around the long flexor tendons of index; 4. Flexor tendons of index; 5. First lumbrical; 6. Long flexor tendon of thumb; 7. Abductor pollicis brevis; 8. First metacarpal. (**f**) Transverse section of left thumb seen from the front: 1. Tendon of the long extensor inserted within the articular capsule of MP joint; 2. Tendon of the short extensor; 3. Articular capsule of MP; 4. Lateral sesamoid (anterior in neutral position); 5. Tendon of the long flexor within a fibrous sheath fixed to the intersesamoid ligament; 6. Medial sesamoid (posterior in neutral position). (**g**) Dorsal view of left hand: 1. Extensor carpi radialis brevis (third metacarpus); 2. Extensor carpi radialis longus (second metacarpus); 3. Short extensor of thumb; 4. Long extensor of thumb; 5. Radial artery going into the palm; 6. First dorsal interosseous. (**h**) Resection of the first dorsal interosseous: 1. Retinaculum of extensors; 2. Second metacarpus; 3. Adductor pollicis (medial sesamoid)

as we said before a rigid structure is more fragile than a deformable structure.

4.3 Humero-Radioulnar Complex

It concerns two important functional systems: the opening/closing brachio-antibrachial compass and the axial rotation of the hand [23–27].

4.3.1 The Junction Humerus/Ulna

It is a 1-DF joint of trochlear type with the olecranon bone block which limits the excursion joint to about 150° of flexion from the bone locking extension. It should be noted that the groove of the trochlea is in a spiral shaped. In full extension this creates a lateral offset which amplitude is normally greater in women and is a

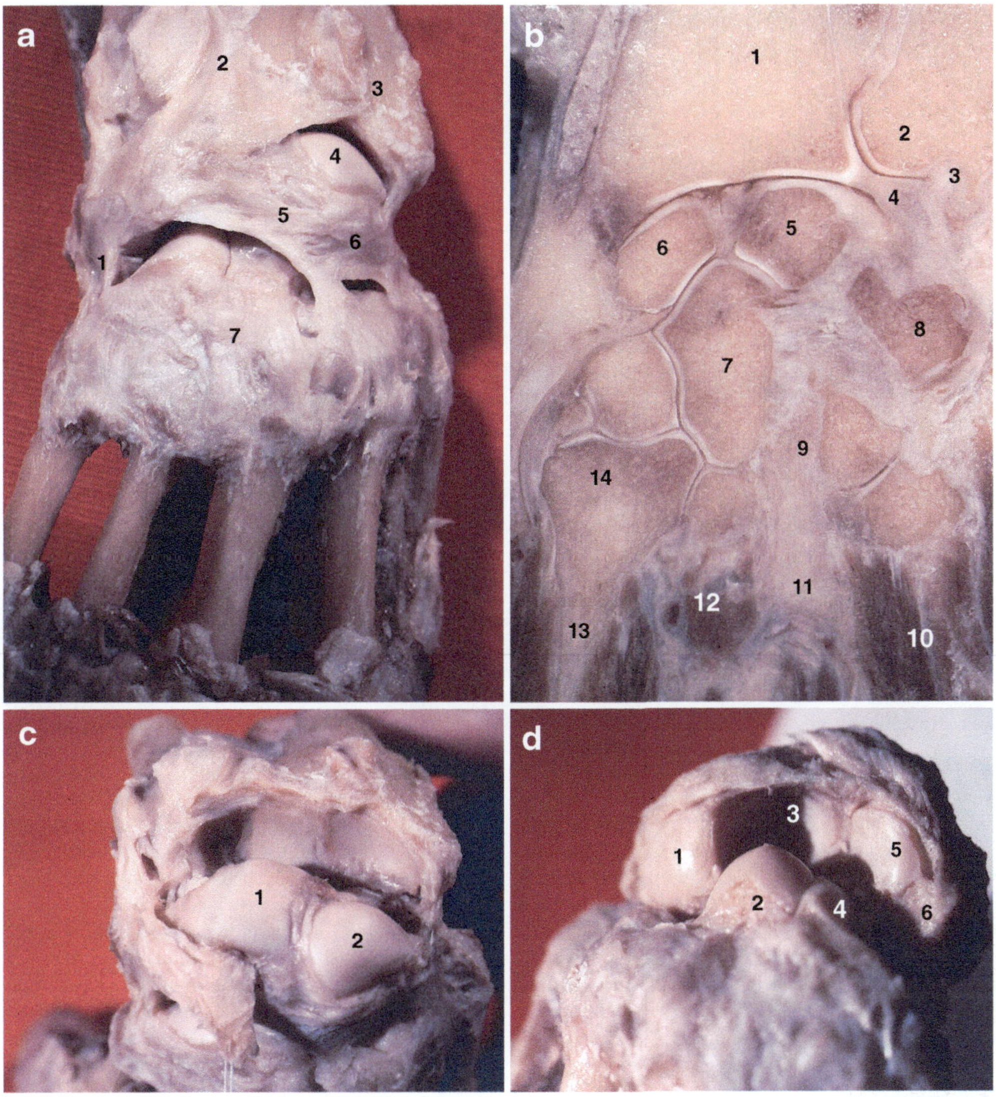

Fig. 4.10 The wrist joint. (**a**) Dorsal view of a right wrist: 1. Ulnar collateral ligament; 2. Dorsal radio-carpal ligament; 3. Styloid process of radius; 4. Scaphoid bone; 5. Dorsal intercarpal ligament; 6. Radial collateral ligament; 7. Distal row of carpus. (**b**) Transverse section of carpus: 1. Distal extremity of radius; 2. Head of ulna with the inferior radio-ulnar joint; 3. Styloid process of ulna; 4. Triangular fibro-cartilage ligament (disc); 5. Lunate; 6. Scaphoid; 7. Capitate; 8. Triquetrum; 9. Hamate; 10. Fifth metacarpus; 11. Fourth metacarpus; 12. Third metacarpus; 13. Second metacarpus; 14. Trapezoid. (**c**) Anterior view of the proximal row after section of the articular interline: 1. Scaphoid; 2. Lunate. (**d**) View of the distal row: 1. Scaphoid tubercle; 2. Capitate with its sagittal crest; 3. Lunate; 4. Hamate; 5. Triquetrum; 6. Ulnar collateral ligament

reliable sexual sign. Lower humeral is lopsided forward which facilitates bending.

Mobilisation of the elbow is a good example of the technical constraints imposed by the operating mode of muscles. We have seen that a muscle cannot shorten over one-third of its length. Therefore the elbow flexors cannot be used at their maximum mechanical efficiency which is to be inserted distally at hand. It is necessary to fix backward the tendons of these muscles in the direction of the axis of the elbow, which reduces their mechanical effectiveness.

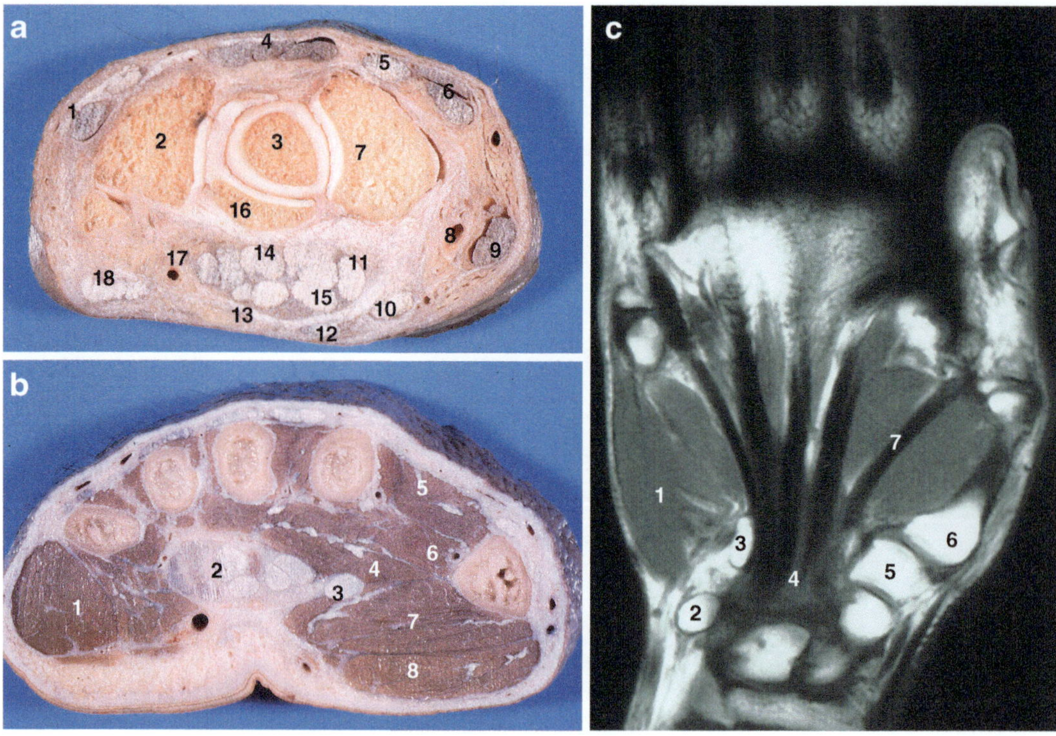

Fig. 4.11 Transverse sections of wrist and MRI of palm. (**a**) Transverse section of a left carpus seen from above: 1. Tendon of the ulnar extensor of carpus; 2. Triquetrum; 3. Capitate; 4. Tendons of fingers extensors; 5. Tendon of index extensor; 6. Tendon of long extensor of thumb; 7. Scaphoid; 8. Radial artery; 9. Tendons of short extensor and long abductor of thumb; 10. Tendon of flexor carpi radialis; 11. Tendon of long flexor of thumb; 12. Tendon of palmaris longus; 13. Transverse carpal ligament (retinaculum flexorum); 14. Flexor tendons within the carpal tunnel; 15. Median nerve; 16. Lunate; 17. Ulnar artery; 18. Tendon of flexor carpi ulnaris. (**b**) Transverse section of left palm seen from behind: 1. Hypothenar eminence muscles; 2. Distal part of the carpal tunnel; 3. Tendon of the long flexor of thumb; 4. Opponens pollicis; 5. First dorsal interosseous muscle; 6. Adductor pollicis; 7. Flexor pollicis brevis with its two parts; 8. Abductor pollicis brevis. (**c**) MRI of the palm: 1. Muscles of hypothenar eminence; 2. Pisiformis; 3. Hamulus of hamate; 4. Carpal tunnel; 5. Trapezium; 6. Base of first metacarpal with trapezo-metacarpal joint; 7. Long flexor tendon of thumb

This reflects a poorly studied phenomenon, which is to double the bending component of brachial muscles: polyarticular biceps and monoarticular brachialis by a symmetrical forearm component. For economical reasons, this component should intervene only when necessary, that is, when the strength is greater than the weight of the segment. It is imperative to automate this process for the reasons—always the same—of the ignorance of the pilot. Therefore it is the synergistic contraction of brachioradialis and biceps/brachialis that contracts automatically when applying a resistance at the end of the forearm. This is also a test of the functional validity of the C7 root.

According to the presence of two bones in the forearm, it is logical to interest both for bending. The short and stocky monoarticular brachialis attaches to the coronoid process of the ulna. The polyarticular biceps—connected to the scapula by two separate long and short tendons—attaches to the tuberosity of the radius, promoting its axial rotation and making it supinator. It should be noted that it is not the most active supinator and it creates one torque with the supinator muscle especially active when the forearm is in semiflexion. The supinator wrapped around the upper third of the radius is the main supinator, regardless of the position of the elbow.

Another important detail is the lacertus fibrosus which is a triangular tendinous sheath detached from the tendon of the biceps to spread on the origin of epitrochlear muscles. In this way, the traction of the biceps that can be very important is distributed intelligently to the medial side of the forearm.

The opposite movement, the extension of the forearm, is made by the powerful brachial triceps with its long polyarticular part fixed under the humeral glenoid with above and outside the tendons of the polyarticular biceps. The anconeus muscle is a dependence of the triceps and should be considered as an active ligament protecting the elbow joint and mainly the radioulnar.

4.3.2 Functional Pronosupination

This pronosupination gives to the hand some very useful and very used positions. The technical problem is to allow the hand to put its palm either upward (or front), favourable for supporting objects (*supinare*) or downward (or back) for taking objects (*pronare*). The amplitude of the rotation is of about 170°.

The solution cannot be a true axial rotation of the radius for mechanical strength reasons. The objective is to rotate the hand knowing that the carp has the same width as the forearm.

4.3.2.1 The Intelligent Solution Is to

- Put a wheel on radius at the top part of the elbow
- Fix the carpus set on a pallet placed on the inferior wide radius, the movable bone (ulna being blocked in its trochlea)
- Operate an axial rotation at both ends with inversion of the articular surfaces:
  - at the top, round radial head turning in the radial notch of the ulna in which it is fixed by the powerful annular ligament and rotating also below the capitulum of the humerus (condyle);
  - at the bottom, round ulnar head around which rotates the radius in the ulnar notch attached to ulnar styloid process by the triangular ligament, playing the role of an intra-articular disc. This articular system can be called "double reversed trochoid" resulting in a tilting movement in X of the radius on the ulna, greatly enhanced by the reverse curvature of the two bones. The pronators place themselves logically at the top: pronator teres from the epicondyle and the coronoid process to the side face of the radius and down and pronator quadratus fixed transversely on the distal quarter of radius and ulna

The supinators are the torque biceps/supinator. The brachioradialis is the muscle of the intermediate position putting the hand at 90°, between pronation and supination, a position widely used and in relation with the fixation of the brachioradialis on the styloid process of the radius.

The role of forearm interosseous membrane enhancing the mechanical resistance of the two zeugopod bones should not be underestimated. It is also interesting to note that the elbow representing the middle joint of the upper limb undergoes significant torsional stresses (like in the use of a tool-style handle of heavy hammer type or tennis racket), which are offset by the pronosupination. This is not the case for the lower limb which does not have this type of joint and resists, thanks to the original anti-torsion system of the knee.

4.4 Cleido-Scapulohumeral Complex

It seems convenient to group in the same functional study the scapulohumeral joint with 3 DF that supports the three-dimensional spatial orientation of the humeral lever and the suspension of the entire upper limb and shoulder girdle to cervicothoracic axial skeleton.

4.4.1 Cervicothoracic Suspension

It is made in the most mobile possible manner with only one sternoclavicular joint of low volume, the functional fixation being provided mainly by muscles. This is the fundamental difference with the lower limb that needs stability above all.

4.4.2 Scapulohumeral Joint

Along with this principle of maximum mobility, it has [28–31]:

- a lax capsule with low ligament reinforcements: superior, medial and inferior glenohumeral ligaments;
- a spherical humeral head of 6 cm in diameter, larger than the femoral head and representing more than a third of a sphere (155° to arc);
- a small scapular glenoid reception cavity almost flat of about 6 cm^2 inclined obliquely forward and upward, which does not allow such a retentive match like in the hip joint, even if the labrum, a fibrous rim around its circumference, contributes to enlarge the reception cavity in a viscoelastic manner (Fig. 4.12);
- everything is under one rounded osteofibrous roof, the coracoacromial arch, formed by two bone processes: at the back the spatula-shaped acromion extending outside the spine of the scapula and in the front, the coracoid process, semi-flexed finger-shaped, sagittally aligned below the clavicle with which it is joined by the two conoid and trapezoid coracoclavicular ligaments. Both bone structures are linked together by the strong coracoacromial ligament

4.4.3 Mobilising and Fixing Muscles

Mobilisation and fixation of the complex can be described in the two following groups.

4.4.3.1 The Scapulohumeral Muscle Group

This group comprises the muscles attaching scapula and humerus [32]:

The Rotator Cuff

They are the muscles surrounding closely the humeral head and forming a sort of cap called rotator cuff due to the strong rotational component of those muscles: behind on both sides of the dorsal scapular spine are the supraspinatus and infraspinatus muscles (lateral rotation) which join the teres minor and on the anterior side of the scapula the subscapularis muscle (medial rotation).

The Deltoid Muscle

It is a triangular muscle covering the joint and giving the external modelling side of the shoulder. It consists of three parts: anterior attached to the lateral third of the clavicle, median fixed to the acromion and posterior to the dorsal spine of the scapula.

It has a fascicular innervation, which means that it receives from the circumflex nerve separate branches for the anterior and middle and for the posterior bundle, that shrinks only if we add a retropulsion against resistance during abduction. Its total contraction makes it the main abductor muscle of the arm and its attachment to a ridge placed at half the humeral shaft gives a large lever arm.

It has several technical problems that it is important to well understand [33].

- Firstly, to lift up the arm in abduction from its starting position (arm along the body), the initial tendency of the deltoid is to pull the humerus upward. Fortunately, there is the additional supraspinatus muscle willing horizontally, fixed to the greater tubercle of the humerus and which pulls off the humerus for abduction. A serous bursa below facilitates sliding.

  The notion of *starter* played by the supraspinatus which contracts undoubtedly together with the deltoid is now rightly rejected. But the important biomechanical fact is that a solid stop is mandatory above the humerus in order to force the humeral head to rotate for abduction by the action of the deltoid. This stop is formed by the supraspinatus tendon with its enhanced and

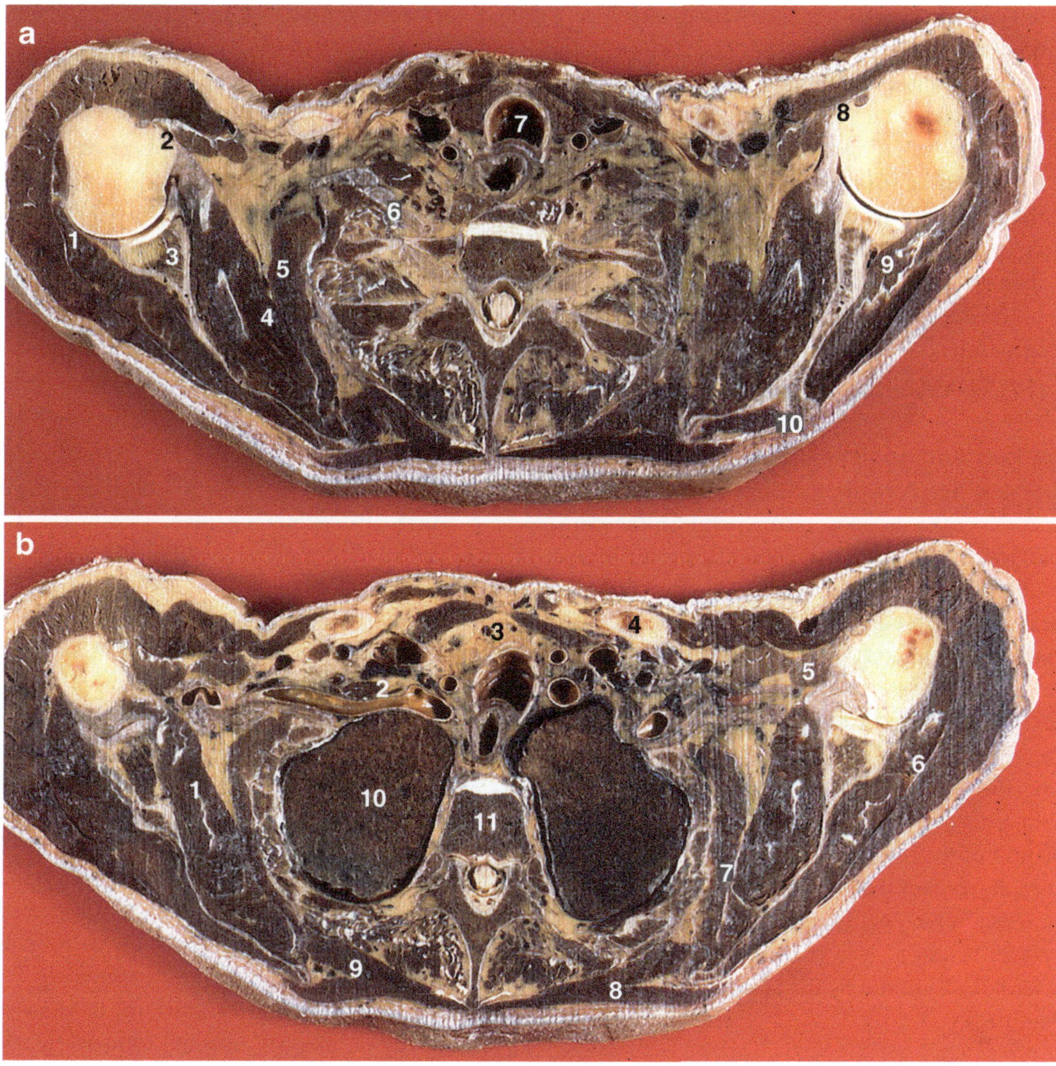

Fig. 4.12 Horizontal sections of superior part of the trunk seen from above. (**a**) Section of the glenohumeral joints: note the oblique position of the scapula and the posterior orientation of the head of humerus: 1. Deltoid muscle; 2. Humeral minor tubercle; 3. Glenoid cavity of scapula; 4. Subscapularis; 5. Serratus anterior; 6. Brachial plexus at the superior part of scalene triangle (anterior and middle scalene muscles); 7. Trachea with oesophagus behind and thyroid gland in front; 8. Tendons of subscapularis fixed on minor tubercle and of long head of biceps within the intertubercular groove; 9. Infraspinatus; 10. Spine of scapula. (**b**) Section below the joints: 1. Subscapularis; 2. Subclavian artery crossing the inferior part of scalene triangle; 3. Group of arterial branches of aorta; 4. Clavicle; 5. Pectoralis minor and major; 6. Infraspinatus muscle; 7. Serratus anterior; 8. Trapezius; 9. Rhomboid; 10. Superior lobe of left lung; 11. Thoracic vertebra

penniform appearance, doubled in the front by the intracapsular biceps tendon that effectively completes the stop effect. This also contributes mostly to the required humeral head centring.

Again, pathology provides convincing arguments: during the rupture of the supraspinatus and/or biceps, the patient may do abduction of the arm only by running laterally the upper limb passively in order to

engage the lateral component of abduction of deltoid and overcome the lack of stop by compressing the humeral head against the glenoid cavity.
- Secondly, what has been said about the limited potential for muscle shortening applies to the abduction of the arm. The third length of deltoid fibres alone can put the arm at no more than 90° and even less, considering the weight of the upper limb. This is seen clearly in cases of paralysis of the trapezius and we will discuss this later.

The Two Antagonistic Brachial Muscles

They are fixed on the glenoid: below, the long triceps; above, the long biceps; and on the coracoid process, the short biceps. The long biceps tendon has a long walk between the two crests of the humerus tubercular, which is closed by a transverse ligament with an inner sliding synovial system. Then it binds very near to the upper part of the glenoid after entering the capsule. Therefore, as explained above, the tendon will play the role of a powerful active ligament maintaining the humeral head. Indeed, the contraction of the biceps flexes the arm, but tends to push the humeral shaft upward. This action is seen, as already mentioned, in the case of tendon rupture, resulting in a rise of the humeral shaft nevertheless limited by the acromiocoracoid arch [34–36].

4.4.3.2 The Scapulo-Spinal Muscle Group

The scapula is a flat bone free of skeletal articulation and only maintained and mobilised by muscles. It realises with the clavicle a triangle of suspension of the upper limb to the axial skeleton with a forearm attached to the sternum by a 2-DF joint allowing limited movements of forward/backward and up/down. The joint is small but solid with a meniscus inside and very strong ligaments, which does not preclude its frequent traumatic dislocation. Similarly the clavicle is secured to the scapula by a powerful coracoclavicular ligament divided into two entities: the trapezoid ligament and medially the conoid ligament.

The Arm Abduction Beyond 90°

This technical problem is responsible for many terminologies appeared in the literature by creating confusion, and it is important to stay closer to the anatomical reality described from living, being in a natural vertical posture [37–41].

The scapula and its clavicular suspension arm in this steady state, the upper limb along the body with the hand in middle position and not supine (palm forward as in the classic position of anatomical description), is not in a frontal but in an oblique plane along the rounded chest wall. Therefore, the coracoid process is like a flexed finger sagittally aligned. Its summit is palpable below the clavicle, and it is the start of a muscular V with coracobrachialis muscle on lateral side and pectoralis minor on the medial side. Thus the glenoid cavity is in an oblique position directed toward the front and the humeral head looks back. Obviously rotators of the rotator cuff can significantly alter this position.

As the deltoid has limited shortening, the solution to raise the upper limb above the head is to change the position of the scapula by tilting it upward. This tends to horizontalise the glenoid cavity, which puts it in a better position to withstand mechanical load on the humerus. Some people talk about moving a bell, but it seems more appropriate to speak about a rocking motion with bascule and counter bascule.

Indeed, the rocking motion in bascule is provided by a pair of muscle formed by the trapezius muscle and serratus anterior, and the counter bascule by the couple rhomboid on the back and pectoralis minor on forward, the only active muscle able to put down the shoulder girdle.

Some important anatomical details of these two antagonistic systems are described below.

The Trapezius

It is also called cucullaris because of its resemblance with hooded monks. It has a rhombic shape with firstly an upper lateral part attached to the occipital bone, secondly a middle lateral part attached over the superior lateral third of the clavicle and of the superior edge of the spine of the scapula, and thirdly, an inferior part. Medially it attaches to the spinous processes of C1 to T11 with a tendinous lozenge in the middle.

Some misinterpretations must be permanently erased:

- firstly, there is no reason to speak about three beams that could contract separately, because unlike the deltoid, the trapezius muscle has a global innervation activating the whole muscle;
- secondly, its vectorial representation should be in the form of a broken line with an upward component of elevation, a slash materialised by the lever of the dorsal spine of the scapula and inferior oblique component to the spine that explains its function of tilting the scapula with elevation of the clavicle. The sternocleidomastoid muscle that attaches to the upper medial third of the clavicle does not have a sufficient lever arm to intervene in the rocking motion;
- thirdly, the attachment to the cephalic mass has no effect on this one, because the trapezius is essentially a shoulder muscle. As age increases, it becomes less and less easy to raise one's arm and the atrophy of the cephalic part of trapezius allows to see better behind it the semispinalis capitis looking like two strong vertical fleshy columns, also very well visible in children.

The Serratus Anterior

It is a muscle with a wide range of fingering from the first to the ninth rib, with a fixation on the medial border of the scapula, which is applied against it. It seems necessary to stop talking about a scapulothoracic articulation or syssarcosis, to designate the connective slip plane on the side of the chest under the serratus anterior. The words "articulation" "joint" should be exclusively used when there is a contact between two skeletal pieces.

When scapula is blocked in position, it is a powerful respiratory inspiration muscle and its action to pull forward the medial border of scapula is a strong component of the rocking motion of the scapula to the elevation of the upper limb, in synergy with trapezius and deltoid. Note that this muscle provides a wide elastic suspension of the anterior train of quadrupeds with a narrower vertical scapula and a well-developed and centred thorn.

The Rhomboid

It is artificially separated into small originated from C6/C7 and large T1/T4. It is an oblique muscle which approximates the medial border of the scapula with the midline of the spinous processes. When the girdle is positioned in the passive or active bascule, it actively reduces the scapula toward the midline with the pectoralis minor lowering the girdle and putting it in neutral position (counter bascule). The active joint of both scapulas is made by a combined action trapezius/rhomboid whose resultant action is horizontal, given by the inverted obliquity of both muscles.

The Levator Scapulae

It requires a special mention because it suspends the upper corner of the scapula from the posterior tubercles of transverse processes of C1 to C4 and participates to the scapula rocking motion. This is a neck muscle that is easily palpable. The levator scapulae is frequently activated, which might contribute to originate some specific myalgia that it is necessary to properly diagnose and treat.

The Powerful Adduction Strap

It is formed by three muscles (the big three):

- on the anterior side, the pectoralis major attached to the two-thirds of the medial clavicle, the anterior portion of the sternum and the upper part of the aponeurosis of the recti muscles;
- in the middle the teres major coming directly from the lower part of the lateral edge of the scapula;
- posteriorly, the latissimus dorsi from the spinous processes of T7 to the sacrum.

The three elements of this strap are attached to the upper third of the humerus on two vertical lateral and medial tubercular ridges. There is a kind of U-shaped folding of concentrating muscle bundles of muscle fibres of both anterior and posterior broad muscles prior to bone fixation.

It raises the question for the need of such a powerful strap to put the humerus in adduction. The answer lies in the possible use of this strap for climbing. This requires raising the entire body by adduction of the humerus being vertical head down pushing it closer to the centreline. These muscles are highly developed among weightlifters and trapeze artists.

Incidentally the latissimus dorsi by its large dorsal medial rotation component provides access to the posterior thoracic skin *(scratching)*. It is also the one that is highly stressed in wheelchair propulsion by providing a thrust extension on wheels, which creates more and more shoulder pathological problems in paraplegic people.

In total, the scapulohumeral complex provides direction and broad movement of the upper limb in space by combining the 2 DF of girdle and the 3 DF of humeral joint. The mechanical strength of the assembly is essentially a problem of muscles.

Conclusion

The complexity of the construction of the upper limb is at the maximum of its functional capabilities. Man clearly shows his mental superiority over the animal, and all the refined technologies he has created owe mainly to the technical organisation of the hand, this extraordinary versatile tool.

References

1. Rabischong P (1971) Les Problèmes Fondamentaux du Rétablissement de la Préhension. Ann Chir 25 (19–20):927–933
2. Steindler A (1955) Kinesiology of the human body. Charles C Thomas, Springfield
3. Rabischong P (2012) Anatomie Compréhensive du Membre Supérieur. In: Bonneau D, Vautravers P, Hérisson C (eds) Médecine du Sport et Thérapies manuelles. Sauramps Médical, Montpellier
4. Tubiana R (2000) Anatomy of the hand and upper limb. In: Tubiana R, Amadio PC (eds) Medical problems of the instrumentalist musician. Martin Dunitz, London, pp 5–53
5. Tubiana R (1981) Architecture and function of the hand. In: Tubiana R (ed) The hand, vol 1. WB Saunders, Philadelphia, pp 493–498
6. Rabischong P (2000) Command and control of hand functions. In: Tubiana R, Amadio PC (eds) Medical problems of the instrumentalist musician. Martin Dunitz, London, pp 91–120
7. Moberg E (1958) Objective methods for determining functional value of sensibility in hand. J Bone Joint Surg Br 40B:3
8. Moberg E (1960) Evaluation of sensibility in the hand. Surg Clin North Am 40(2):357–362
9. Dellon AL (1990) The sensational contributions of Erik Moberg. J Hand Surg Br 15(1):14–24
10. Mansat M (1977) Anatomie topographique et chirurgicale du plexus brachial. Rev Chir Orthop 63:20–26
11. Samii M (1990) Special considerations regarding the treatment of brachial plexus lesions. In: Samii M (ed) Peripheral nerve lesions. Springer, Berlin
12. Narakas A (1978) Surgical treatment of traction injuries of the brachial plexus. Clin Orthop Relat Res 133:71–90
13. Bonnel F, Allieu Y, Bruner P (1980) Anatomical and surgical principles of the brachial plexus in newborn children. Int J Microsurg 2(1):12–15
14. Seddon HJ (1954) Peripheral nerve injuries. Medical Research Council Special Report 282. Her Majesty's Stationery Office, London
15. Sunderland S (1968) Nerves and nerve injuries. Livingstone, London
16. Bonnel F, Rabischong P (1976) Contribution à l'étude de la systématisation du plexus brachial. In: Actualités en Rééducation Fonctionnelle et Réadaptation, Masson
17. Rabischong P, Peruchon E, Pech J (1977) Is man still the best robot? In: Proceedings of the VIIth international symposium on industrial robots, pp 49–57
18. Landsmeer JMF (1955) Anatomical and functional investigations of the articulation of the human finger. Acta Anat 25(suppl 24):1
19. Landsmeer JMF (1990) Introduction à la morphologie fonctionnelle du doigt. Ann Kinésithér 17(5):203–210
20. Landsmeer JMF (1976) Atlas of anatomy of the hand. Churchill Livingstone, Edinburgh
21. Pieron AP (1973) The mechanism of the first carpometacarpal joint. An anatomical and mechanical analysis. Acta Orthop Scand Suppl 148:1–104
22. Stack HG (1973) The palmar fascia. Churchill Livingstone, Edinburgh
23. Iselin M, Iselin F (1967) Traité de chirurgie de la main. Flammarion Médecine Sciences, Paris
24. Tubiana R (1981) The hand, vol 1. Saunders, Philadelphia
25. Tubiana R (1997) La Main. Anatomie fonctionnelle et examen clinique. Masson, Paris
26. Tubiana R (1980) Traité de chirurgie de la main, Tome 1 Anatomie, physiologie. Masson, Paris

27. Tubiana R, McCullogh CJ, Masquelet AC (2004) Voies d'abord chirurgical du membre supérieur. Elsevier Masson, Paris
28. Basmajian JV (1978) Muscles alive, 4th edn. Williams and Wilkins, Baltimore
29. Inman VT (1944) Observations of the function of the shoulder joint. J Bone Joint Surg 26A:1–30
30. Souquet R (1983) Syndromes canalaires du membre supérieur. Monographies du GEM Expansion Scientifique Française
31. Codman EA (1934) The shoulder. Thomas Todd, Boston
32. Fischer LP, Carret JP, Gonon GP et al (1977) Etude cinématique des mouvements de l'articulation scapulo-humérale. Rev Chir Orthop 63(suppl II):108–112
33. Leroux JL, Micallef JP, Blotman F et al (1992) Shoulder abduction analysis in normal and pathological conditions using the Elite System. Eur J Phys Med Rehabil 2:89–95
34. Lucas D (1973) Biomechanics of the shoulder joint. Arch Surg 107:425–432
35. Poppen NK, Walker PS (1976) Normal and abnormal motion of the shoulder. J Bone Joint Surg Am 58:195–201
36. Saha AK (1983) Mechanism of shoulder movements and a plea for the recognition of "zero position" of gleno-humeral joint. Clin Orthop 173:391–393
37. Gagey OJ, Bonfait H, Gillot C (1988) Anatomie fonctionnelle et mécanique de l'élévation de l'épaule. Rev Chir Orthop 74(3):209–218
38. Antony NT, Keir PJ (2010) Effects of posture, movement and hand load on shoulder activity. J Electromyogr Kinesiol 20:191–198
39. Ebaugh DD, Spinelli BA (2010) Scapulothoracic motion and muscle activity during the raising lowering phases of an overhead reaching task. J Electromyogr Kinesiol 20:199–205
40. Kaplan EB (1953) Functional and surgical anatomy of the hand. Lippincott, Philadelphia
41. Wynn Parry CB (1966) Rehabilitation of the hand, 2nd edn. Butterworths, London

Eye Mobility

5

Abstract

Vision is a complex function that allows the identification of shape, colour and movement of objects. With two eyes, only one image is perceived in 3D: it is stereoscopy. This requires a perfect fusion of the two retinal images made by the optic chiasma and the retinal macula—a high-resolution point placed along the optical axis. Eye movements have to be executed around the geometric centre of the eyeball explaining why a protraction force given by the superior oblique muscle has to be used as a counterpart of the retrotraction force produced by the four recti muscles. The images are guiding the muscular dynamic coordination and are managed at the level of the superior colliculus representing the mandatory functional link between visual images and eye movements. The tectospinal neural pathway is the activating system of the eye/head servo-mechanism, allowing the movement of both eyes and head in case of large visual explorations. The oculomotor nerves (III, IV, VI) are purely motor nerves requiring a proprioceptive anastomosis with the sensitive nerve of the territory: the trigeminal nerve (V), which has for this reason the longest brainstem nucleus.

5.1 Introduction

Vision is a sensory programme and we have already largely spoken about it in Chap. 2. Still, we have to recall that the vision programme allows the identification of shapes and colours of objects in visual space, as well as of their movements.

In addition, we have two eyes and see only one single image, which is the principle of stereoscopy. A stereoscopic vision involves a lateral shift of the two retinal images and especially their perfect superposition, pixel by pixel.

This technical problem is solved by two specific complex structures. The first is the optic chiasm that crosses retinal fibres at 50 % in humans and primates. They then go to the integrator of the two half-retinal images, the lateral geniculate nucleus with six layers, and then project through the optic radiations to the primary visual cortex of the calcarine fissure (area 17) of the occipital lobe (Fig. 5.1). This configuration is

P. Rabischong, *Comprehensive Anatomy of Motor Functions*, DOI 10.1007/978-3-319-04169-8_5,

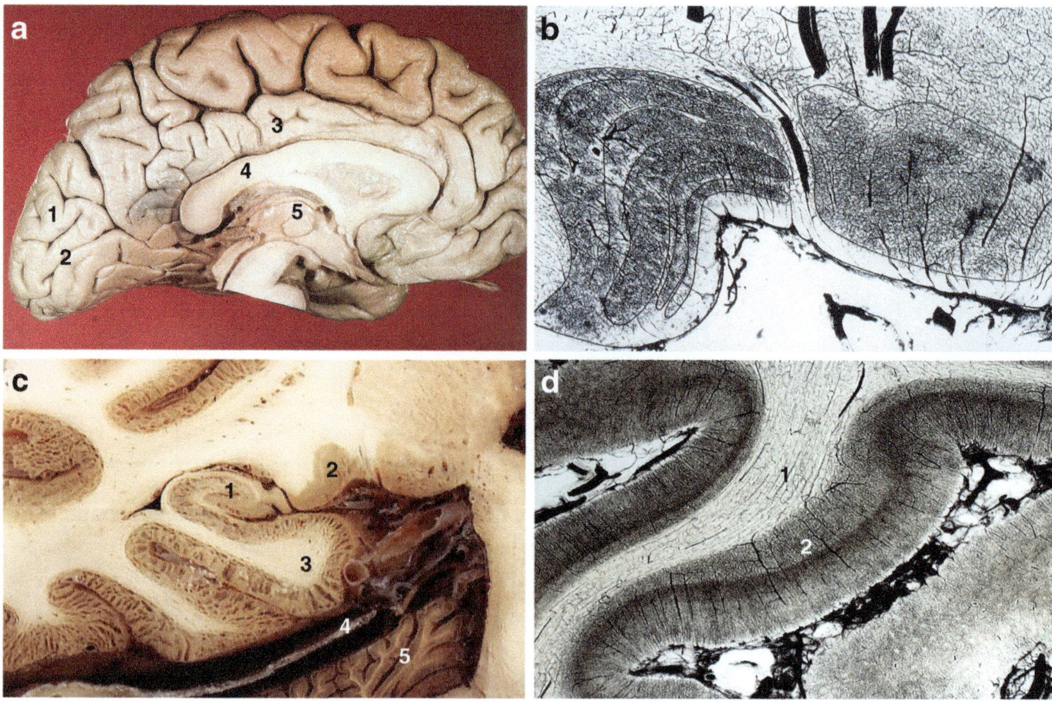

Fig. 5.1 Visual cortex and lateral geniculate nucleus (LGN). (**a**) Medial aspect of the interhemispheric section: 1. Cuneus of occipital lobe; 2. Calcarine fissure; 3. Cingular gyrus; 4. Corpus callosum; 5. Third ventricle. (**b**) Vascular microinjection (H. Duvernoy): the six layers of the NGL (three for each retina). (**c**) Coronal section of temporal lobe seen from behind: 1. Hippocampus; 2. LGN; 3. Gyrus parahippocampi; 4. Tentorium cerebelli; 5. Cerebellum. (**d**) Microinjection of vessels of primary visual cortex (H. Duvernoy): 1. White matter; 2. Gray matter with Gennari stria

such that each cerebral hemisphere sees a visual hemifield.

But this is still not sufficient to guarantee a perfect fusion of images. There is a high-resolution point in the retina (135,000 cones per mm^2) placed in the optical axis and that is responsible for 80 % of vision: the macula with the fovea, a small crater enhancing the precision of point setting, which gives a perfect centring of images and therefore stereoscopy.

One technical problem is the difficult implementation of the coupling of the two eyes, which may not be perfect at birth. If, subsequently, there is a shift of the two optical axes that cannot be retrieved, it is anticipated in the construction plan to provide a correction to avoid the very debilitating diplopia thanks to corticoretinal fibres described by Mawas which are capable of inhibiting one macula.

5.2 Three Specific Features Are Important to Understand

5.2.1 A Broad Visual Space Exploration

It can be obtained by moving the two eyeballs along two main axes, vertical and horizontal. In order to achieve this, it seems logical to introduce two pairs of rectus muscles: superior/inferior for the vertical axis and medial/lateral for the horizontal axis of visual exploration. The precise functional coupling of the two globes for stereoscopic vision imposes that eyeball rotations be perfectly centred on the centre of rotation of the globes [1–3]. But, during their contraction, the recti muscles exert a backward traction (Fig. 5.2).

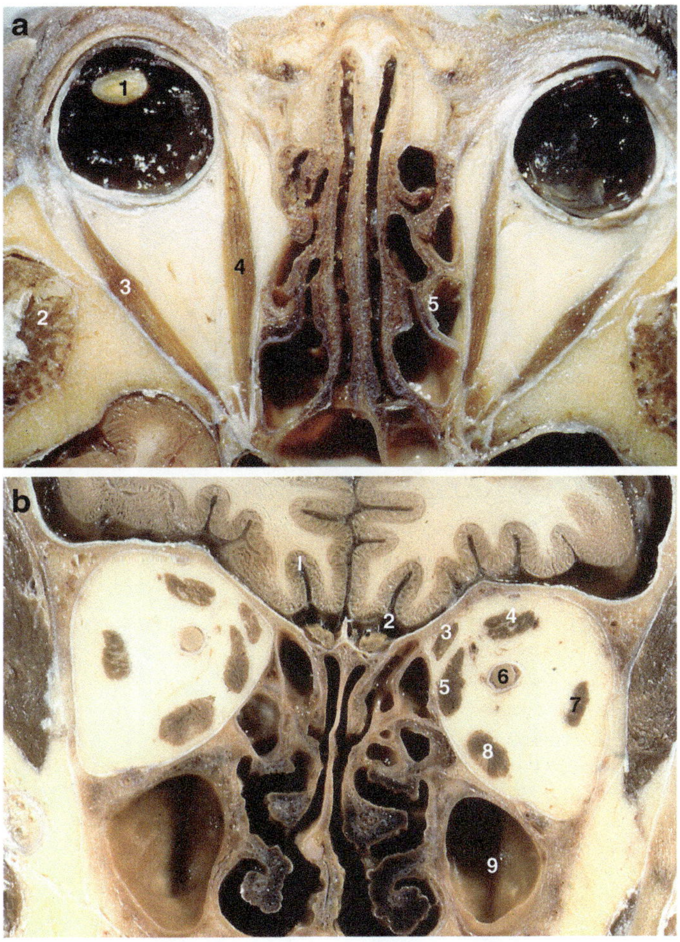

Fig. 5.2 Anatomical sections of orbital cavities. (**a**) Axial section: 1. Lens (back post mortem position); 2. Temporal muscle; 3. Lateral rectus muscle; 4. Medial rectus muscle; 5. Ethmoidal cells. (**b**) Coronal section: 1. Orbital part of the frontal lobe; 2. Olfactory tract within the olfactory groove; 3. Superior oblique muscle; 4. Superior rectus and levator palpebrae superioris; 5. Medial rectus; 6. Optic tract; 7. Inferior rectus; 8. Lateral rectus; 9. Maxillary sinus

To counter this retrotraction, a first idea is to limit the degrees of freedom of the globes by placing them in a fibrous sac, the Tenon's capsule, and sheath of the bulb that acts as a gliding space. This continues backward through the sclera and the sheath of the optic fascicle and forward in the episcleral space. On the sides, the capsule is fixed by fibrous fins to the sidewalls of the orbit.

However, this passive limitation must be coupled with an active compensation by an opposing force to the protraction. This justifies the necessary presence of the superior oblique muscle, which reflects on a fibrous pulley placed in the superomedial angle of the orbit (Figs. 5.3 and 5.4). From there, the tendon fibres are spread in a wide fan on the superior aspect of the globe passing under the superior rectus. The contraction of this superior oblique muscle obviously pushes forward the eyeballs, compensating the protraction of the recti. This ingenious anatomical solution could probably explain why the superior oblique muscle has a particular nerve, the trochlear nerve (IV). It is the only cranial nerve to emerge back near the superior colliculus, which is precisely the centre of the mandatory junction between visual images and eye movements [4, 5].

So, a dynamic stabilisation of the globes is given by a dynamic equilibrium of seven muscle forces: the four recti and the two superior and inferior obliques. They allow ocular coupled movements in all directions:

- vertically up and down: supraduction and infraduction;

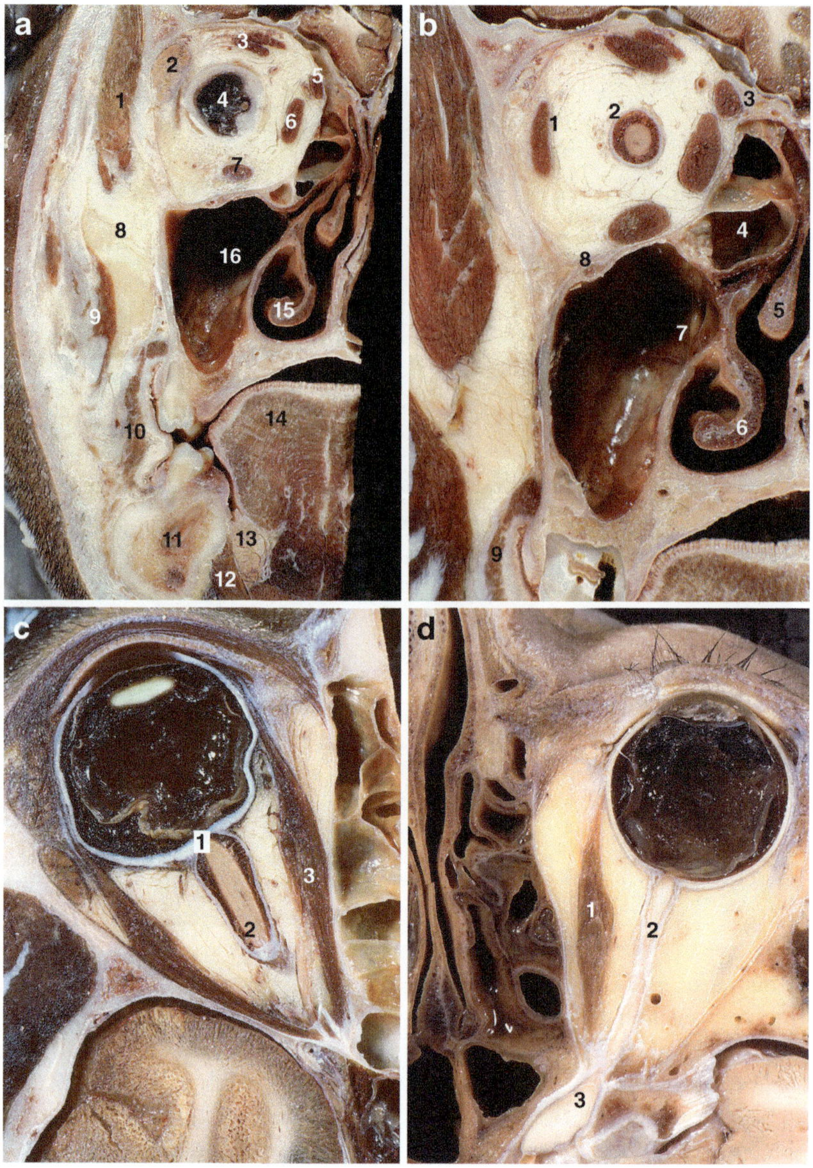

Fig. 5.3 Coronal and axial sections of the orbits. (**a**) Coronal section of right face at the level of the posterior part of eye ball: 1. Anterior part of temporal muscle; 2. Lacrimal gland; 3. Levator palpebrae superioris and superior rectus; 4. Posterior part of eye ball with the optic papilla; 5. Superior oblique; 6. Medial rectus; 7. Inferior rectus; 8. Fat pad in front of the temporal muscle tendon (Bichat); 9. Masseter; 10. Buccinatorius; 11. Mandible with inferior dental canal; 12. Mylo-hyoid muscle; 13. Sublingual salivary gland; 14. Tongue; 15. Inferior turbinate; 16. Maxillary sinus. (**b**) Coronal section at the middle part: 1. Lateral rectus; 2. Optic fascicle with its meningeal sheaths well visible in this young fellow. That prove the cortical diencephalic origin of the nervous retina and the externalization of the optic pathway which is not a cranial nerve (ophtalmencephalon); 3. Posterior ethmoidal canal (nerve and vessels); 4. Ethmoidal cells; 5. Middle turbinate; 6. Inferior turbinate; 7. Natural drainage hole of the maxillary sinus; 8. Infra-orbital canal (nerve from V2 and vessels); 9. Buccinatorius. (**c**) Left axial section on a young fellow: 1. Optic papilla; 2. Optic fascicle with meningeal space and central optic artery; 3. Medial rectus and its nerve. (**d**) Right axial section on an elderly: 1. Medial rectus; 2. Optic fascicle without meningeal space; 3. Optic canal

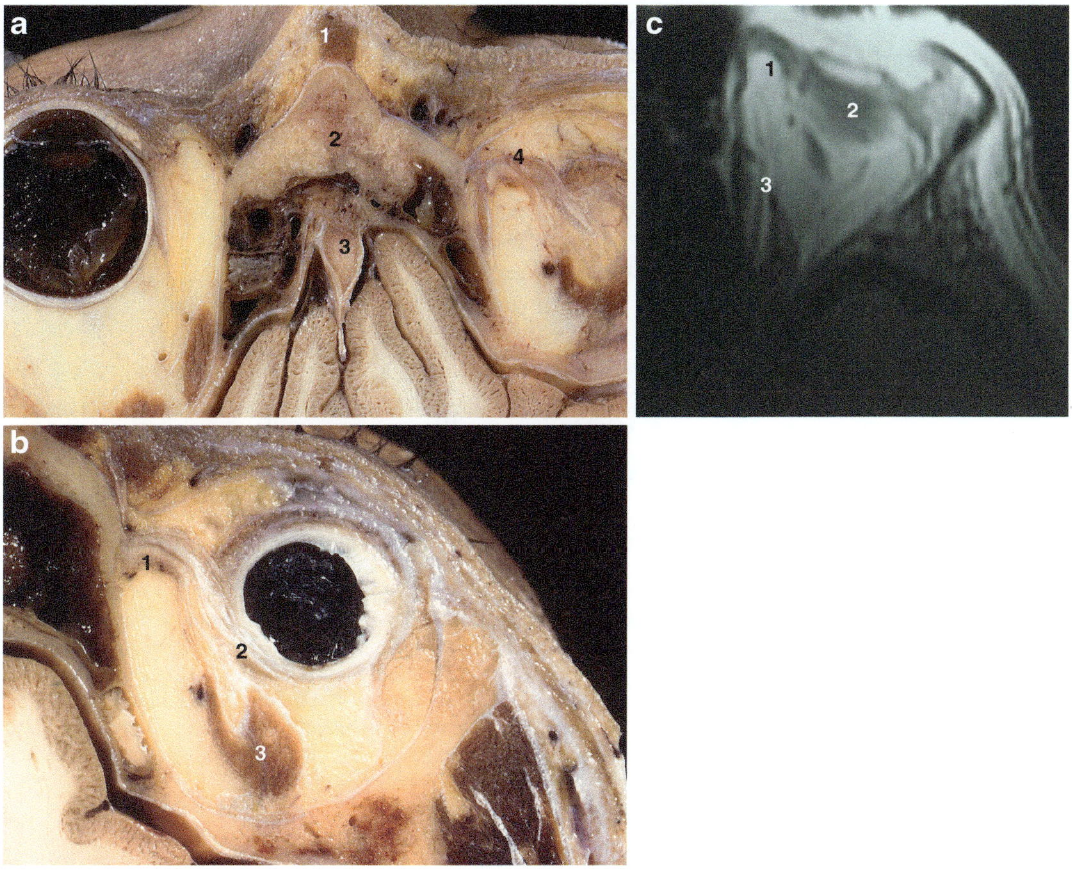

Fig. 5.4 The oblique superior muscle. (**a**) Axial section of the orbits: 1. Procerus muscle; 2. Fronto-nasal spine; 3. Crista galli (ethmoid bone and falx cerebri insertion); 4. Pulley (trochlea) for the reflection of the superior oblique muscle tendon. (**b**) Axial section of a right orbit: 1. Pulley; 2. Tendon on the medial and posterior part of the eye ball; 3. Superior oblique muscle. (**c**) MRI in axial section of right orbit: 1. Pulley; 2. Tendon thrown back on the eye ball; 3. Superior oblique muscle

- horizontally right and left: abduction and adduction;
- obliquely: intorsion and extorsion.

In addition, by decoupling their bilateral action and due to the power of the foveal fixation, they can converge or diverge when fixing a point moving forward and backward in the front of the nose. Any paralytic or paretic muscle causes misalignment with a more or less important diplopia.

5.2.2 Central Control of Eyeball Movements

The important point to remember is that the images are guiding all the coordination of the ocular movements. In fact, the visual pathways have two different paths: a first track, retino-geniculostriate, ending in the visual occipital cortex, and the second track, retino-colliculo-cortical, which from the Y ganglion cells of the

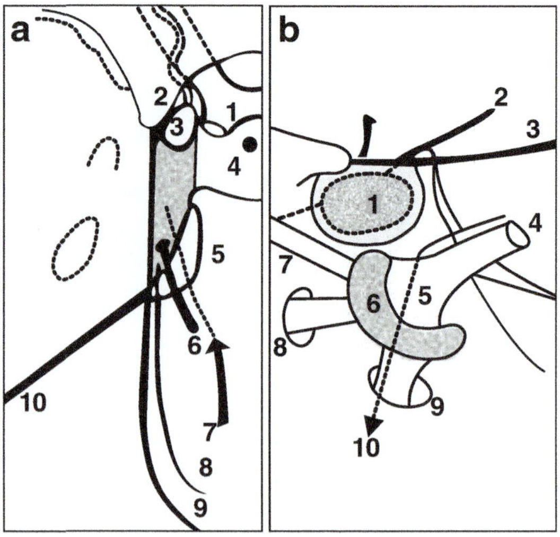

Fig. 5.5 Drawing of cavernous region and trigeminal cavum. (**a**) Superior view: 1. Optic chiasm; 2. Anterior clinoid process; 3. Internal carotid artery emerging from the cavernous space; 4. Hypophysis; 5. Dorsum sellae with the posterior clinoid process shifted medially for 3 mm creating the dural roof of the cavernous sinus; 6. Ocular motor nerve (III); 7. Abducens nerve (VI) passing through the Dorello canal at the petrous bone apex; 8. Trochlear nerve (IV) following the superior border of the tentorium; 9. Fixation of the tentorium on the superior border of the petrous bone; 10. Incisura tentorii for the brain stem. (**b**) Lateral view of the cavernous region and trigeminal cavum: 1. Hypophysis within sella turcica; 2. Ocular motor nerve (III); 3. Superior border of the tentorium fixed on anterior clinoid process; 4. Sensitive root of trigeminal nerve (V); 5. Triangular plexus; 6. Trigeminal ganglion (Gasser); 7. Ophtalmic nerve (V1) in the superior orbital fissure; 8. Maxillary nerve (V2) in the foramen rotundum; 9. Mandibular nerve (V3) in the foramen ovale; 10. Motor root of trigeminal nerve (masticatory nerve) following the V3

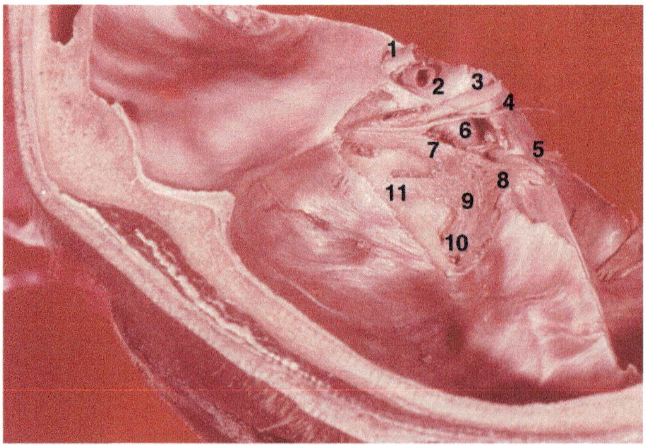

Fig. 5.6 Oculomotor and left trigeminal nerves. Dissection of the left trigeminal cavum (resection of the dura) in the middle cerebral fossa: 1. Optic fascicle; 2. Internal carotid artery; 3. Ocular motor nerve (III); 4. Trochlear nerve (IV); 5. Abducens nerve (VI); 6. Parkinson triangle with the internal carotid artery; 7. Ophtalmic nerve (V1); 8. Sensitive root of the trigeminal mixed nerve (V); 9. Triangular plexus and trigeminal ganglion; 10. Mandibular nerve (V3) and motor root (masticatory nerve); 11. Maxillary nerve (V2)

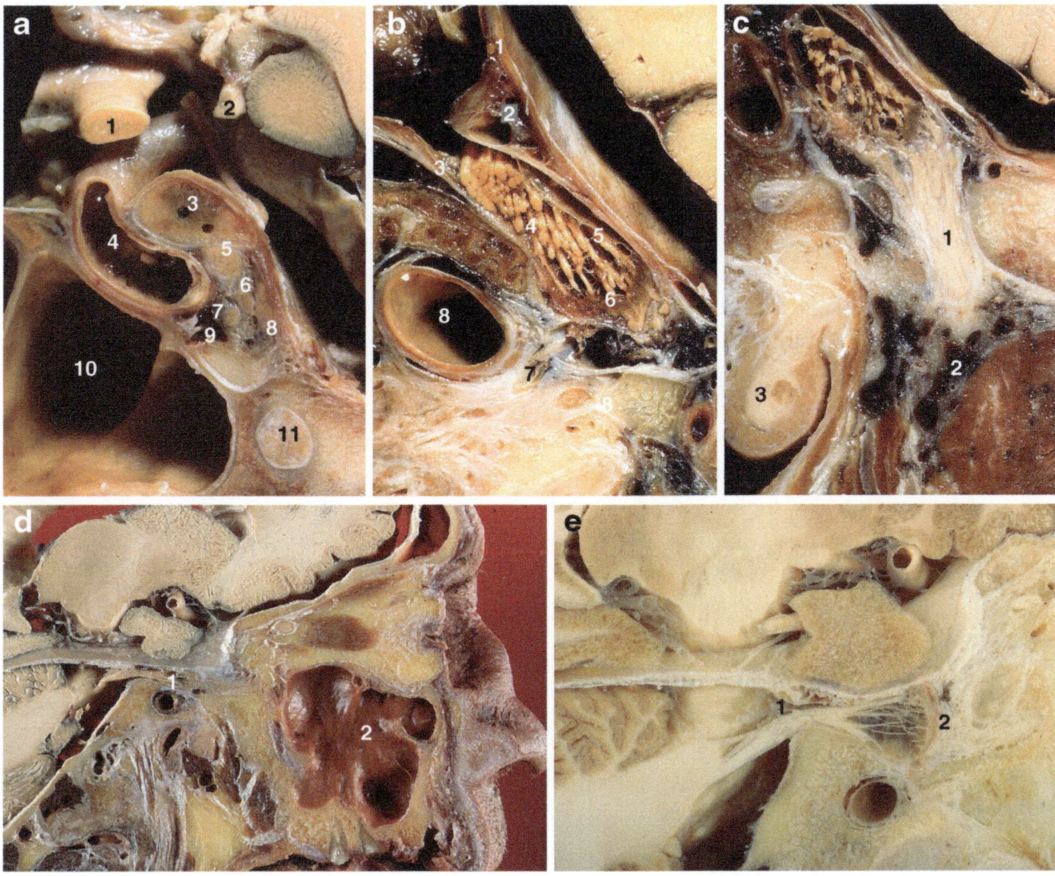

Fig. 5.7 Oculomotor and trigeminal nerves. (**a**) Coronal section of the anterior part of the left cavernous sinus seen from the front: 1. Optic fascicle and optic chiasm; 2. Olfactory tract; 3. Anterior clinoid process; 4. Anterior loop of the internal carotid artery; 5. Ocular motor nerve (III); 6. Trochlear nerve (IV); 7. Abducens nerve (VI); 8. Ophtalmic nerve (V1); 9. Anterior part of the venous cavernous sinus; 10. Sphenoid sinus; 11. Maxillary nerve (V2). (**b**) Coronal section of the posterior part of the cavernous region: 1. Superior border of the tentorium cerebelli with a little groove for the trochlear nerve; 2. Inferior petrous sinus; 3. Motor root of the V; 4. Sensitive fibres of the triangular plexus; 5. Arachnoid membrane; 6. Trigeminal ganglion in the bottom of the cavum; 7. Petrosus major nerve (parasympathetic from VII); 8. Internal carotid artery within its bony canal. (**c**) Coronal section of the oval foramen: 1. Mandibular nerve (V3); 2. Pterygoid venous plexus; 3. Valve of the tuba auditiva. (**d**) Parasagittal section of right hemiface: 1. Trigeminal cavum and internal carotid artery; 2. Maxillary sinus. (**e**) Sagittal section: 1. Trigeminal nerve; 2. Trigeminal ganglion and triangular plexus with atrophy of the nervous fibres (elderly person)

retina, is articulated with layers V and VI of the superior colliculus and with the roof of the midbrain where the visuomotor junction occurs. It also receives connections from the frontal premotor cortex (area 8) placed in front of the motor precentral gyrus, being the seat of the centre for foveal fixation with a strong participation of the caudate nucleus.

The tectospinal pathway that connects with the oculomotor nuclei III and VI on both sides starts from colliculus and mesencephalic tectum and then descends to the cervical spinal cord where it articulates with the motor centres of the rotator muscles of the head (SCM, splenius muscles and cardan muscles). This pathway is the major player of the eye/head servomechanism that moves eyes and head together in tracking moving objects, which was described in the chapter devoted to cervical spine.

It may also explain the ocular saccades that are quick jerks of bulb recall (mostly vertical but also lateral) [6]. That can be observed on a subject watching a passing landscape (such as in a train).

They are not consciously perceived and can also be recorded in the functional exploration of nystagmus, in connection with disturbances of the vestibular nuclei and medial cerebellum (fastigial nucleus).

The pilot of the machine is obviously ignorant of all these technicalities but he can consciously and voluntarily use the powerful foveal fixation we talked about earlier. Choosing to watch a particular area of the environment is a seemingly simple decision but is the technical basis of many life tasks such as reading and writing. The premotor area centre of the foveal fixation also explains the classical conjugate deviation of head and eyes observed in stroke and in hemiplegia.

5.2.3 Ocular Cranial Nerves

It should be noted that all the nerves of the extraocular muscles, the oculomotor nerve (III) for superior, inferior and medial recti and inferior oblique, the abducens nerve (VI) for lateral rectus and the trochlear nerve (IV) for the superior oblique, are purely motor. But their wealth in muscle spindles imposes a proprioceptive sensory component [7–9] which is given by the first branch of the trigeminal nerve (V1) (Fig. 5.5). It joins the motor nerves in the lateral wall of the cavernous sinus (Figs. 5.6 and 5.7). The proprioceptive neurons are not placed in the trigeminal ganglion but directly into the trigeminal nuclei of the brainstem. This is why it is the longest brainstem nucleus of all the cranial nerves with three parts: spinal, pontine and mesencephalic. The same process of anastomosis with the trigeminal nerve (lingual nerve) applies to the purely motor hypoglossal nerve (XII).

Conclusion

In the absence of instructions for use given by the constructor of this complex machinery, we can discover in this amazing eye mobility function a logic construction plan and all the mandatory central and peripheral nervous connections, allowing us to understand their why and how, which is precisely the objective of the comprehensive approach defined initially.

References

1. Büttner-Ennever JA (1988) Neuroanatomy of the oculomotor system. Elsevier, Amsterdam
2. Gunny R, Yousry TA (2007) Imaging anatomy of the vestibular and visual system. Curr Opin Neurol 20 (1):3–11
3. Krauzlis RJ (2005) The control of voluntary eye movements: new perspectives. Neuroscientists 11 (2):124–137
4. Leigh RJ, Zee DS (2006) The neurology of eye movements, 4th edn. Oxford University Press, New York
5. Rub U, Jen JC, Braak H et al (2008) Functional neuroanatomy of the human premotor oculomotor brain stem nuclei: insights from post-mortem and advanced in vivo imaging studies. J Comp Neurol 187 (2):167–180
6. Sparks DL (2002) The brain stem control of saccadic eye movements. Nat Rev Neursci 3(12):952–964
7. Winckler G (1956) L'innervation proprioceptive des muscles extrinsèques du globe oculaire chez l'Homme. CR 43e RéunionAss Anat, Lisbonne, pp 848–857
8. Eberhorn AC, Horn AK, Fischer P et al (2005) Proprioception and palisade endings in extraocular eye muscles. Ann NY Acad Sci 1039:1–5
9. Samii M, Janetta PJ (eds) (1981) The cranial nerves. Springer, New York

Facial Mimic

6

Abstract

Some striated muscles are included within the skin of the facial mask. They are responsible for specific movements of the two palpebral openings of orbits and of the mouth. They are activated by the facial nerve (VII), but their coordination is made by the skin innervated by the trigeminal nerve (V). The technical problem of the superior orbital floor is linked to the trophicity of the cornea, which has to be perfectly transparent and, hence, cannot receive a vascular network. This explains the need to have a double trophic liquid space: one on the internal side given by the aqueous humour permanently secreted and resorbed in the anterior chamber and the other one on the external surface, in contact with air. This arrangement justifies the precise and efficient organisation of the lacrimal system, which must maintain a liquid film on the cornea by moving periodically the eyelids with the orbicular muscle (blinking) and also to evacuate within the nasal fossa the tears permanently produced by the lacrimal gland. Some anatomical specifications of the Riolan and Horner muscles are the right technical answer to the problem. The mouth is an orifice bounded by two lips encircled by a complex "star-shaped" muscle allowing it to move in different directions for mimic and phonation. The facial nerve has in fact two parts in its brainstem nucleus corresponding to these two functions, which explains the possibility of complete or incomplete facial palsy.

6.1 Introduction

The facial skin mask has, included in its deep layer, striated muscles so called skinny muscles or platysma, allowing its animation focused on three holes: the two palpebral openings of orbits and the oral orifice.

It is important to note that the facial muscles have no muscle spindles nor proprioceptive corpuscles; as since they are inserted within the skin, it is then the skin sensitivity which provides their functional coordination through the trigeminal nerve (V). Truncal anesthesia performed by a dentist changes the sensitivity

P. Rabischong, *Comprehensive Anatomy of Motor Functions*, DOI 10.1007/978-3-319-04169-8_6,

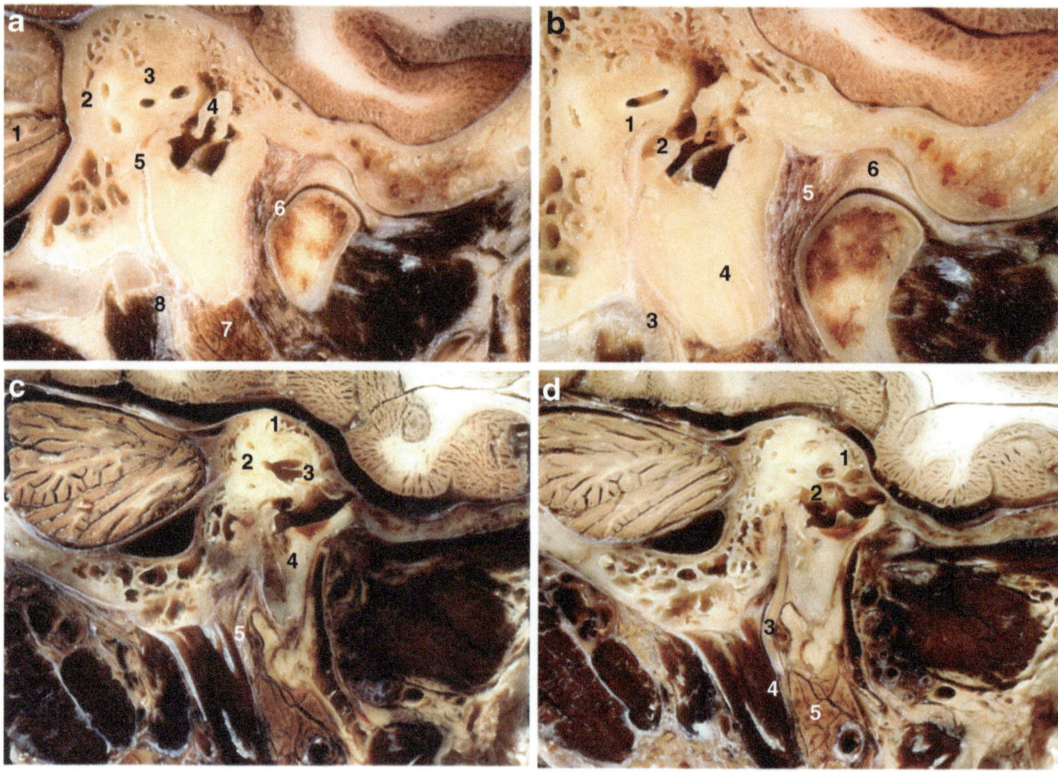

Fig. 6.1 The extra-temporal facial nerve (VII). (**a**) Sagittal section of a right ear seen from outside: 1. Cerebellum; 2. Posterior semi-circular canal (SCC); 3. Lateral SCC; 4. Incudo-mallear joint; 5. Descending portion of the facial canal; 6. TM joint; 7. Parotid gland; 8. Facial nerve emerging at the stylomastoid foramen. (**b**) Serial section inside the previous one: 1. Facial nerve elbow below the lateral SCC; 2. Chorda tympani between malleus and incus; 3. Facial nerve and stylomastoid artery; 4. Mastoid process; 5. Fibrovenous retrocondylar pad; 6. Meniscus of TMJ. (**c**) Sagittal section at the level of utricular macula: 1. Superior SCC; 2. Posterior SCC; 3. Utricular macula; 4. Tympanal bone (vaginal crest); 5. Digastric posterior belly nerve. (**d**) Sagittal section 3 mm inside the previous section: 1. Geniculate ganglion; 2. Stapes; 3. Facial nerve entering within the parotid gland; 4. Digastric posterior belly; 5. Parotid gland

of the face that can induce in the patient a fake feeling of paralysis. But the numbness felt is an impression fortunately transient due to a loss of facial expression.

This also explains why the mixed facial nerve (VII) of the second visceral arch becomes a purely motor nerve when it arrives within the face after innerving the stylohyoid muscle and the posterior belly of the digastric muscle (Figs. 6.1, 6.2 and 6.3). We must emphasise the importance for surgeons who can, for facemask resuscitation in case of facial palsy, successfully use a suture with the hypoglossal nerve (XII), a purely motor one [1–7]. Sutures between nerves of identical structure (purely sensory or purely motor) have always better clinical outcomes.

6.2 Facial Superior Orbital Floor

The first major technical problem of the construction of the facial superior orbital floor is the cornea, the window of the eye and also, by its two radius of curvature, the first dioptre of the optical system. It must necessarily be perfectly translucent and transparent to ensure excellent image capture. This excludes for its trophicity the presence of a vascular network disrupting the vision significantly.

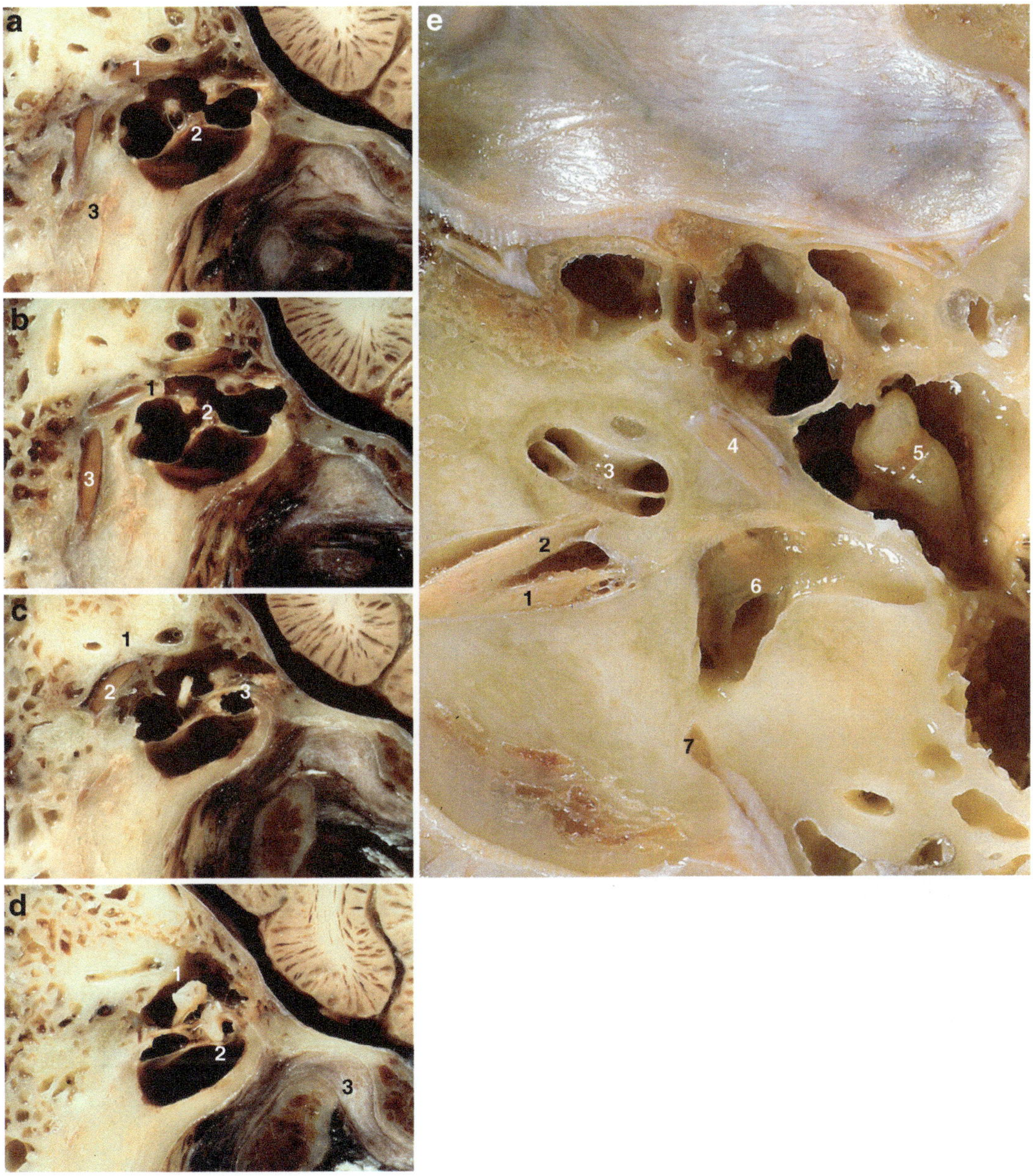

Fig. 6.2 The intrapetrous facial nerve. (**a**) Serial anatomical sagittal sections of a right ear (from outside to inside): 1. Facial nerve in its tympanic portion; 2. Tympanic membrane; 3. Posterior canal of chorda tympani. (**b**) Successive section: 1. Stapedial muscle; 2. Stapes and its tendon; 3. Facial nerve in its descending mastoid canal. (**c**) Successive section: 1. Lateral SCC; 2. Elbow of facial nerve; 3. Chorda tympani. (**d**) Successive section: 1. Lateral SCC; 2. Anterior process of malleus; 3. Meniscus of TM joint. (**e**). Axial section of a right ear seen from above: 1. Vestibular nerve; 2. Cochlear nerve; 3. Cochlea; 4. Facial nerve in its tympanic portion; 5. Incudo-mallear joint; 6. Vestibule and utricular macula; 7. Endolymphatic canal and its ampulla

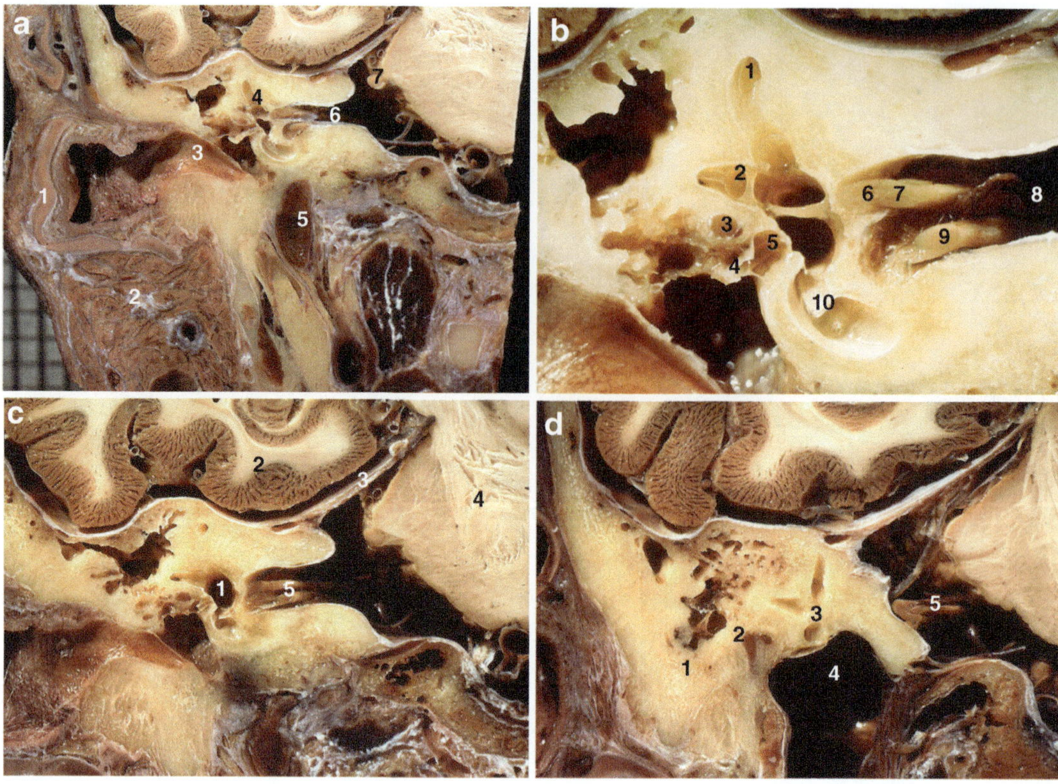

Fig. 6.3 Serial anatomical coronal sections of a right ear seen from the front. (**a**) Note that the roof of the meatus acusticus externus is at the same level as the floor of the meatus acusticus internus: 1. Cartilage of the meatus acusticus externus; 2. Parotid salivary gland; 3. Posterior part of the meatus acusticus externus with tympanic membrane; 4. Cochleo-vestibular complex; 5. Internal carotid artery; 6. Meatus acusticus internus; 7. Trigeminal nerve (V). (**b**) Details of the cochleo-vestibular complex: 1. Superior SCC; 2. Ampulla of the lateral semicircular canal; 3. Facial nerve; 4. Posterior branch of stapes; 5. Oval window with plate of stapes; 6. Vestibular nerve (VIII); 7. Facial nerve; 8. Labyrinthic artery; 9. Cochlear nerve; 10. Spiral lamina of the first roll of cochlea. (**c**) Posterior section: 1. Vestibule with lateral and superior SCC; 2. Temporal lobe of brain; 3. Tentorium cerebelli; 4. Pons; 5. VII and VIII nerves. (**d**) Posterior section: 1. Mastoid process; 2. Facial nerve in its descending canal; 3. Posterior part of the lateral and superior SCC; 4. Jugular vein bulb; 5. Facial nerve within the groove formed by cochleovestibular nerve (VIII)

6.2.1 Layers of the Cornea

It has several layers from surface to depth:

- a stratified squamous epithelium with three cell seats: superficial, intermediate and basal;
- a nerve plexus layer dependent of the ophthalmic branch of the trigeminal nerve (V1) and limited by the anterior limiting lamina of Bowman;
- a connective tissue stroma rich in mucopolysaccharides in which are keratocytes;
- the limiting posterior of Descemet;
- finally a pavement endothelium.

It continues directly with the sclera at the limbus level and with the bulbar conjunctiva. Its thickness is variable and in the order of 1 mm or more.

6.2.2 Trophicity of the Cornea

Technically, it is necessary to ensure the trophicity of this avascular membrane by maintaining a fluid trophic space on both sides:

- on the internal side, the anterior chamber with a large space is located in front of the lens and iris diaphragm. The aqueous humour is placed inside. It is a transparent liquid, similar to cerebrospinal fluid, secreted continuously by the ciliary processes of the ciliary body and

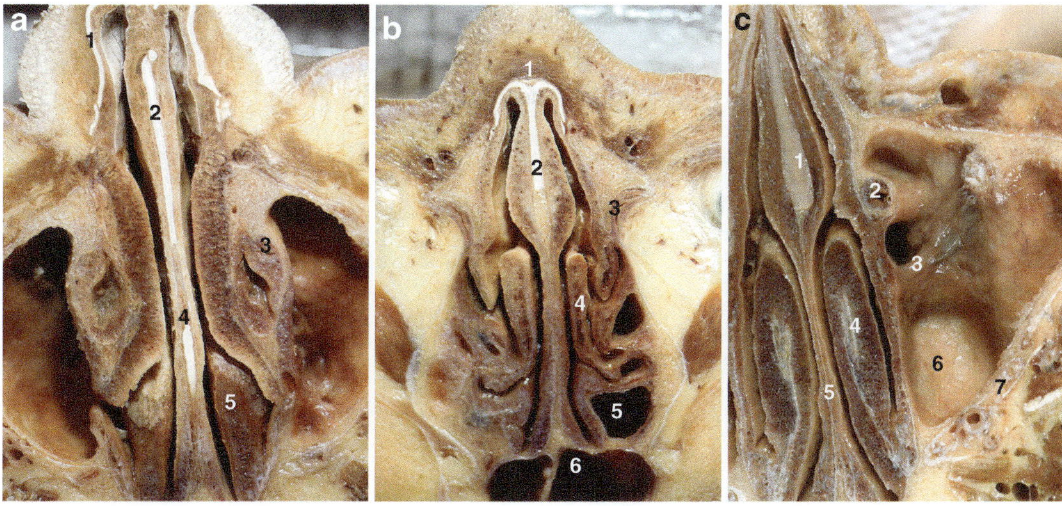

Fig. 6.4 Serial axial sections of nasal fossa seen from below. (**a**) Inferior section: 1. Lateral cartilage; 2. Septal cartilage; 3. Lacrimo-nasal duct; 4. Olfactory part of the nasal septum; 5. Middle turbinate. (**b**) Superior section: 1. Nasal muscle; 2. T cartilage (union of lateral and septal cartilage); 3. Lacrimal duct; 4. Superior part of the middle turbinate; 5. Posterior ethmoidal cell; 6. Sphenoid sinus. (**c**) Middle section on another piece: 1. Septal turbinate (Sulsenti); 2. Lacrimo-nasal duct; 3. Mark of the suborbital canal; 4. Middle turbinate; 5. Septal olfactory thin mucosa; 6. Roof of maxillary sinus; 7. Pterygopalatine artery and parasympathetic nerves

continuously absorbed by the venous sinus of the sclera containing the fine Schlemm's canal at the iridocorneal angle. The lack of absorption of liquid by pinching the angle and reduction in size of the canal creates glaucoma, phenomenon of hypertension within the inextensible eyeball causing, as in brain hydraulic disorders, nerve damage in the retina, a major cause of blindness in adults;

- on the external side, the solution is more complex because the cornea, looking to the outside world, is exposed to air. It is therefore necessary to maintain a fluid film on the corneal surface without disrupting vision. This is done by the extraordinary lacrimal system that of course does not exist in fishes that are totally and permanently immersed in water.

It consists of a gland logically placed at the upper side corner of the orbit, the lacrimal fluid flowing in the superior fornix of the bulbar conjunctiva. The permanent secretion is controlled by post-ganglionic parasympathetic fibres of the pterygopalatine ganglion activated by the facial nerve and conveyed by the lacrimal branch of the trigeminal nerve, which receives the vegetative communicating branch of the zygomatic nerve on the side wall of the orbit (Figs. 6.4 and 6.5).

This creates a lacrimal lake placed between bulbar and palpebral conjunctiva that is needed to spread periodically onto the anterior surface of the cornea. To do this, the curtain eyelid, formed by both the upper and lower eyelids containing tarsi with a fibro-cartilaginous structure and moored to medial and lateral walls of the orbit by palpebral ligaments, serves not only to cover the cornea with an opaque curtain but also as "windshield"-style brush. The windshield of modern cars has a geometric shape with asymmetrical curves and the wipers are fitted with springs to ensure a perfect touch of the brushes onto the surface of the windshield.

Technically, the problem is the same for the curtain eyelid. The motor system is the orbital orbicularis muscle, large circular intradermal patch applied to the contours of the orbit for its orbital part and on eyelids for its palpebral part. This part closes the curtain eyelid by a blink symmetric movement fully automated and performed almost unconsciously many times a day. A total voluntary contraction of the muscle completely closes the lid aperture.

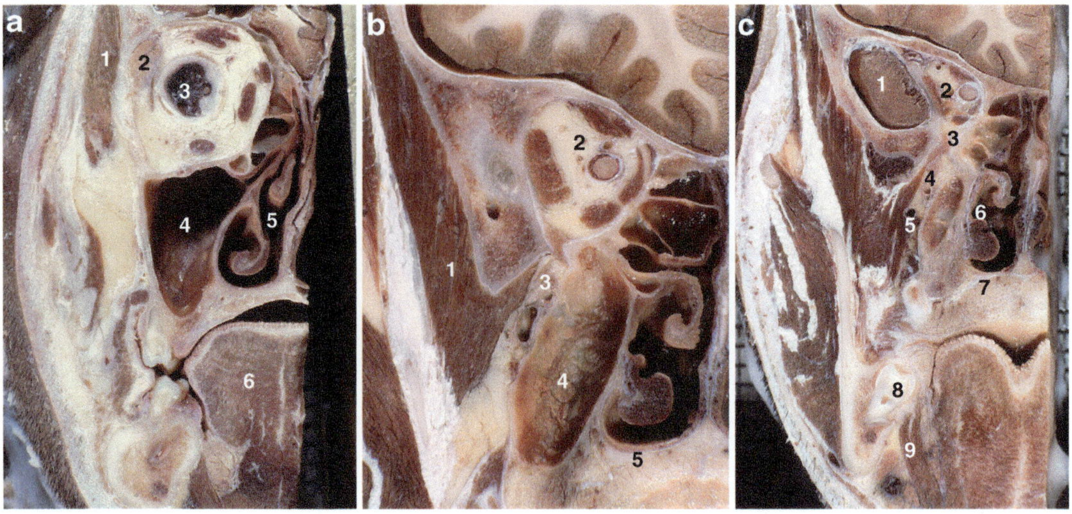

Fig. 6.5 Parasympathetic pterygo-palatine ganglion of VII (tear and mucosa secretion). (**a**) Coronal section of a right hemiface seen from the front: 1. Temporal muscle; 2. Lacrimal gland; 3. Posterior part of the eye ball; 4. Maxillary sinus; 5. Nasal fossa; 6. Tongue. (**b**) Coronal section of the posterior part of the maxillary sinus: 1. Temporal muscle; 2. Parasympathetic ciliar ganglion (III) for iris contraction; 3. Anterior part of the pterygo-maxillary fossa with pterygopalatine ganglion and maxillary artery; 4. Posterior part of the maxillary sinus. (**c**) Posterior coronal section: 1. Apex of temporal lobe of brain; 2. Apex of orbit; 3. Inferior fissure of orbit; 4. Pterygopalatine parasympathetic ganglion (lacrimal ganglion of VII); 5. Maxillary artery; 6. Middle meatus below middle turbinate; 7. Horizontal part of palate; 8. M3 within the mucosa; 9. Mylohyoid muscle

Careful analysis showed the presence of fine fibres in the muscle lid margin representing preciliar and retrociliar portions described by Riolan. This allows its application as close as possible, to the convex surface of the cornea and to spread a liquid film on it. It explains first the permanent tearing in patients with trachoma, a viral disease that destroys the lid margin and put away lacrimal lake from the lacrimal points of the eye lid and, also, one of the major complications of facial nerve paralysis eliminating the automated blinking ensuring the maintenance of lacrimal fluid film. This creates a likelihood of trophic keratitis which can lead to blindness due to corneal clouding by desiccation. To avoid this, it is sometimes needed to suture both eyelids (blepharorrhaphy).

6.2.3 Evacuation of Tears

The problem of the evacuation of tears secreted continuously remains. The most logical and discreet solution is to use nasal fossa (Fig. 6.6). Two small holes, lacrimal points, offset by 2 mm on the medial upper and lower lid margins and applied to the lacrimal lake mark the opening of lacrimal canaliculi. After an elbow and a straight path, they flow into the lacrimal sac secured by a doubling of the medial palpebral ligament on the lacrimal groove of the lacrimal bone. The bag continues by the nasolacrimal duct in the vertical direction down to the inferior meatus of the inferior turbinate (Fig. 6.7).

The technical originality of the system is the mandatory pump effect from the lacrimal lake carried out by the lacrimal part of the orbicularis muscle, the Horner muscle. Being applied on the lacrimal sac and canaliculi, its contraction—together with all the orbicularis during blinking—compresses them to prevent an eventual reflux into the canaliculi and creates the lacrimal lake aspiration during a relaxation phase of the contraction (Fig. 6.8). To complement the muscle action, some vertical fibres described by Gerlach are pulling the top of the bag to promote the pumping effect.

This ingenious system can work for a lifetime with billions of blinking and an impressive reliability. Technical sophistication of details such as shift of lacrimal points without which the pump

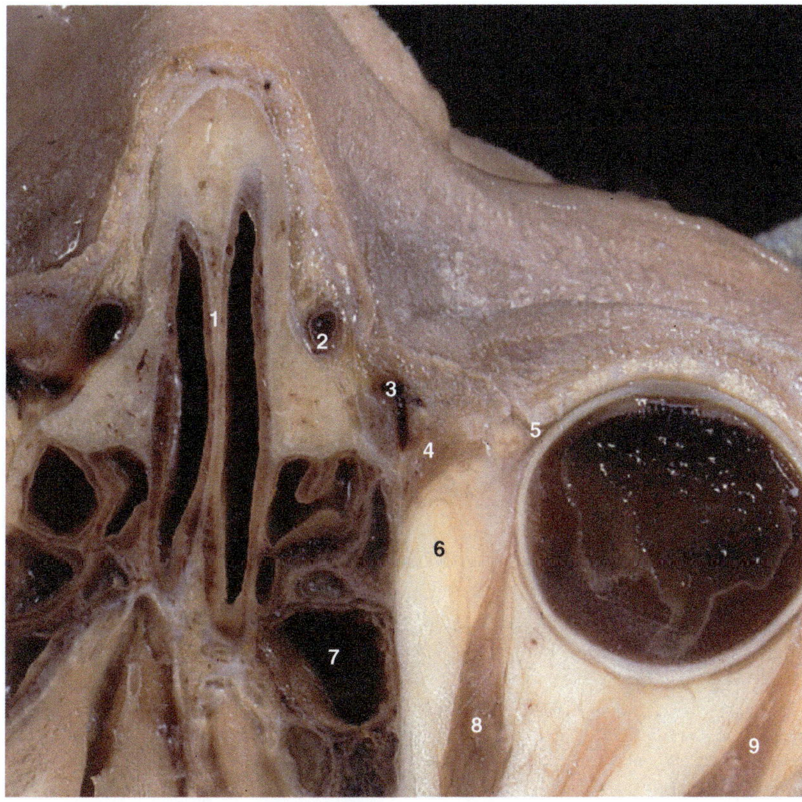

Fig. 6.6 Lacrimal sac and Horner muscle. 1. Olfactory mucosa on nasal septum; 2. Facial vein; 3. Lacrimal sac; 4. Lacrimal part of the orbicularis muscle (Horner muscle) pushing on the lacrimal sac during blinking and creating a pump effect to suck up the tears within the conjonctival lacrimal lake; 5. Conjonctival cul-de-sac; 6. Orbital fat tissue; 7. Posterior ethmoidal cell; 8. Medial rectus; 9. Lateral rectus

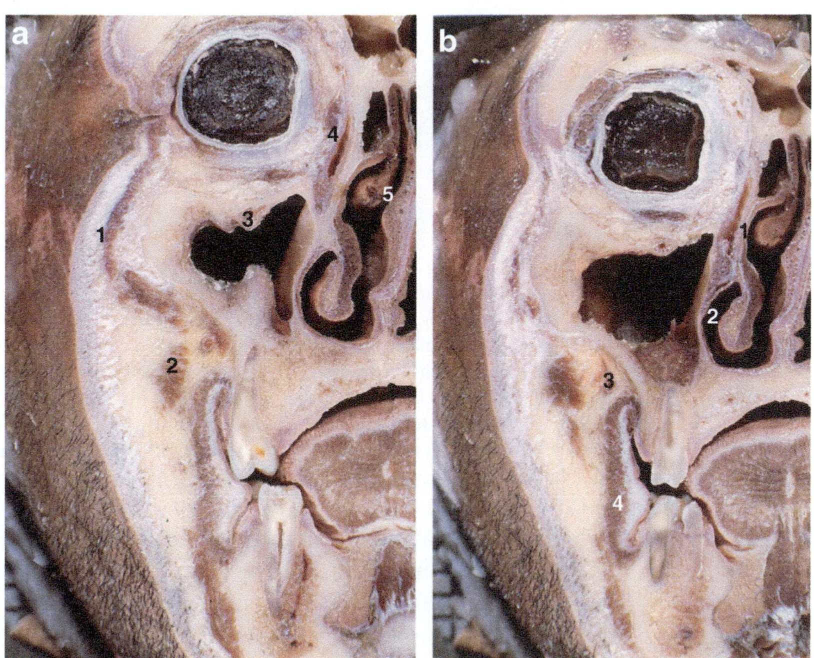

Fig. 6.7 The lacrimonasal duct to drain off tears. (a) Coronal section of a right hemiface seen from the front: 1. Orbicularis oculi muscle; 2. Zygomatic muscle; 3. Suborbital artery and nerve (V2); 4. Lacrimonasal duct and lacrimal part of the orbicular muscle (Horner muscle); 5. Middle turbinate. (b) Serial coronal section behind the previous: 1. Lacrimonasal duct; 2. Inferior meatus below the inferior turbinate where the canal ends; 3. Facial artery; 4. Buccinatorius

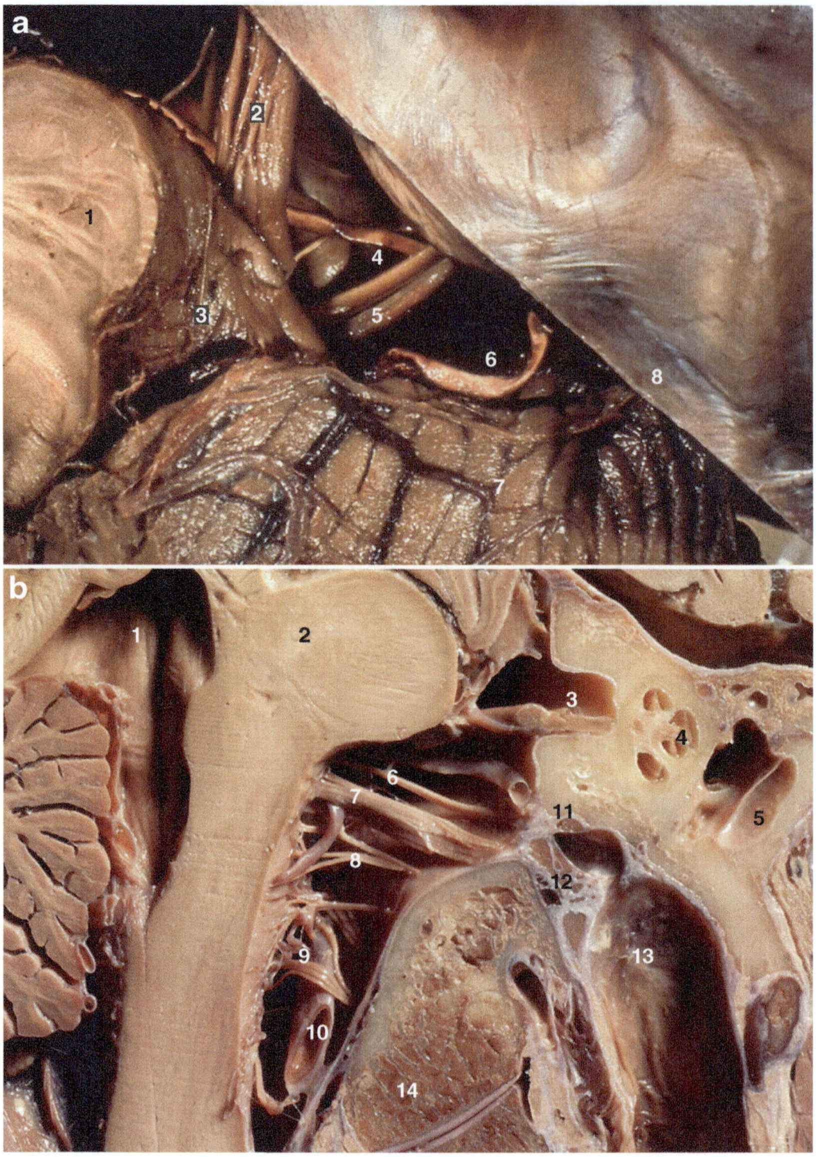

Fig. 6.8 The facial nerve entering within the temporal bone. (**a**) Axial section of the cerebello-pontine angle: 1. Pons; 2. Sensitive root of the trigeminal nerve (V) with motor root on the medial side; 3. Middle cerebellar peduncle; 4. Facial nerve (VII); 5. Cochleo-vestibular nerve (VIII); 6. Anterior inferior cerebellar artery (AICA); 7. Right cerebellar hemisphere; 8. Transsected dura of the tentorium cerebelli. (**b**) Oblique parasagittal section of a left ear: 1. Fourth ventricle; 2. Pons; 3. Meatus acusticus internus with VII and VIII nerves; 4. Cochlea; 5. Meatus acusticus externus with a part of the tympanic membrane; 6. Glosso-pharyngeal nerve (IX); 7. Vagal nerve (X); 8. Accessorius nerve (XI); 9. Hypoglossal nerve (XII); 10 Vertebral artery; 11. Superior ganglion of IX; 12. Superior and inferior ganglions of X; 13. Jugular vein; 14. Occipital condyle (C0)

effect would be greatly diminished is an argument for eliminating all forms of randomness in the construction of this vital apparatus for the maintenance of the cornea and, hence, the vision.

What is also surprising is that the wink is bilateral, synchronous and automated with a variable periodic rate from one subject to another. But blinking with only one eyelid is also possible, for various reasons, which all have to do with the desire to send a signal to another human. This mixture of automatic and voluntary actions can be frequently found and we

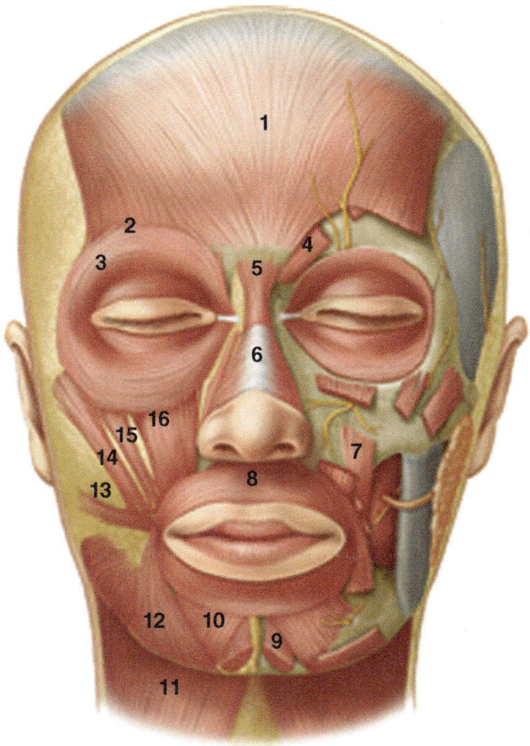

Fig. 6.9 Facial muscles (after Totrtora). 1. Occipitofrontal muscle (frontal part); 2. Orbicularis oculi (pars orbitalis); 3. Orbicularis oculi (pars palpebralis); 4. Corrugator supercilii; 5. Procerus muscle; 6. Nasalis; 7. Levator anguli oris; 8. Orbicularis oris; 9. Mentalis; 10. Depressor labii inferioris; 11. Platysma; 12. Depressor anguli oris (triangular muscle); 13. Risorius; 14. Great zygomatic; 15. Small zygomatic; 16. Levator labii superioris alaeque nasi

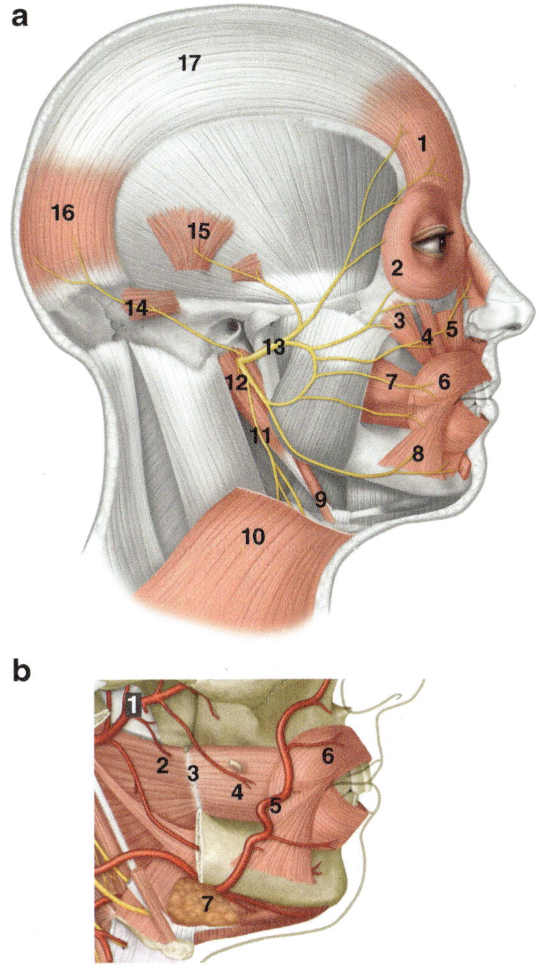

Fig. 6.10 Profile view of the facial mask. (**a**) Profile view: 1. Occipitofrontal muscle; 2. Orbicularis oculi; 3. Great zygomatic; 4. Small zygomatic; 5. Levator labii superioris; 6. Orbicularis oris; 7. Buccinatorius; 8. Depressor anguli oris; 9. Stylohyoid muscle; 10. Platysma; 11. Posterior belly of digastric muscle; 12. Facial nerve (VII), cervicofacial branch; 13. Facial nerve, temporofacial branch; 14. Posterior auricular muscle; 15. Superior auricular muscle; 16. Occipitofrontal muscle (occipital part); 17. Epicranial fascia. (**b**) Inferior part of the face after resection of the masseter muscle 1. Maxillary artery; 2. Superior constrictor muscle; 3. Pterygopalatine raphe; 4. Buccinatorius muscle; 5. Facial artery; 6. Orbicularis oris; 7. Submandibular gland

saw a good example of it in the part related to breathing.

One can also understand why the facial nerve (VII) has two functionally distinct parts in its brainstem central nucleus: a superior part, which is the basis of the vitality of the cornea by controlling the symmetrical automated eyelid blinking, and a lower part operating in a more conscious and voluntary way because of its involvement in the complex mobilisation of mouth orifice [8].

6.3 Facial Inferior Buccal Floor

The single oral orifice is at the level of the face mask bounded by the two upper and lower lips registering at their contact a horizontal slot more or less thick and long, limited laterally by the two labial commissures. They contain in their deep part one orbicularis muscle made by thick circular fibres surrounding the buccal cavity, on which are fixed other facial muscles (Figs. 6.9 and 6.10):

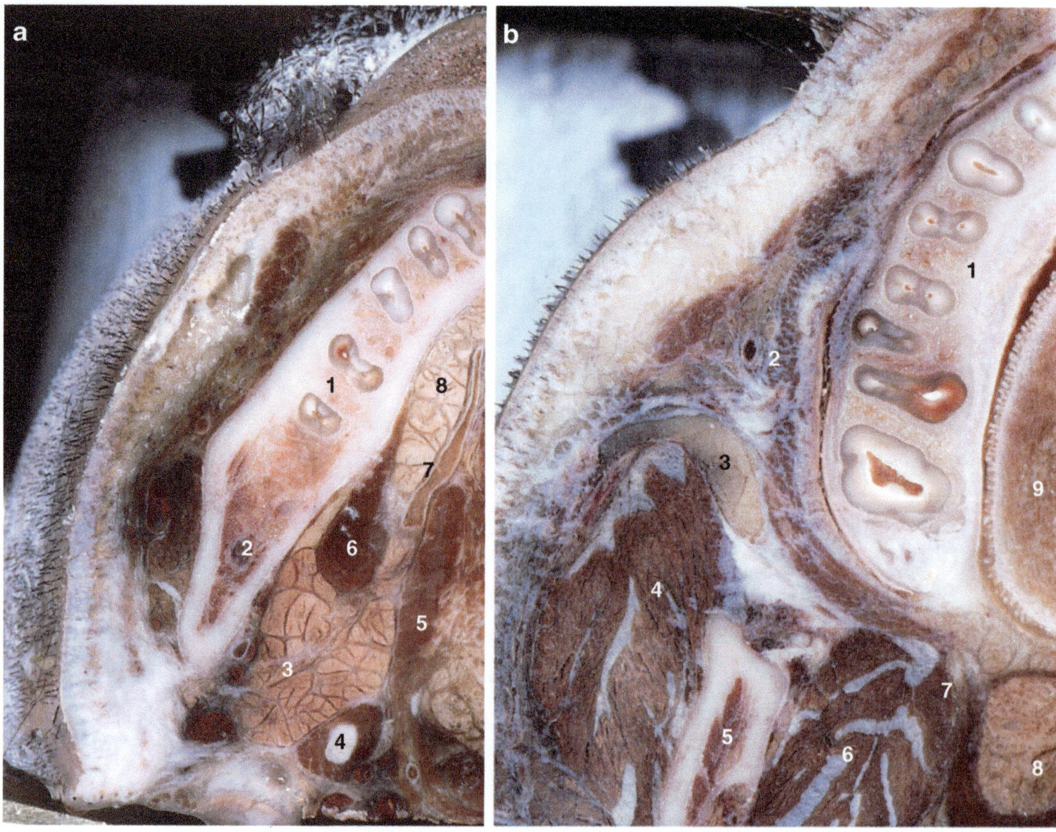

Fig. 6.11 Axial anatomical sections of a left hemiface seen from above (young person). (**a**) Section of the corpus mandibulae: 1. Alveolar part; 2. Inferior dental canal; 3. Submandibular gland; 4. Digastric muscle; 5. Hyoglossal muscle; 6. Mylohyoid muscle; 7. Wharton canal from the submandibular gland; 8. Sublingual gland. (**b**) Section of the maxilla: 1. Superior alveolar arcade; 2. Buccinatorius; 3. Fat pad (Bichat); 4. Masseter; 5. Mandibular ramus; 6. Medial pterygoid; 7. Bucco-pharyngeal raphe; 8. Palatine tonsil; 9. Superior part of the tongue

- myrtiform or lower of the nasal septum at above and middle;
- small zygomatic from the crest of the zygomatic malar to upper lip;
- major zygomatic fixed to the malar on top and with the angle of the mouth in bottom;
- risorius arranged transversely from the fascia of the masseter to the labial commissure;
- canine from the canine fossa of the maxilla to the commissure;
- superficial and deep reliever of superior lip;
- triangular lips from mandible to the commissure;
- square chin or quadratus labii inferioris from the mandible to the lower lip.

Thus, the orbicularis oris muscle is surrounded by a kind of muscle in the form of a star going in all directions and allowing various deformations of the mouth orifice, which constitute the highly suggestive mimic expression of internal emotions variable according to the different cultures (joy, sadness, doubt, anger, ...) and also the kisses up to the true anastomosis (etymologically mouth to mouth) [9]. But this voluntary deformation has a major role in the articulated language, as we shall see later.

In addition, the buccinator muscle (from *buccina*, the horn) is also fixed on the lateral angle of the mouth, starting from the pterygomandibular raphe which attaches behind to the constrictor of the pharynx. It forms the muscle lining of the cheek and may pressurise the air in the oral cavity which is the basis of the set of wind instruments (Fig. 6.11).

Finally, the facial palsy (VII) may be incomplete or complete affecting the orbital floor with more or less deficiency of eyelid closure with the risk of trophic keratitis and/or the oral floor with lowering of the commissure on the paralysed side that creates a significant disfigurement [10–12]. Multiple treatments are often disappointing because the complete atrophy of the muscles is irreversible. But the paralysis called a frigore reversibly affects the trunk of the facial nerve in the petro-mastoid canal and may regress completely after a variable period of time.

It should also be noted the fibrillary tremor of Sedan which is a series of clonic contractions of the orbicularis oculi muscle and eventually other facial muscles (painful twitch). It can become intolerable requiring surgical intervention initiated by Janetta in order to remove, by an interposition of a fragment of muscle, the mechanical beats of an artery on the facial nerve trunk causing the twitching.

References

1. Terzis JK, Konophaos P (2008) Nerve transfers in facial palsy. Facial Plast Surg 24:177–183
2. Samii M (1984) Facial nerve grafting in acoustic neurinoma. Clin Plast Surg 11(1):221–225
3. Campers A, Socolovsky M (2007) Facial reanimation by means of the hypoglossal nerve: anatomic comparison of different techniques. Neurosurgery 61(suppl 3):41–49
4. Yetiser S, Karapinar U (2007) Hypoglossal-facial nerve anastomosis: a meta analytic study. Ann Otol Rhinol Laryngol 116:542–549
5. Kunert P, Podgorska A, Bartoszewicz R et al (2011) Hemihypoglossal-facial nerve anastomosis for facial nerve palsy. Neurol Neurochir Pol 45 (5):452–460
6. Fish U (1981) Surgery of Bell's palsy. Arch Otolaryngol 107:1–11
7. Samii M, Gerganov V (2013) Surgery of cerebellopontine lesions. Springer, Heidelberg
8. May M (1986) The facial nerve. Thieme Stratton, New York
9. Morris JS, Friston KJ, Büchel C et al (1998) A neuromodulatory role for the human amygdala in processing emotional facial expressions. Brain 121 (1):47–57
10. House JW, Brackman DE (1985) Facial nerve grading system. Otolaryngol Head Neck Surg 13:146–147
11. Linstrom CJ (2002) Objective facial motion analysis in patients with facial nerve dysfunction. Laryngoscope 112:1129–1147
12. Volk GF, Pantel M, Guntinas-Lichius O (2010) Modern concepts in facial nerve reconstruction. Head Face Med 6:25

Masticatory Function 7

Abstract

The masticatory function is realised by the temporomandibular joint (TMJ) activated by masticatory muscles: mandibular strap and temporal range for the closure, and lateral pterygoid for the opening. The biomechanical organisation of the somato-gnathic complex depends on the type of feeding explaining its different types in animals. Human beings are intended to use all kinds of food and their mouth is also designed for articulated language. This justifies the discordant TMJ with a meniscus and the particular protraction movement of the condyle during the aperture of the mouth given by the lateral pterygoid muscle which is fixed on both the mandible and the meniscus. It is a tendency to functionally link mandibular and upright postures, and as a hypothesis, it can be suggested to find in the thalamic organisation some similarities in the management of these two bilateral stabilisations very often asymmetrical (dominance and laterality).

7.1 Introduction

The somato-gnathic complex comprises a certain number of organs around the oral cavity whose main function is chewing (or manducation). This concerns not only solid foods but also words (phonation) with the major role of the lingual modulator of sounds and cleaner of mouth toward the pharynx.

The mandible is the mobile bone of the complex. It is articulated to the skull base by the temporomandibular joint (TMJ). It is set in motion by the masticatory muscles responsible for: mouth opening by the lateral pterygoid muscle combined synergistically to the digastric one and mouth closure by the mandibular strap formed by the masseter and the medial pterygoid and by the temporal range providing selective clamping of the dental arches (Fig. 7.1).

On the embryological plan, the whole face with the three floors (orbital, maxillary and mandibular) derives from the first visceral arch colonised by the trigeminal nerve (V). The second visceral arch gives the hyoid bone – supporting the tongue and the larynx – which is linked to the mandible: (i) by the muscles of the floor of mouth (mylohyoid and anterior belly of the digastric belonging to the first arch); (ii) by the posterior belly of the digastric; and (iii) by the stylohyoid attached to the styloid process

P. Rabischong, *Comprehensive Anatomy of Motor Functions*, DOI 10.1007/978-3-319-04169-8_7,

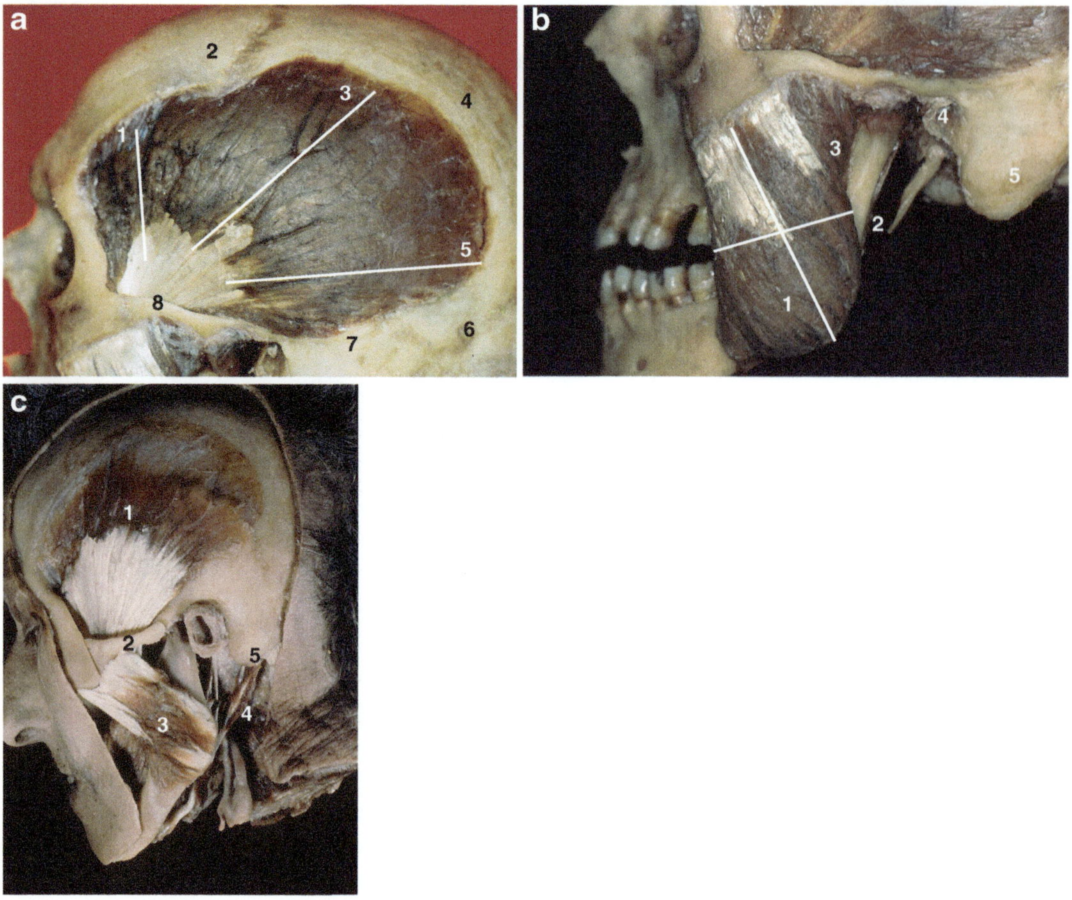

Fig. 7.1 Left temporal and masseter muscles. (**a**) Dissection of temporal: 1. Vertical anterior part; 2. Frontal bone; 3. Oblique middle part; 4. Parietal bone; 5. Horizontal inferior part (antagonist of the lateral pterygoid); 6. Occipital bone; 7. Temporal bone; 8. Zygomatic arch. (**b**) Dissection of masseter: 1. Superficial part almost perpendicular to the occlusal line; 2. Styloid process; 3. Deep part; 4. Meatus acusticus externus; 5. Mastoid process. (**c**) Dissection of an hemiface: 1. Temporal with its tendon; 2. Zygomatic arch; 3. Masseter; 4. Posterior belly of digastric

(territory of the second arch colonised by the facial nerve, VII). This explains the multiple functional synergies between the mandible and larynx.

The comparative anatomy has shown clearly the morphological and functional specialisation of stomato-gnathic complex depending on the type of feeding in different species. Therefore, the masticatory apparatus and teeth are preset accordingly.

Ruminants feed on plants, and their entire digestive tract from the mouth to the anus is built based on this type of food. They have cylindrical transverse condyle, allowing a lateral displacement of the jaw adapted for grinding and finishing plants from merycic rumen, the first gastric pouch.

Carnivores chase their *mobile provender* made mainly of meat. They need a strong and stable clamp jaw with a strong and retentive closing system on the prey that is realised by the canines offset hook, which is possible with a 1-DF TMJ hinge and very powerful temporal muscles. A lagomorph, like a rabbit, performs rapid jaw shear movements through a condyle lying in anteroposterior axis with pterygoid and masseter muscles responsible for the powerful alternative mandibular movement. The hen, known by its absence of teeth, compensates for this disadvantage by the use of a gizzard mill effective for grinding grain.

Finally, man is undoubtedly an omnivore and the structure of his digestive tract and glands proves it. Its TMJ allows all movements of specialised types briefly described before: opening, closing, antetraction, retrotraction and diduction.

It seems interesting to analyse in a comprehensive vision the original solution provided to the technical problem of eating [1].

7.2 Movements of the TMJ

They are numerous and conditioned by the configuration of the TMJ and the muscular actions.

7.2.1 Joint

This is probably the one throughout the body that is the most sought, in chewing food and in phonation. It must withstand very important occlusive forces and allow multidirectional movements, involving rotation and translation. Therefore, all joint surfaces are covered with fibrocartilage and contact zones of hyaline cartilage [2]. The following two original technical features can be described.

7.2.1.1 The Articular Contact Surfaces

On the mandible, there is the condylar process with its head or articular condyle placed above the neck and, on the temporal bone, the articular tubercle above extending forwardly the mandibular or glenoid fossa. It is therefore a question to have two convex cartilage surfaces by nature discordant in contact.

Both mandibular condylar heads form an angle of their transverse portion of 140° which is open toward the front with individual variations. In addition, the mandibular neck is slightly inclined backward and the head is flexed forward to be opposed to the tubercle of the temporal. Moreover, when the mandible is in place, and not placed on a table in the position which it used to be when described by ancient anatomists, the mandibular ramus or rising part is vertical and the body of the mandible goes obliquely downward (which is a test of anatomical quality of the drawing of the mandible in place) (Fig. 7.2).

7.2.1.2 The Meniscus

All species with a mandible have a biconcave articular disc (meniscus) with a thick posterior rim which fits exactly into the mandibular fossa in intercuspidation position. This disc creates the connective damping of compression forces during dental clamping. A thinner anterior rim receives the meniscal attachment of the lateral pterygoid muscle. A middle area interposes exactly between the condyle and the articular tubercle.

The articular disc continues back through the capsule that attaches to the neck and with the mandibular periosteum as well. There are thus two separate articular cavities. The first superior one has an S shape extending from the upper portion of the posterior rim of the disc to beyond the articular tubercle on the superior face of the lateral pterygoid meniscal part. The second inferior one extends from the posterior part of the condylar head to its cartilage edge, at the base of the condylar insertion of the lateral pterygoid.

A significant anatomical detail is, behind the disc, the existence of a fibro-venous retrocondylar space, also called posterior connective vascular complex or less precisely bilaminar area (Fig. 7.3). It is made of a fibrous connective tissue in which lies a very rich venous plexus. It occupies the space between the disc overlying the condylar head and the anterior part of the tympanic bone joined to the temporal bone by the petrotympanic fissure of Glaser. This structure plays the role of a rear stop for the mandibular condyle.

This is the appropriate technical solution to the problem of opening the mouth which is done by pulling forward the mandibular condyle. That imperatively requires an extensible and flexible structure filled with venous blood which changes the posterior empty fibrous tissue and make possible the protraction of the condyle. This original feature can be seen on an MRI coronal section

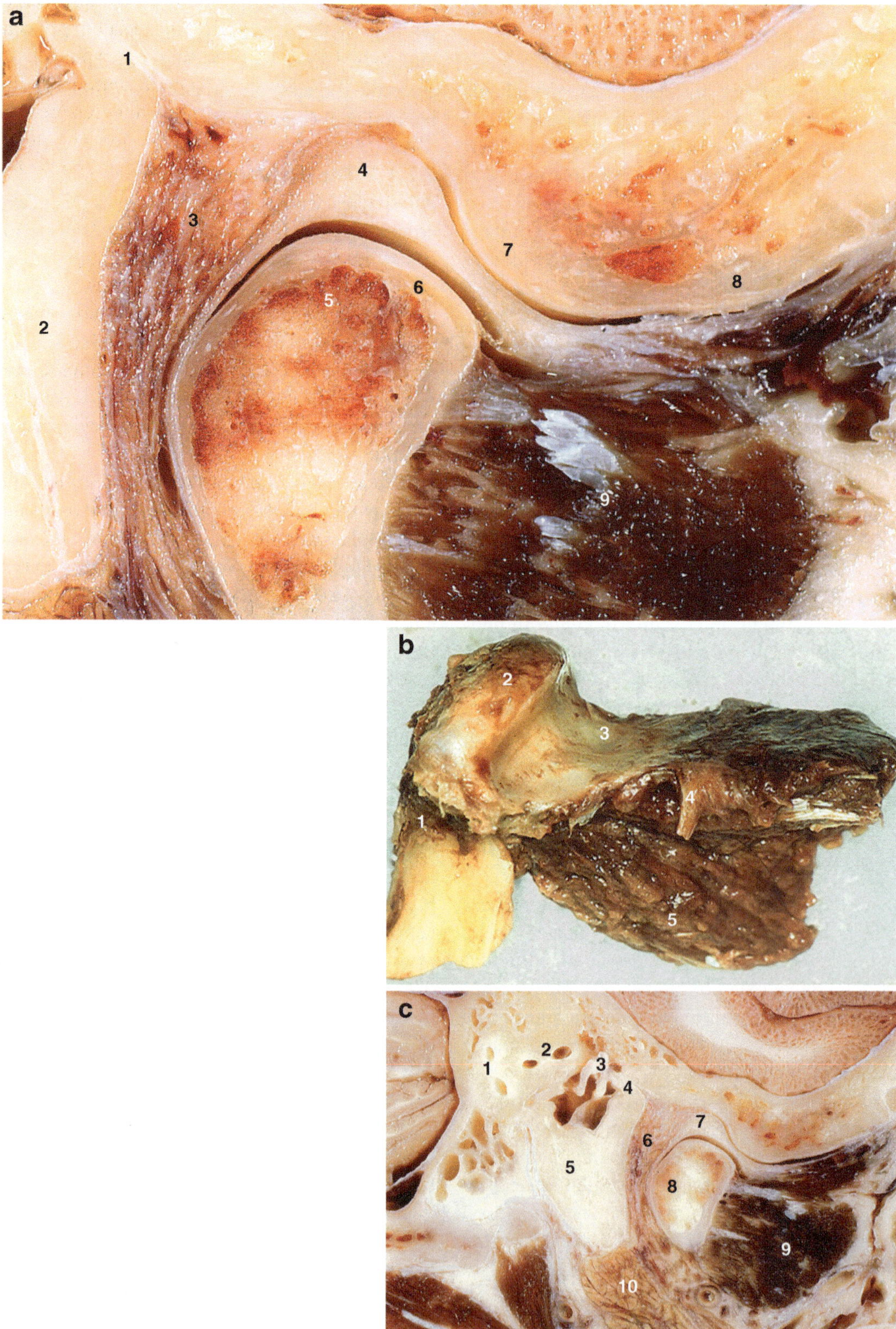
a
1
2
3
4
5
6
7
8
9
b
1
2
3
4
5
c
1
2
3
4
5
6
7
8
9
10

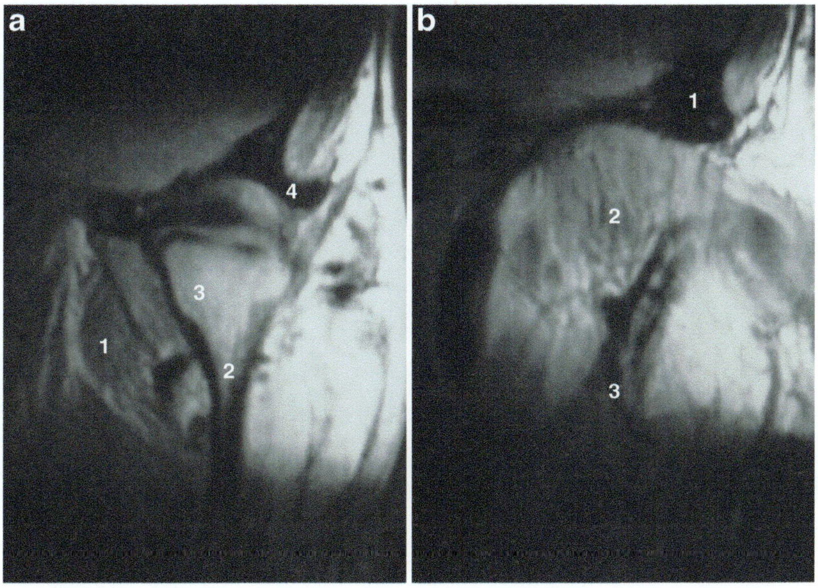

Fig. 7.3 MRI of the right TMJ in coronal sections. (**a**) Mouth closed: 1. Medial pterygoid muscle; 2. Collar; 3. Mandibular condyle; 4. Zygomatic arch. (**b**) Mouth open: 1. Articular tubercle of temporal; 2. The retrocondylar fibro-venous cushion like an open accordion with very visible thin veins; 3. Mandible

and can be compared with an accordion full of blood instead of air.

7.2.2 Joint Mobilisation

It is done by a series of muscles responsible for the different mandibular movements.

7.2.2.1 Closing the Mouth and Dental Clamping Intercuspation

This is the active part of the food chewing, which is carried out by two systems.

The Mandibular Strap

It is formed by two symmetrical components, outside the masseter muscle [3, 4] and inside the medial pterygoid. One should note the oblique orientation of the two muscles pulling on the mandibular body perpendicular to the occlusal line [5]. This solves the problem of the need for cusp compression forces correctly oriented to avoid lateral stresses that may cause dental malposition. It is obvious that the mechanical perfection is impossible to imagine, which justifies and maintains the orthodontist's work.

The Temporal Range

This muscle, located on the side face of the head, has actually a fan shape with anterior muscle fibres, middle vertical and posterior horizontal. It is also innervated by three separate nerve

Fig. 7.2 Temporo-mandibular joint (TMJ). (**a**) Sagittal section: 1. Fissura tympanosquamosa; 2. Tympanal bone; 3. Retrocondylar fibro-venous cushion (bilaminar zone); 4. Posterior thick part of meniscus within the mandibular fossa of the temporal bone; 5. Mandibular condyle; 6. Inferior articular cavity of the condylo-meniscal joint; 7. Articular tubercle of temporal bone; 8. Superior articular temporo-meniscal cavity; 9. Lateral pterygoid muscle (LPM). (**b**) Dissection of the disc and lateral pterygoid muscle: 1. Collar; 2. Posterior part of the disc covering the mandibular condyle; 3. Anterior part of the disc and superior articular cavity; 4. Meniscal superior head of LPM; 5. Inferior part of LPM fixed on fovea pterygoidea below the condyle. (**c**) Sagittal section of TMJ: 1. Posterior SCC; 2. Lateral SCC; 3. Incudomallear joint; 4. Tympano-squamous fissure; 5. Vaginal crest of tympanic bone; 6. Retrocondylar fibro-venous cushion; 7. Meniscus; 8. Mandibular condyle; 9. LPM; 10. Parotid gland

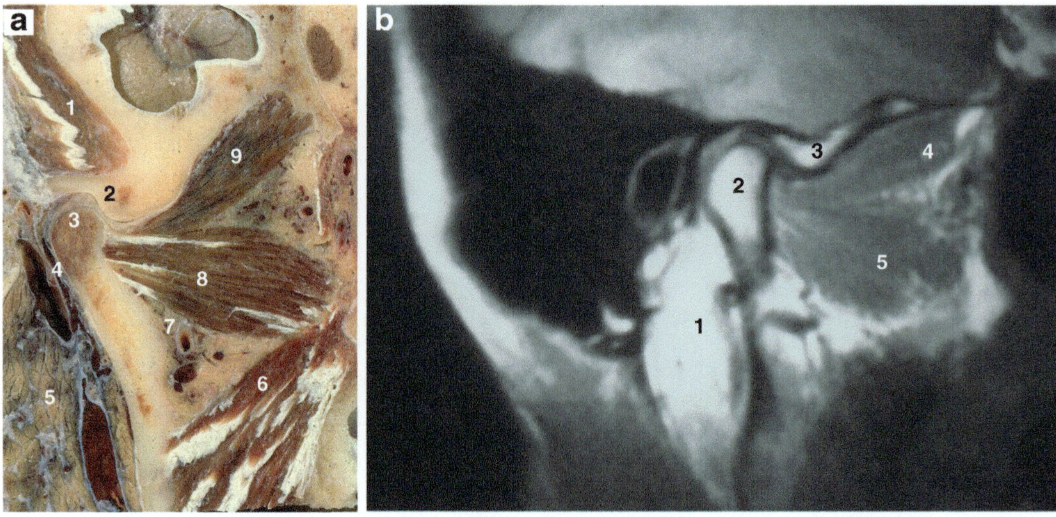

Fig. 7.4 Oblique sagittal sections along the 45° axis of the lateral pterygoid muscle (LPM). (**a**) Anatomical section: 1. Temporal muscle; 2. Articular tubercle of temporal bone; 3. Articular disc (note its S shape); 4. Mandibular condyle with fovea pterygoidea for the inferior beam of LPM; 5. Parotid gland with venous plexuses; 6. Medial pterygoid muscle; 7. Maxillary artery; 8. Inferior mandibular beam of LPM; 9. Superior meniscal beam of LPM. (**b**) MRI sagittal section of LPM made along its 45° axis: 1. Parotid gland; 2. Mandibular condyle; 3. Articular tubercle of temporal; 4. Superior part of LPM; 5. Inferior part of LPM

branches: anterior, middle and posterior deep temporal nerves. It is therefore typically a fascicular innervation creating a fascicular contraction for varying dental tightening from front to back. It is easy to realise and feel it ourselves. The horizontal posterior muscular beam is antagonist of lateral pterygoid in bringing actively back the mandibular head, by pulling on the coronoid process.

The tendon of the temporal muscle is literally embedded in the coronoid process. The aponeurosis of the muscle is a tough fibrous sheath that attaches to the zygomatic arch below and to the long curved frontoparietal superior line (temporal line) above. A functional fat pad, called ball of Bichat, is at the front of the temporal tendon playing the role of a cushion pad.

7.2.2.2 Active Aperture of the Mouth

This movement is dependent on the lateral pterygoid muscle which is not accessible to palpation as some still believe (Fig. 7.4). It is composed of two upper and lower portions [6–8]. To be understood, it must be examined through its sagittal and axial anatomical sections.

On a sagittal section made along its main axis placed at 45° to the sagittal plane, we can clearly see the top beam obliquely upward and the lower beam obliquely downward. Both lateral and medial pterygoid form the Z image that we sometimes find in muscular planes.

The upper beam binds mainly on the front part of the articular disc and the upper part of the pterygoid fossa. This avoids exclusive traction on the disc that may detach or dislocate it forward (Fig. 7.5).

The lower beam attaches to most of the pterygoid dimple and moves toward the side face of the lateral pterygoid process where it binds (Fig. 7.6). This allows to understand that when opening the mouth, the mandibular head does not rotate around its geometric centre as in carnivores, but instead it is towed forward with an axis obliquely downward, forcing it by abutting on the temporal articular tubercle to pivot in a combined movement of translation and rotation with a slight lowering of the mandibular head (Fig. 7.7).

In patients who underwent surgical removal of the larynx with resection of part of the floor of

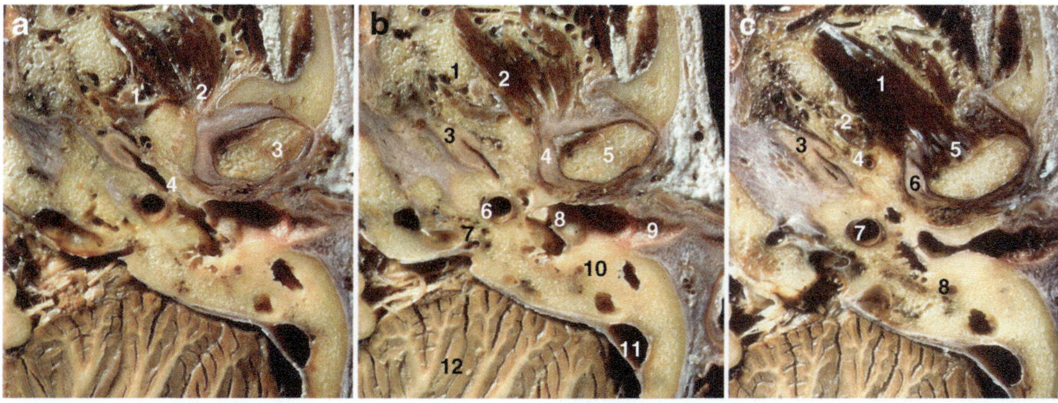

Fig. 7.5 Axial sections of right TMJ seen from above. (**a**) Superior section: 1. Mandibular nerve (V3); 2. Meniscal fixation of the lateral pterygoid muscle (LPM); 3. Mandibular condyle; 4. Bony canal of tuba auditiva (protympanum). (**b**) Middle section: 1. Mandibular nerve (V3); 2. Superior meniscal beam of LPM; 3. Tuba auditiva; 4. Meniscal attachment; 5. Mandibular condyle; 6. Internal carotid artery; 7. Cochlea; 8. Tympanic membrane; 9. External auditory meatus; 10. Facial nerve (VII); 11. Venous sigmoid sinus; 12. Cerebellum. (**c**) Inferior section: 1. Inferior mandibular beam of LPM; 2. Mandibular nerve (V3); 3. Tuba auditiva; 4. Middle meningeal artery; 5. Fovea pterygoidea for attachment of inferior beam of LPM; 6. Thick medial part of disc; 7. Internal carotid artery; 8. Facial nerve in its descending canal

the mouth, the mouth opening is always possible by lateral pterygoid action. But under normal conditions, there is a torque of the mandibular beam formed by the lateral pterygoid pulling forward and the anterior belly of the digastric pulling backward. This can be observed during mouth opening against resistance.

7.2.2.3 Protraction, Retrotraction and Diduction movements

It is the lateral pterygoid muscle that is responsible for active protraction and the horizontal posterior bundle of the temporal muscle for retrotraction, with a posterior end-stop limited by the tympanal bone and an anterior one limited by the joint capsule.

On an axial section, the biomechanical compromise of lateral pterygoid is clearly visible with a 45° intermediate direction between anteroposterior traction for the mouth opening and transverse lateral displacement (lateral excursions) of the mandible (Figs. 7.8 and 7.9). In bilateral synergy, both muscles open the mouth, and in alternating unilateral action, they allow the diduction movement that cannot be exactly transversal, as in ruminants with transverse cylindrical mandibular condyle. In fact the lateral translation is combined with a certain degree of rotation of the contralateral condyle.

It is interesting to note that, in the closing tightening dental maximum intercuspal, all masticatory muscles contract including muscles of the opening. Paradoxically, this outbreak muscle increases the dental compressive forces but creates a protective mechanism of the articular disc against over constraints. In fact, the upper beam of the lateral pterygoid which provides an S-shaped active strap pulling the front portion of the disc and generating a down decompression force reducing the upward compression force generated by mandibular head. However, it is possible to observe cracking discs or ejections by mechanical hyperpressure.

7.3 Control of Mandibular Posture

Posture is by definition a steady state of a body segment. This implies a control system to

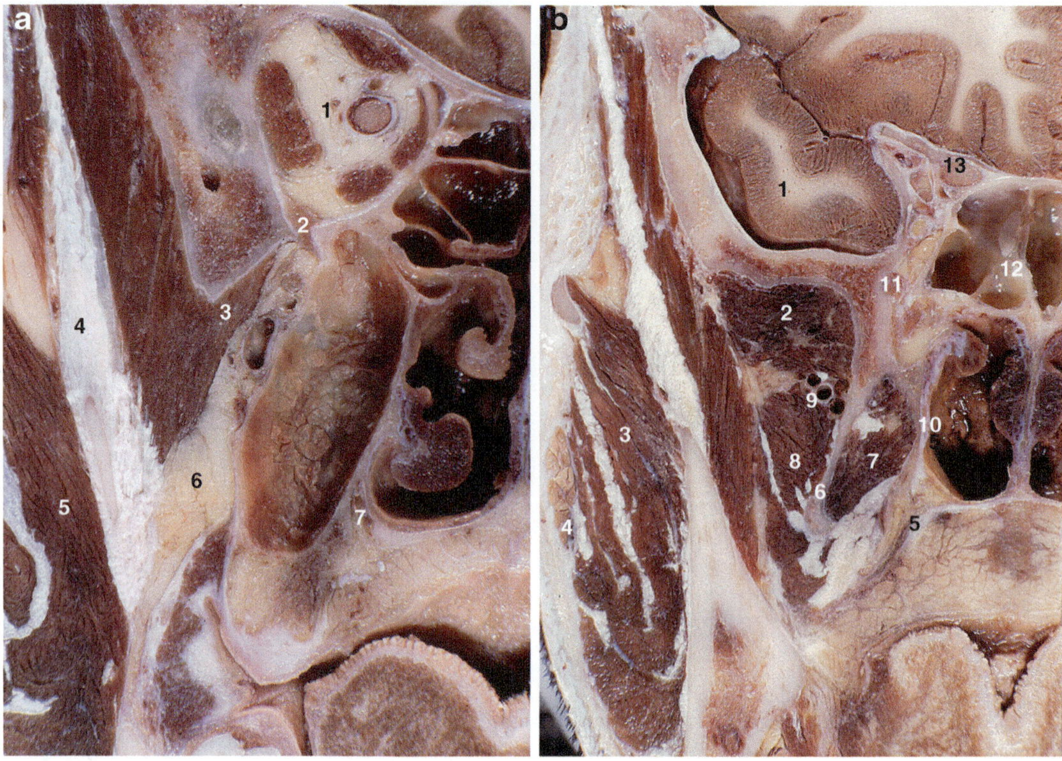

Fig. 7.6 Masticatory muscles on serial coronal sections seen from the front. (**a**) 1. Apex of orbit with optic fascicle and ophthalmic artery; 2. Inferior orbital fissure; 3. Medial beam of the temporal; 4. Tendon of the temporal muscle; 5. Masseter; 6. Fat pad; 7. Inferior palatine vessels and nerves. (**b**) Section behind: 1. Temporal lobe; 2. LPM superior beam; 3. Masseter; 4. Stenon canal of parotid gland; 5. Reflection around the hamulus of medial lamina of pterygoid process of the tendon of the tensor veli going on the superior aspect of velum; 6. Lateral lamina of pterygoid process; 7. Medial pterygoid muscle; 8. LPM inferior beam; 9. Maxillary artery; 10. Medial lamina of pterygoid process; 11. Vidian nerve; 12. Sphenoid sinus; 13. Optic fascicle and ophthalmic artery within the optic canal

maintain the posture that uses neural circuits, operating either under conscious and voluntary control or under an automatic mode.

7.3.1 Reminder of the Basic Principles of Motor Control

Man is technically a very complex machine mostly driven by an ignorant and even those who know do not function better than the ones who know nothing about biology. In other words, the main keyword of the motor organisation is automation.

This presupposes two levels of control. The first level is the voluntary and conscious decision for actions where the command language is expressed in terms of movements—and not of muscles—that it is not needed to know to function perfectly.

The second level is the unconscious executive level where the language is the analytical processing of all the information from the biological sensors and giving orders to all effector muscles in a precise spatiotemporal sequence with correction of errors in real time by the inferior olive of the brainstem. Some more complex motor programmes learned through education such as cycling or playing piano can be stored in the subcortical areas of the basal ganglia central representing a real motor library.

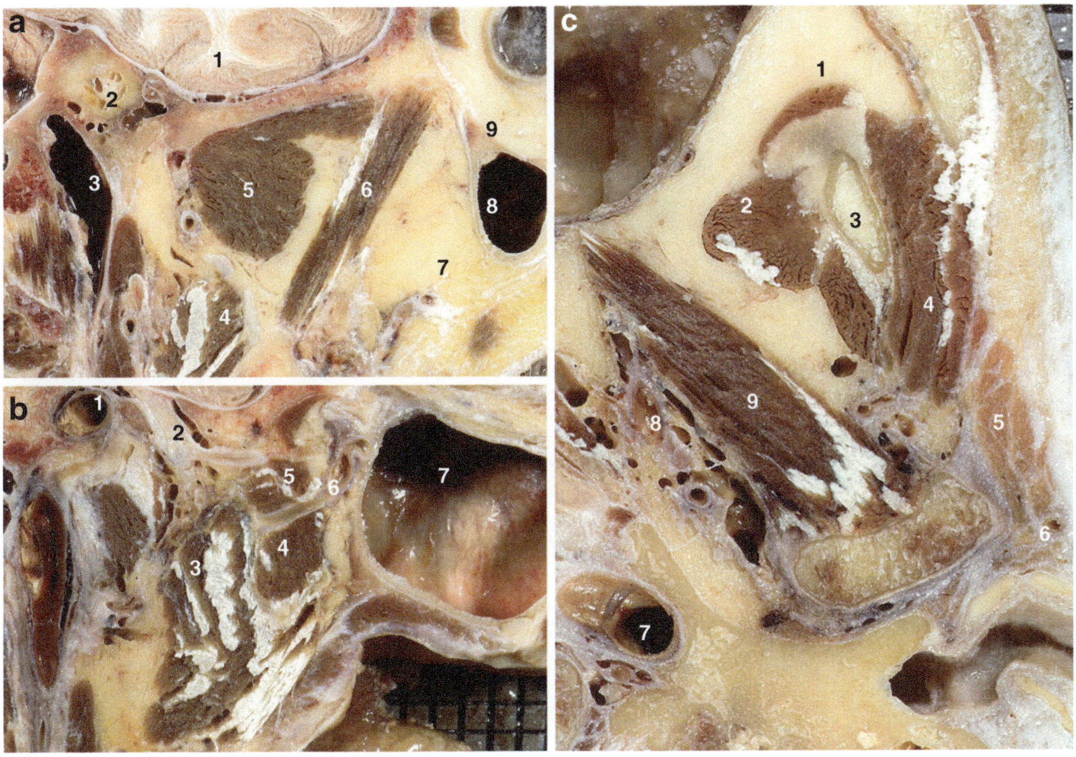

Fig. 7.7 Sagittal and axial sections of the right infra-temporal fossa. (**a**) Lateral sagittal section: 1. Temporal lobe of brain within the middle cerebral fossa; 2. Cochlea; 3. Internal jugular vein; 4. Medial pterygoid muscle; 5. Inferior LPM; 6. Medial beam of temporal muscle; 7. Fat pad (boule de Bichat); 8. Maxillary sinus; 9. Orbit fat. (**b**) Medial sagittal section: 1. Internal carotid artery; 2. Mandibular nerve (V3) emerging from the foramen ovale; 3. Medial pterygoid muscle; 4. Inferior LPM; 5. Superior LPM; 6. Maxillary artery; 7. Maxillary sinus. (**c**) Axial section: 1. Fat pad (Bichat); 2. Medial beam of temporal; 3. Coronoid process and temporal tendon; 4. Masseter; 5. Parotid gland; 6. Superficial temporal artery; 7. Internal carotid artery; 8. Maxillary artery; 9. Inferior beam of LPM

The central nervous system must control 600 muscle actuators, viscoelastic, non-reversible and non-linear by means of a motor keyboard within the area 4 of the precentral gyrus of the frontal lobe (motor homunculus of Penfield) completed with a sensitive keyboard involving all areas of sensitive skin areas along the 1,2,3,5 postcentral gyrus of the parietal lobe. There is therefore a stato-kinetic cortical control loop which can be actuated by a cerebro-cerebello-cerebral interconnection, the action decision being elaborated in premotor areas (Fig. 7.10).

The important point is to understand that two information are required for motor control, as was clearly demonstrated in robotics: firstly, the state of the motors perceived by the brain through the muscle spindles and Golgi organs, identifying the mechanical stiffness of the muscles and not their length during their three possible states: relaxed, contracted or stretched; and secondly, the angles of body segments for position and movement control.

But the joint goniometry cannot be given by muscles that are viscoelastics and redundants, with no central library of muscle inserts for geometric computation nor by ligaments that are indicators of joint excursion limit.

The goniometer is actually the skin covering all body segments, which, by its mechanical deformation during movement or posture, transmits by the sensors described by Ruffini as a stretching information to the brain, enriched and refined by the vision.

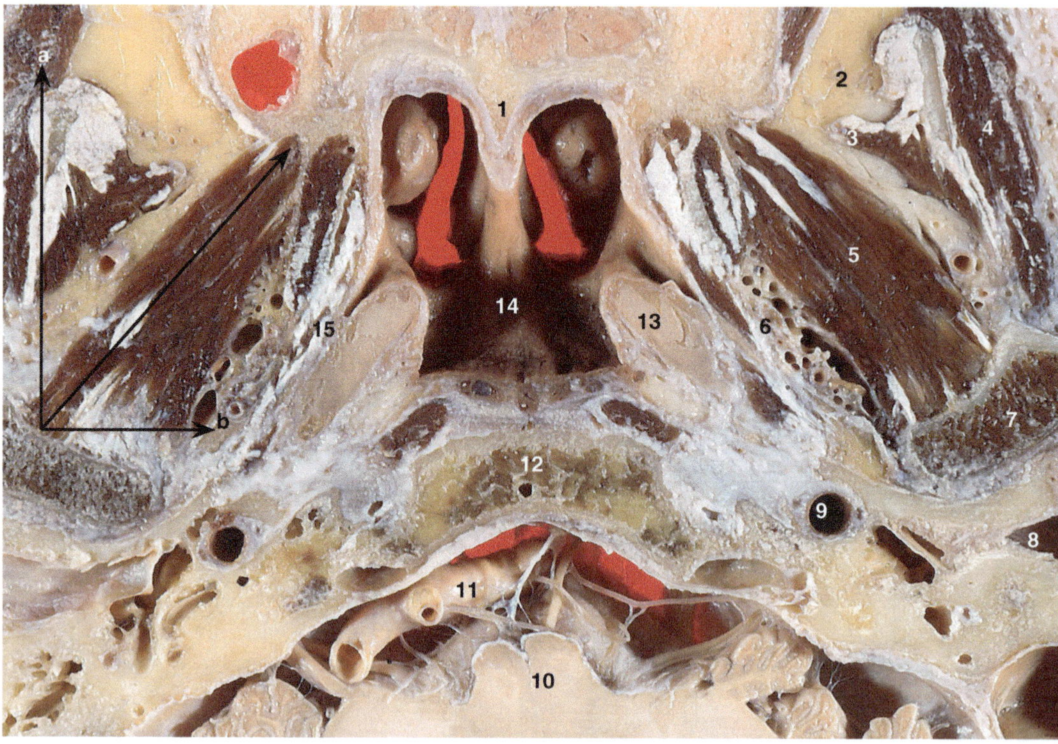

Fig. 7.8 Axial section seen from below: the biomechanical compromise of LPM: (***a***) *Anterior traction on mandibular condyle for opening the mouth by bilateral synergic action;* (***b***) *Lateral movement (diduction) by unilateral action with rotation.* 1. Posterior palatine spine; 2. Fat pad (boule de Bichat); 3. Medial beam of temporal muscle; 4. Masseter; 5. Inferior beam of LPM; 6. Mandibular nerve (V3); 7. Oblique orientation of condyle (140° angle); 8. Meatus acusticus externus; 9. Internal carotid artery; 10. Medulla oblongata; 11. Junction of the two vertebral arteries giving the basilar trunk; 12. Clivus; 13. Fibro-cartilaginous valve of tuba auditiva at the pharyngeal ostium; 14. Rhinopharynx; 15. Tensor veli

Muscles are consciously perceived only when they are painful. Therefore, cutaneous and visual inputs allow us to consciously exist in a three-dimensional space (static and dynamic body scheme).

7.3.2 Mandibular Position and Movement

From this short reminder, it can be concluded that the mandible is constantly sough and may adopt variable postures from the muscle relaxation to the strong contraction often indicative of individual mental states as stress, aggression, pain, anguish, ... [10].

Bruxism is the intercusp rubbing, movement usually involuntary, and is particularly observed in children. These postures are not consciously identified by muscular sensations almost non-existent if the muscles are not sore, but by skin of the face and neck. The face is richly innervated by the three branches of trigeminal nerve (V) corresponding to three floors and the cervical plexus for the neck. The facial nerve (VII), as we have seen, is the motor nerve of platysma muscles that animate the facial mask and its conscious perception is given by the trigeminal nerve. It is the same for TMJ which under normal conditions is dimly perceived. But in the case of capsular or disc disturbances, mobilisation can become painful creating a specific syndrome requiring

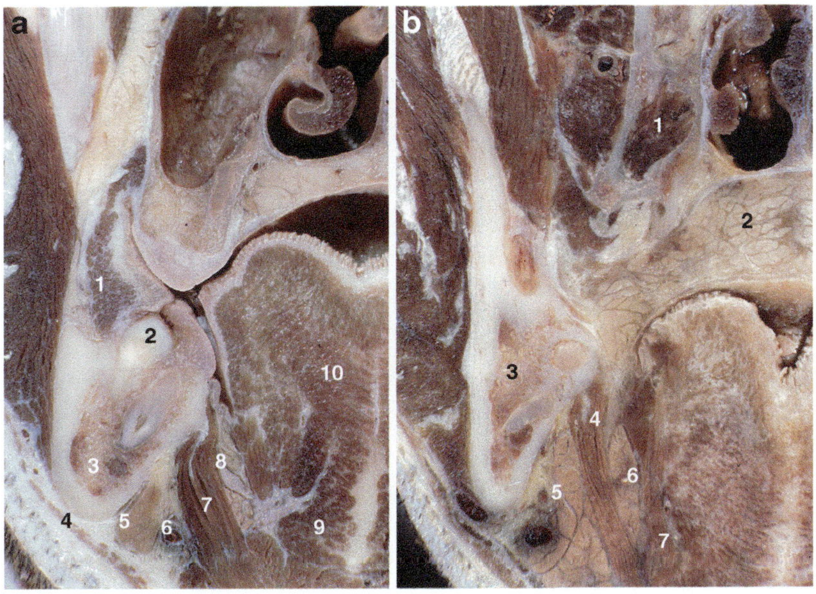

Fig. 7.9 Coronal sections of right inferior hemiface. (**a**) Section of posterior part of mandibular corpus: 1. Buccinatorius; 2. Third molar below the mucosa; 3. Inferior dental canal; 4. Platysma; 5. Submandibular gland; 6. Submental artery; 7. Mylohyoid muscle; 8. Sublingual gland; 9. Genioglossal muscle; 10. Tongue. (**b**) Section of the mandibular ramus: 1. Pterygoid fossa; 2. Palate; 3. Mandible; 4. Mylohyoid muscle; 5. Submandibular gland and facial vein; 6. Sublingual gland and Wharton canal; 7. Hyoglossal muscle

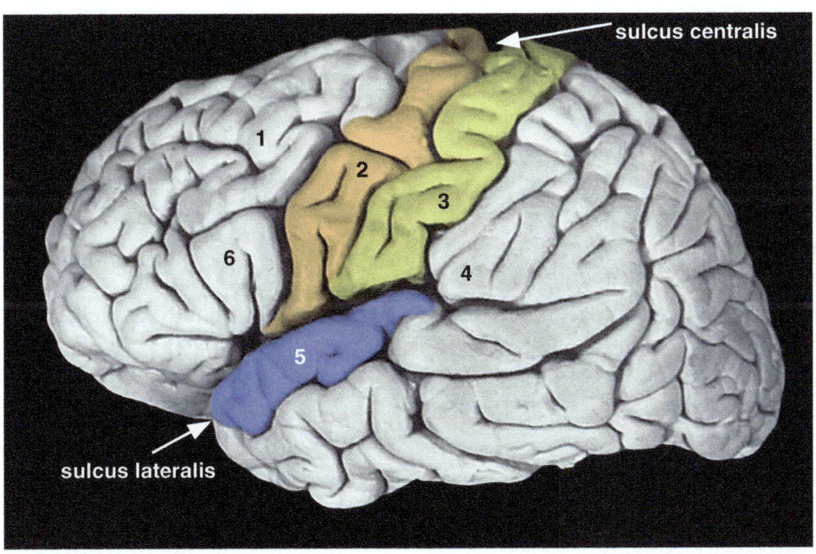

Fig. 7.10 Central cortical command of mobility and stato-kinetic loop. 1. Premotoric area (conscious decision of action); 2. Motor keyboard of area 4; 3. Sensitive keyboard of areas 1,2,3,5; 4. Superior temporal gyrus for speech understanding (Wernicke area 22); 5. First temporal gyrus for hearing; 6. Inferior part of F3 for articulated language (Broca areas 44 and 45)

appropriate orthodontic or surgical or manual treatment.

7.3.3 Mandibular Posture and Upright Posture

Posturology has made huge progress since the initial work of Baron and invaded specialised consulting offices. Interpretation of measurements needs a more scientific deepening to be indisputable as we mentioned earlier.

It should be remembered that the technical problem of postures is mainly related to maintaining a steady state of axial and symmetrical anatomical structures. Upright posture is highly automated by the postural servomechanism with several inputs: visual, cutaneous (dynamic plantar force plate), vestibular with threshold and tendino-muscular with the myotatic reflex responsible for the left/right spinal balance. The system is unstable and requires redundant permanent dynamic corrections by impulses with anticipation. It is the same situation with the mandibular posture stabilised by symmetrical muscles. However, we know that the symmetry of axial structures is not perfect: we have one eye bigger than the other, two slightly different hemifaces, a foot longer than the other one, ... The chewing often occurs asymmetrically with a favourite side which also may vary depending on the condition of dental arches or in case of unilateral dental pain. This frequent asymmetry of masticatory activity also generates an asymmetric swallowing. The facial mask animated by bilateral muscles is often slightly asymmetric, which can be visible during phonation.

So, given some technical similarities between mandibular and vertical postures, one may wonder whether there is any possible justification for associating them or not; for instance, by prescribing a stabilogram in patients with TMJ disturbances. One can formulate some assumptions about this.

All synchronised left/right system imposes an initiator or controller side triggering the synchronous sequence on two neural centres. This is the phenomenon of dominance or laterality that applies to all symmetrical systems with synchronous speed.

The comprehensive solution to this problem is sought in the organisation of the central nervous system where the management of motor automatisms is dependent of the brain stem and cerebellum with a first-level analysis in spinal cord. Regarding the mandible and the face in general, trigeminal nerve of the first visceral arch, as we said before, has the longest central nucleus of all cranial nerves partially because it processes the proprioceptive contingent of all pure motor nerves: III, IV, VI and XII. It is also the sensitive part of the skin-muscle close loop control of position and movement of masticatory muscles. It also intervenes in the regulation of the facial mask thanks to the connection between trigeminal spinal nucleus (pars interpolaris) and motor nucleus of the facial nerve (VII). From there, the important trigeminothalamic pathway with its two dorsal and ventral parts receives fibres from the dorsal spinal column. Those fibres convey information from Pacini receptors of the plantar force plate (posture balance) and Ruffini of the periarticular skin (goniometry) and after relay in gracile and cuneate nuclei with a connection to the cerebellum participate in the formation of medial lemniscus.

It is therefore at the thalamus level before the cortical projection that it is possible to think about an interference between different topographical information converging in. The thalamic ventrolateral nuclei have a ventral posterior part (VP) called ventrobasal complex with the ventral posterior lateral anterior nucleus (VPLa) which receives the fibres of the conscious stato-kinetic sense of the medial lemniscus and the posterior part (VPLp) receiving the epicritic tactile sensitivity. In fact, it is acceptable to talk about interference between different sensory inputs concerning different postural functions.

Finally, this functional similarity of mandibular and vertical postures would authorise in a certain way their rapprochement but further research is needed to validate this assertion. It is obviously very difficult to have a holistic perception of this neural complexity and thus should be wary regarding too reductionist interpretations that lead to dangerous pedagogical dogmatism.

However, we note that, mechanically, the TM joint is a very much solicited joint with a mode of posture depending on many external factors due to its strong integration into the body and mind real life. This may explain in its chronic pain a frequent "somatisation" as it is as well the case for knee or spine.

References

1. Rabischong V, Rabischong P (1986) Anatomie fonctionnelle et biomécanique de l'articulation temporo-mandibulaire. Les Questions d'Odonto-Stomatologie 44:327–333
2. Christensen FG (1969) Some anatomical concepts associated with the temporo-mandibular joint. Ann Aust Coll Dent Surg 2:39–60
3. Hellings G, Klineberg L (1983) The masseter muscle: the silent period and its clinical implications. J Prosthet Dent 49(1):106–112
4. Lavigne G, Frysinger P, Lund JP (1983) Human factors in the measurement of the masseteric silent period. J Dent Res 62(9):985–988
5. Orthlieb JD (1983) La courbe de Spee; un impératif physiologique et prothétique. Cah Prothese 44:89–116
6. Duday H, Lapeyre L (1982) La place du ptérygoïdien latéral dans les mouvements de la mandibule. Rev Electroencephalogr Neurophysiol Clin 12: 153–157
7. Grant PG (1973) Lateral pterygoid, two muscles. Am J Anat 138:1–10
8. McNamara JA (1983) The independent functions of the two heads of the lateral pterygoid muscle. Am J Anat 138:197–206
9. Rabischong P, Guerrier Y, Vignaud J (1981) Anatomical bases for the surgical access to the pterygopalatine fossa. Anat Clin 2:209–222
10. Abrams IN (1963) Oral muscles pressures. Angle Orthod 33(2):83–104
11. Gelb H (1983) Clinical management of head, neck and TMJ pain dysfunction. W.B. Saunders, Philadelphia, PA

Swallowing and Speaking

8

Abstract

The pharyngolaryngeal organisation can explain clearly how the two major complex functions of swallowing and speaking can be achieved. The first aspect is the aerodigestive crossway which requires a very precise regulation of both the air flow and the digestive tract, because only air can transit in superior and inferior pharynx and no solid nor liquid food can penetrate within the nasal fossa or the larynx. The reason for this complex organisation has to be found in the fact that the mouth is masticating not only food but also sounds coming from the larynx used to the language. This justifies for safe swallowing a great quantity of specific anatomical features like soft palate, epiglottis, junction of the base of the tongue with the arytenoid pad, mobility of the larynx,... To understand phonation mechanisms, some basic notions of acoustics, phonetics and linguistics are needed. The control of voice is given by the ear and the mandatory equipressure on both sides of the eardrum justifies the complex organisation of soft palate and tuba auditiva. The roles of the tongue and of the vocal cords are also very important to perceive the incredible possibilities of the language, which are directly related to the typically human specific brain cortical areas.

8.1 Introduction

Swallowing and speaking are two important functions using complex automated mechanisms that must be technically precisely understood. The concerned anatomical area is the pharyngolarynx that embryologically drifts from the segmentation of the mesoderm surrounding the foregut, giving rise to **four visceral arches** limited externally by ectoblastic slots and internally by entoblastic pouches [1]. Each arch is colonised by a mixed nerve.

8.1.1 The Mandibular Arch

The first and also the largest one, the mandibular arch, builds the face and therefore most of the splanchnocranium. It has three floors, each colonised by a branch of the trigeminal nerve

P. Rabischong, *Comprehensive Anatomy of Motor Functions*, DOI 10.1007/978-3-319-04169-8_8,

(V)—a mixed nerve—with a sensory root and a trigeminal ganglion (of Gasser), and a motor root, the masticatory nerve. Each of its branches carries parasympathetic ganglions originating from other nerves (indicated in parentheses):

- on V1, the ciliary ganglion (from III) for iridian sphincter;
- on V2, the pterygopalatine ganglion (from VII) for lacrimal and mucous glands;
- on V3, the otic ganglion (from IX) for parotid gland;
- on V3, the submandibular ganglion (from VII) for submandibular and sublingual salivary glands.

All these ganglions distribute their postganglionic fibres using the branches of the trigeminal nerve.

The upper part of the visceral arch forms the floor of the orbital cavity, and the lower part, the mandible, which is built from the draft of Meckel's cartilage that gives, in the back, the malleus and the incus ossicles of the middle ear. The muscles of this first arch are the masticatory muscles, the floor of mouth, the tensor tympani and the tensor veli.

8.1.2 The Hyoid Arch

The second one, the hyoid arch, gives birth to the styloid process of the temporal bone, the stylo-hyoid ligament, the little horn of the hyoid bone and the half top of it. The muscles of the hyoid arch are the stapedian, the stylohyoid, the posterior belly of digastric and the platysma of the face. The related nerve is the facial nerve (VII)—a mixed nerve—with a sensory root and its geniculate ganglion, and a motor root innervating all the muscles. We must stop talking about VIIbis or intermediate of Wrisberg, which was a mistake of the ancient anatomists, because it is indeed the sensory root of the facial nerve.

8.1.3 The Hyothyroid Arch

The third one, the hyothyroid arch, forms the lower half of the hyoid bone, its great horn and the upper part of the thyroid cartilage. The nerve is the glossopharyngeal (IX)—a mixed nerve—with two roots: firstly, a sensory root with its superior and inferior ganglions for the sensory innervation of the posterior third of the tongue, the mucosa of the promontorium in the middle ear, the palatine tonsils and the upper part of the pharynx, and secondly, a motor root for the following muscles: stylopharyngeal, superior and middle pharynx constrictors and levator veli.

8.1.4 The Thyro-Aryteno-Cricoid Arch

The fourth and last one, the thyro-aryteno-cricoid arch, represents the lower visceral hyo-laryngeal skeleton with the thyroid, arytenoids and cricoid cartilages.

The nerve is the vagus nerve, or pneumogastric nerve (X). Firstly, it has a sensory root and two jugular and plexiform ganglia with a sensitive area covering the mucosa of the larynx by the superior laryngeal nerve, the lower part of the pharynx, the auditory tube and a portion of the skin of the external auditory meatus. Secondly, it shows a motor root visible on the lower lateral side of the medulla oblongata, corresponding to the inferior recurrent laryngeal nerve. This nerve innervates all the intrinsic muscles of the larynx: cricothyroid, posterior and lateral cricoarytenoid, oblique and transverse interarytenoids and inferior thyroarytenoid or vocal muscle.

Below its visceral arch territory, the vagus nerve—also called pneumogastric—becomes a large parasympathetic nerve that descends in the thoracoabdominal cavity to end into the pelvis.

To retain all these anatomical data, it is necessary in a comprehensive approach to identify the technical problems and assess the relevance of the anatomical solution. In that view, the three following points have to be discussed: the aerodigestive crossroad, swallowing and phonation.

8.2 The Aerodigestive Crossroad

On the face, two types of holes are visible:

- firstly, both nostrils marking the beginning of the respiratory airway whose goal is to let

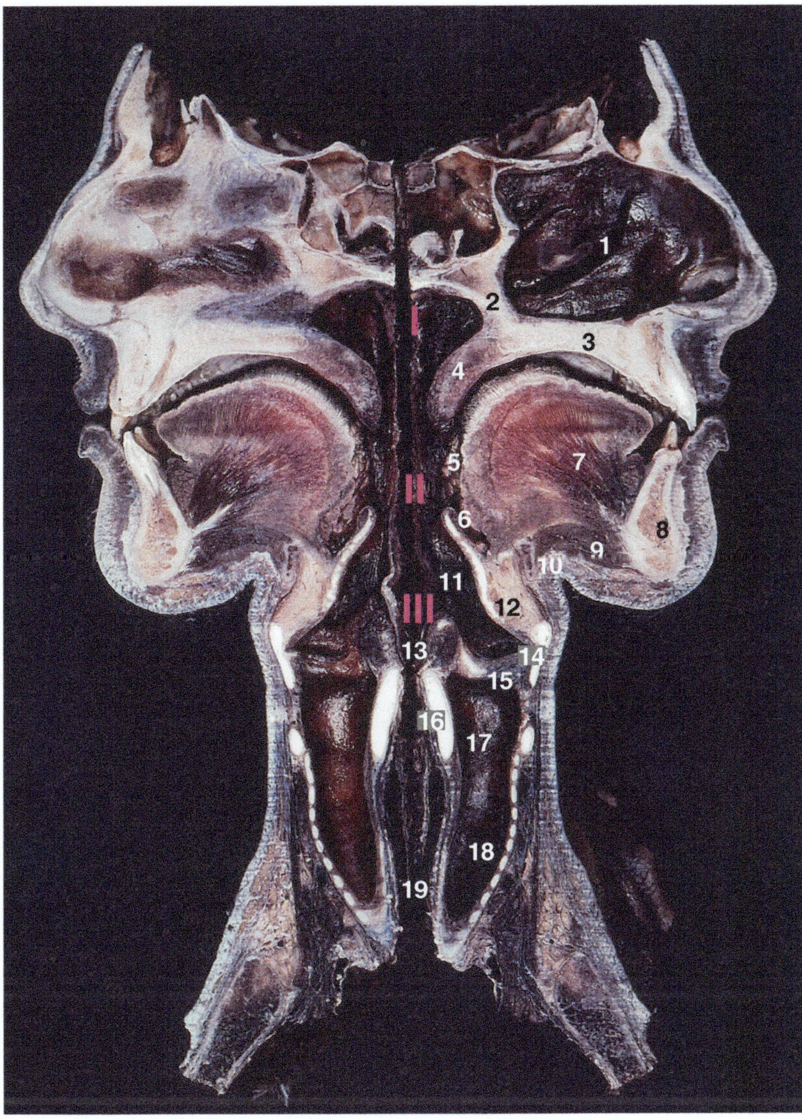

Fig. 8.1 General topography on a double median sagittal section of face. *This section was made on a young medical student of 20 years old, given after death for anatomical studies by his parents*. The three floors of the pharynx: I. Rhinopharynx, II. Oropharynx, III. Laryngopharynx. 1. Nasal fossa; 2. Choane limited by vomer; 3. Hard palate; 4. Velum palatinum; 5. Base of the tongue; 6. Vallecula epiglottica; 7. Tongue; 8. Chin (symphysis); 9. Buccal muscle floor; 10. Hyoid bone; 11. Epiglottis and laryngeal aditus; 12. Pre-epiglottic space (fat tissue); 13. Arytenoid pad (arytenoid cartilages and arytenoid muscles); 14. Thyroid cartilage (anterior part); 15. Glottis (rima glottidis); 16. Cricoid cartilage (posterior lamina); 17. Infraglottic space; 18. Trachea; 19. Oesophagus

enter and exit the air into the nasal cavity, larynx, trachea, bronchi and lungs;

- secondly, placed below, the oral mouth orifice which opens, after the double teeth harrow, into the large dining room whose ceiling is the oral palate covered by the mucosa. Its floor is made by the deformable tongue which represents the beginning of the digestive tract continuing in the oesophagus and the stomach (Fig. 8.1).

The airway accepts and transports air only. The digestive tract lets solids or liquids go through the oesophageal tube after mixing and grinding in the oral cavity.

Technically, it is certain that the airway has exclusive transport of air and therefore cannot accept other materials. It should be logical to separate these two channels, considering that their final destination is totally different. Yet, by observing the region, we can see that the airway is first above within the nasal fossa and then anterior, at neck and chest levels, unlike the digestive tract that comes first in a lower position in the mouth and then is posterior at neck and chest levels. How to explain the crossroad of these two tracts at the level of the pharynx?

One can find the solution in the fact that the mouth dining room does not chew only solid or liquid food, but also sounds coming from the larynx and modulated or shaped by the oral-nasal resonators to generate an articulated speech. It is therefore the phonatory function which determines such a construction pattern.

From this main general principle, we can construct the complete system by introducing the absolute necessity of a strict control of the pharyngeal circulation with a major challenge: no liquid nor solid particles in upper or lower airways made by the reinforced gaping tracheal tube for easy air flow. Eventually, air can go into the oesophagus, which is a supple hose. The problem of the subsequent disposal for the air to go up is related to permissive or restrictive ethnic customs.

The master organ to avoid the digestive flow to enter into the nasal cavity is the soft palate. It can close the nasopharynx by its muscle contractions and thus plays a crucial role in both swallowing and phonation.

8.3 Swallowing

This is a feature that allows sending back in the oesophageal tube the content of the oral cavity. It is absolutely necessary to avoid nasal reflux—exactly what the soft palate does by obstructing the nasopharynx. It should also prevent the introduction of oral contents into the larynx that is exclusively reserved for the passage of air [2–6].

Swallowing is a complex function that requires a full automation and involves a series of phenomena perfectly adjusted.

8.3.1 The Tongue

It occupies the whole oral cavity and is able, by its mobility, to move the content of the cavity backward and to clean its peripheral parts by passing along the dental arches. The floor of the cavity is closed by a muscular diaphragm made of the wide mylohyoid muscle doubled superficially with the anterior belly of the digastric muscle. A contraction lifts all these muscles and pushes the tongue and the buccal content backward. The content falls into the pharynx through a narrow space (isthmus of fauces) bounded laterally by the lower pillars of soft palate, between which are two lymphoid masses, the tonsils. Those vigilant guardians against infectious agents are not always fully effective, but knowing better now the important role of immune lymphoid posterior circle described by Waldeyer leads oral surgeons to remove them much less frequently than before.

8.3.2 The Closure of the Larynx

After the first fast phase and while the soft palate has closed the nasopharynx to prevent ingress of liquid or solid, the larynx also must be closed [7]. This is done by a double movement: climb of the larynx and retropulsion of the vertical base of the tongue which appends to the arytenoid pad.

To understand it, it is needed to remember that the larynx is formed by three articulated cartilage pieces:

- the thyroid cartilage (from the Greek "the shield") shaped like a screen rearwardly open;
- the cricoid cartilage, a circular ring connecting to the trachea, which is a reinforced hose by cartilaginous rings and movable vertically (Figs. 8.2, 8.3 and 8.4);
- the two arytenoid cartilages, placed astride on the upper edge of the cricoid and in a square form. A sagittal process (vocal process) is included in a long muscle called vocal muscle or muscular vocal cord made of fine twisted fibres fixed in the angle of the thyroid shield (Fig. 8.5). The arytenoids can move laterally or rotate on their vertical axis.

This laryngeal set is suspended to the hyoid bone, which is attached to the skull base by the stylohyoid ligament, doubled by the stylohyoid muscle, with a viscoelastic suspension formed by the digastric muscle. Its posterior belly is inserted on the medial side of the mastoid process and, after reflection of its intermediate tendon by the means of a fibrous pulley placed on the middle part of the hyoid bone (little horn), its

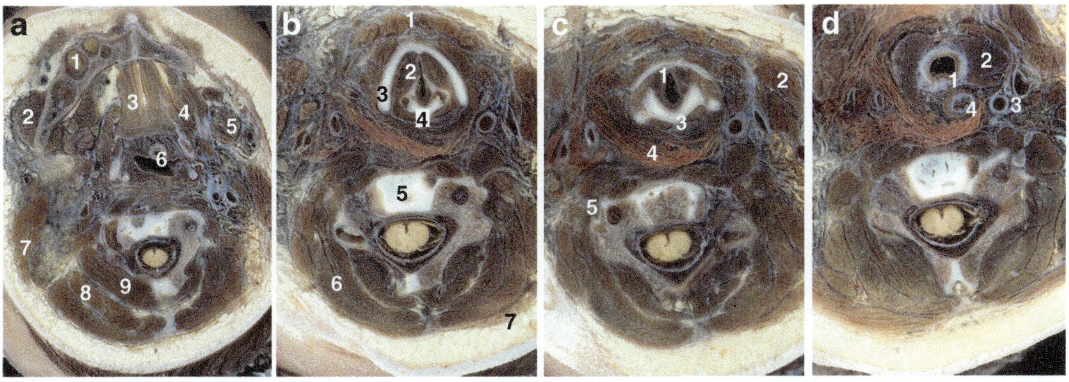

Fig. 8.2 Axial sections of a newborn head seen from below. (**a**) Mandible level: 1. Mandible; 2. Masseter; 3. Geniohyoid muscle; 4. Digastric; 5. Parotid gland; 6. Rhinopharynx with epiglottis; 7. SCM muscle; 8. Semispinalis of the head; 9. Oblique inferior. (**b**) Thyroid cartilage level: 1. Infrahyoid muscles; 2. Vocal cord and vocal process of arytenoid cartilage; 3. Thyroid cartilage; 4. Superior part of cricoid cartilage; 5. Intervertebral disk; 6. Splenius capitis; 7. Trapezius. (**c**) Cricoid cartilage level: 1. Glottis; 2. SCM; 3. Inferior part of cricoid cartilage; 4. Peripharyngeal haematoma; 5. Vertebral artery. (**d**) Trachea level: 1. Trachea; 2.Thyroid gland; 3. Common carotid artery and internal jugular vein; 4. Oesophagus

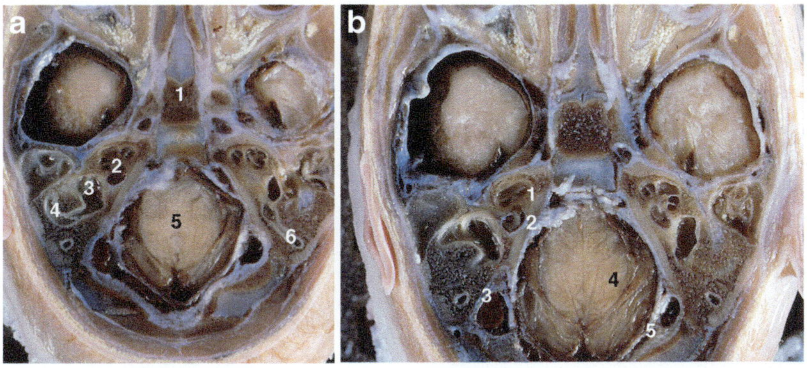

Fig. 8.3 Axial sections of new born skull base. (**a**) Inferior section: 1. Basisphenoid synostosis; 2. Cochlea (same size as adult); 3. Vestibule; 4. Lateral SCC; 5. Brain stem; 6. Posterior SCC (adult size). (**b**) Superior section: 1. Cochlea; 2. Stapes; 3. Superior SCC; 4. Brain stem; 5. Cerebellar tonsils

anterior belly is fixed on the back of the chin. This realises an elastic suspension system that is able to raise the larynx.

The tongue, based on the floor of the mouth, is propelled upward and backward by the mylohyoid diaphragm inserted on the hyoid bone and an oblique line of the internal aspect of the mandible body.

Several important details demonstrate the intelligence and reliability of the whole system.

As we saw, the laryngeal continence is achieved by juxtaposing the base of the tongue and the arytenoid pad formed by the two arytenoid cartilages joined by the thick interarytenoid muscles. This looks like a mouth, and between the lips, there is the interposition of the epiglottis. This soft and flexible fibrocartilage is like a petal with its petiole fixed in the angle of the thyroid cartilage by the thyroepiglottic ligament. It folds down passively and is "stuck" between the two lips. Its front part has the form of a tile, with two oblique slopes, and plays an important role in covering the laryngeal inlet, mainly in diverting the digestive flow laterally into the piriform sinuses, also called pharyngolaryngeal gutters (Fig. 8.6).

It has to be added that the middle part of the larynx is occupied by a sagittal slit bordered by the two vocal muscles, the glottic slit or glottis. It can, by the lateral mobility of the arytenoids,

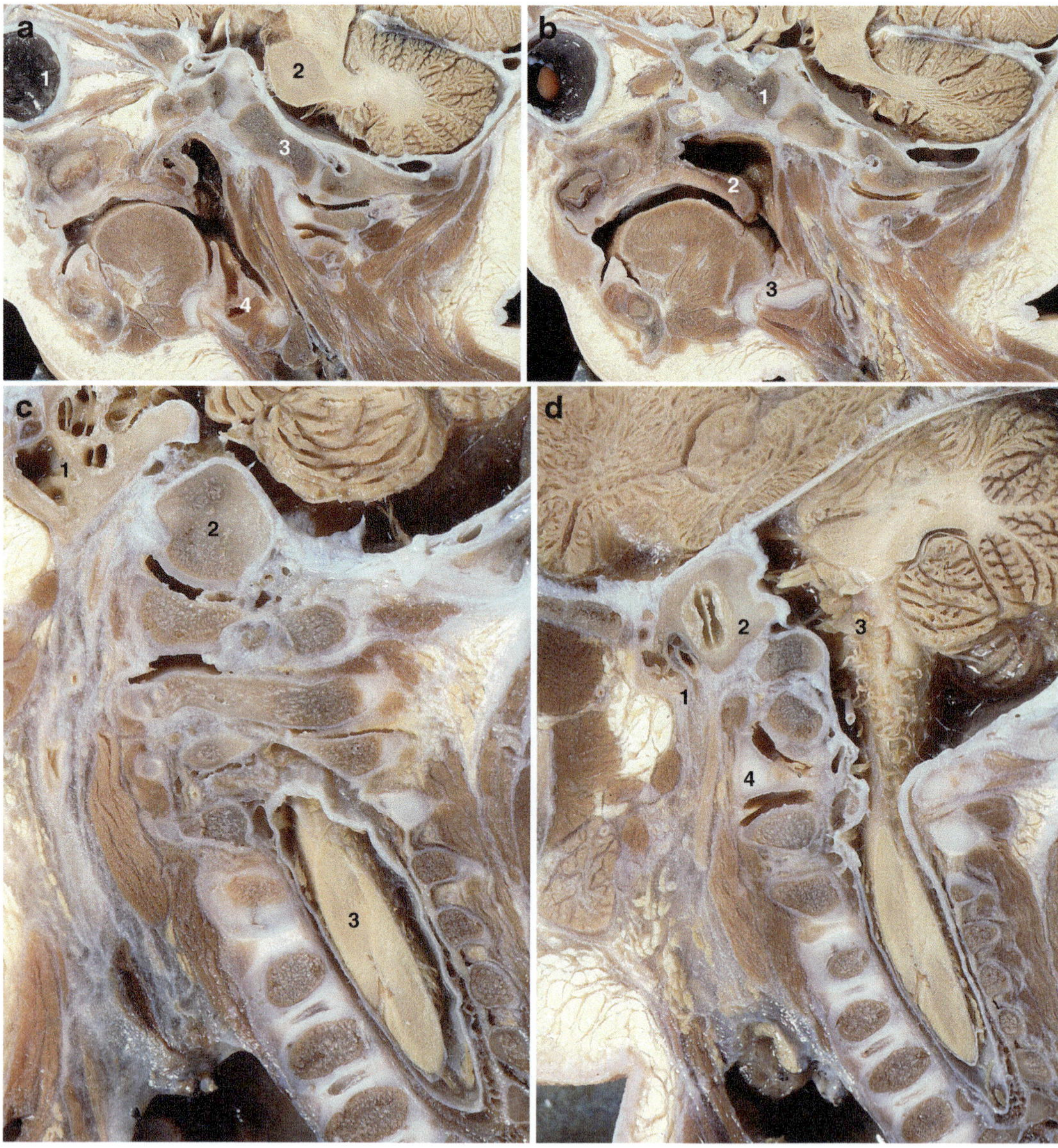

Fig. 8.4 Parasagittal sections of face and cervical spine of a newborn. (**a**) Parasagittal section of skull base: 1. Eye ball, optic fascicle and muscles; 2. Pons; 3. Clivus; 4. Larynx with epiglottis. (**b**) More lateral parasagittal section: 1. Spheno-occipital synostosis; 2. Velum palatinum; 3. Hyoid bone. (**c**) Craniocervical junction: 1. Cochlea (adult size); 2. Lateral occipital; 3. Spinal cord. (**d**) Occipital foramen: 1. Internal carotid artery; 2. Cochlea; 3. Brain stem and cerebellum; 4. C2

either open for breathing by vocal cord abduction activated by the posterior cricoarytenoid muscle or close completely for laryngeal continence by adduction done by interarytenoid muscles.

Finally, the technical secret of this last phase of pharyngeal swallowing is to achieve a lateralisation of the digestive tract (epiglottis and piriform sinuses) and a medialisation of the airway (glottic slit). In that way, the larynx is also a sphincter (Figs. 8.7 and 8.8).

It may happen that this sphincter function has imperfections in the case of "false road" with

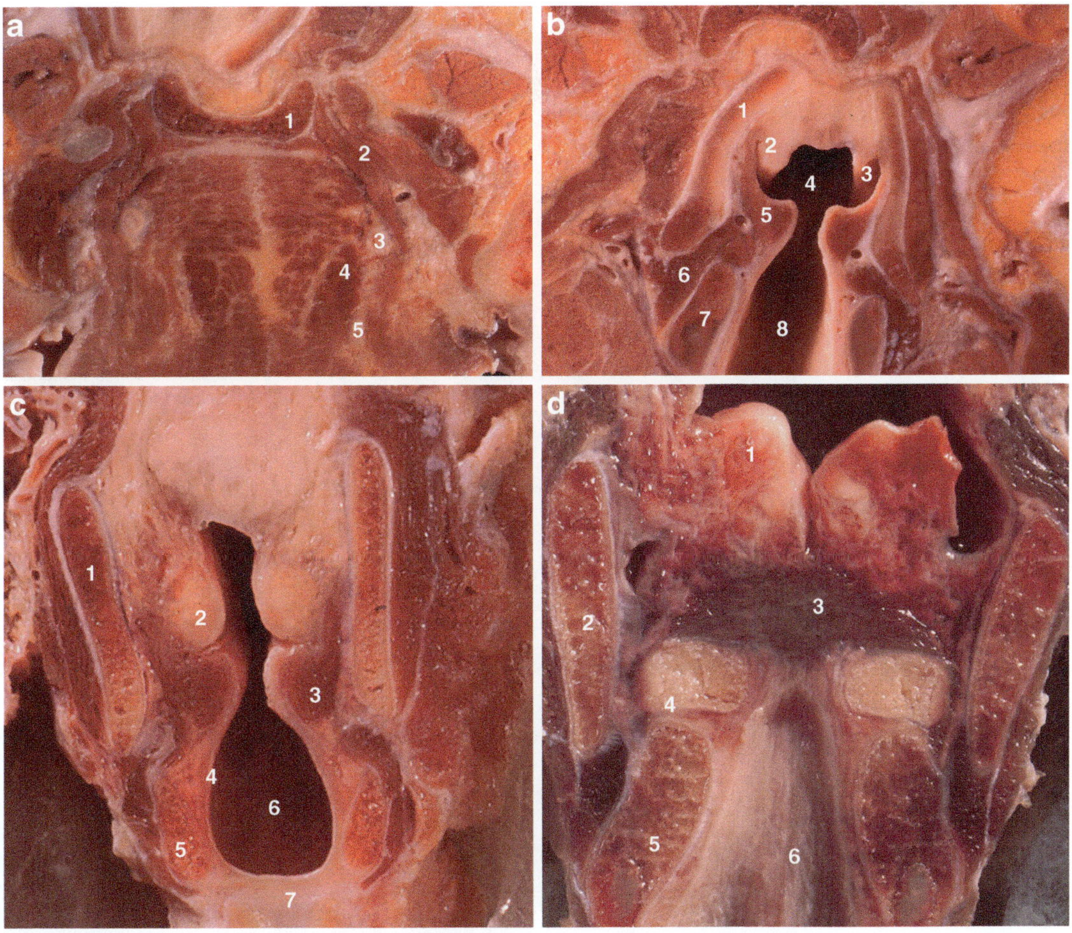

Fig. 8.5 Serial coronal sections of larynx from front to back (fresh cadaver). (**a**) 1. Hyoid bone body; 2. Mylohyoid muscle; 3. Hypoglossal nerve (XII) and lingual artery; 4. Hyoglossal muscle; 5. Lingual nerve (V3). (**b**) 1. Thyroid cartilage; 2. Ventricle fold; 3. Ventricle (Morgagni); 4. Infraglottic space; 5. Vocal cord; 6. Cricothyroid muscle; 7. Cricoid cartilage; 8. Glottic space. (**c**) 1. Thyroid cartilage; 2. Ventricular band; 3. Vocal cord; 4. Elastic conus; 5. Cricoid cartilage; 6. Infraglottic space; 7. Trachea. (**d**) 1. Arytenoid cartilage; 2. Thyroid cartilage; 3. Interarytenoid muscles; 4. Cricoarytenoid joint; 5. Cricoid cartilage; 6. Infraglottic space

laryngeal vestibule intrusion of solid or liquid particles. In this case, an effective cough reflex happens that corresponds to the high sensitivity to foreign bodies of the laryngeal mucosa innervated by the superior laryngeal nerve. For example, banquets often end in joy with many alcoholic toasts. Laughing loudly and drinking at the same time most often end by fits of coughing to clean the laryngeal vestibule.

In another domain, surgery of the larynx, primarily for cancer, has made much progress but remains subject to a number of anatomical constraints:

- to avoid postoperative swallowing problems, it is necessary to preserve the tongue base and a part of the interarytenoid pad;
- the laryngeal cartilages are avascular formations unaffected by the tumoral process that primarily affects the mucosa. In case of resection of the epiglottis, swallowing disorders are minor;
- the glottis represents a vascular and lymphatic barrier differentiating tumours above and below glottis;
- in the case of total resection of the larynx, reconstruction of continuity of the digestive tract is possible, but this is not always the

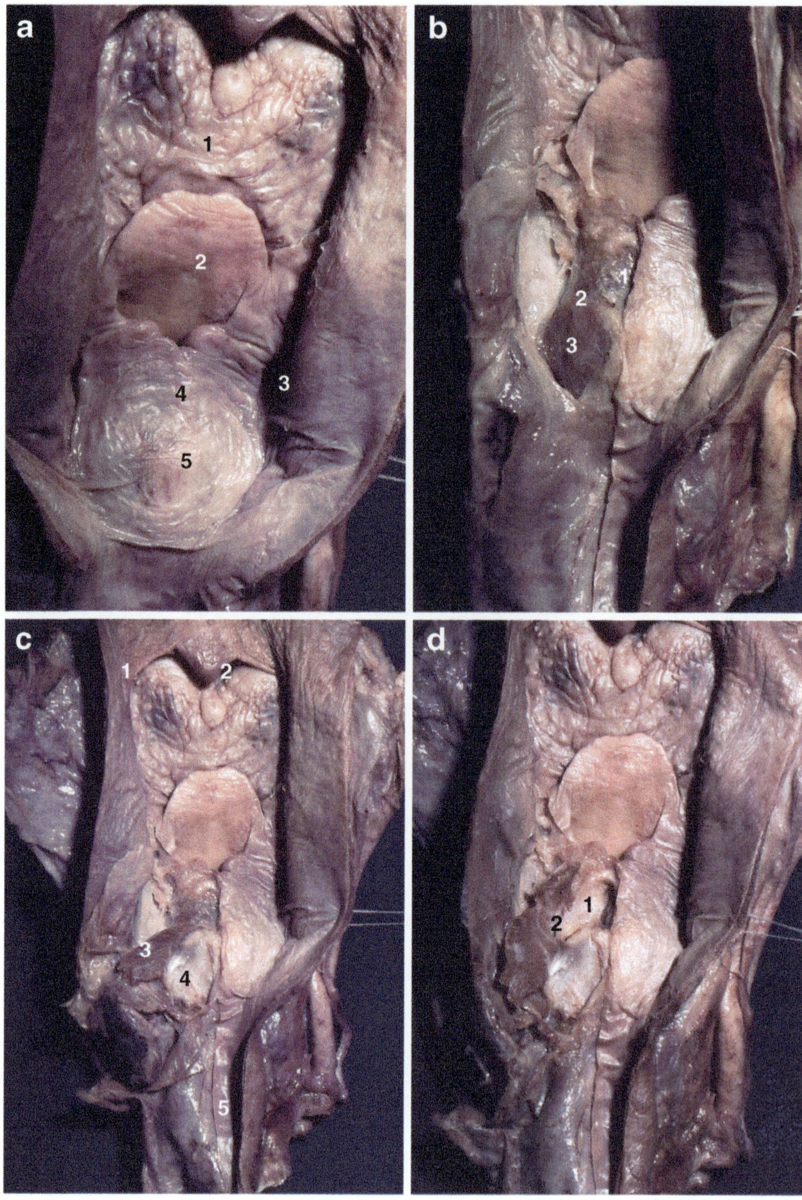

Fig. 8.6 Dissection of posterior muscles of larynx seen from behind. (**a**) Section of pharyngeal constrictors: 1. Vertical basis of tongue; 2. Posterior aspect of epiglottis; 3. Pirifomis sinus; 4. Arytenoid pad; 5. Posterior part of the cricoid cartilage covered by mucosa. (**b**) Median opening of mucosa: 1. Interarytenoid muscles with corniculate tubercle on the top (Santorini); 2. Anastomosis between superior and inferior laryngeal nerves; 3. Posterior cricoarytenoid muscle (posticus). (**c**) Posticus reclined: 1. Lateral border of isthmus of fauces; 2. Uvula; 3. Reclined posterior cricoarytenoid muscle; 4. Posterior aspect of cricoid cartilage; 5. Oesophagus. (**d**) Interarytenoid muscles reclined: 1. Posterior concave aspect of arytenoid cartilage; 2. Cricoarytenoid joint

case for the airway, which requires a stoma by anastomosis of the trachea to the skin. The absence of glottal voice means no voice at all, but it is possible with a specific rehabilitation to restore an "oesophageal voice" by modelling belching sounds in the oral cavity and by the use of an amplifier. The larynx prostheses are for the time being not clinically useful.

8.4 Phonation

Swallowing, practised hundreds of times a day, is a quick and fully automated procedure. But the main function of the larynx is to emit sounds, which will then become phonemes and words by modulation into the supraglottic resonators.

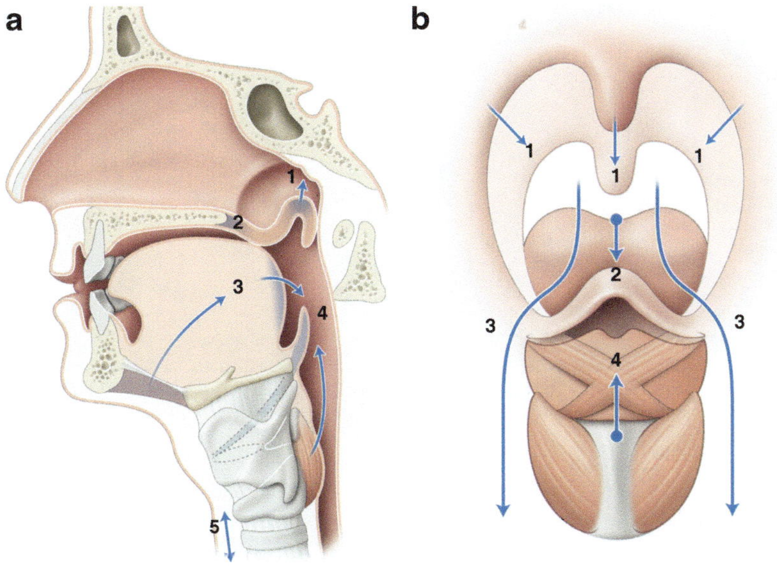

Fig. 8.7 Schematic drawing of swallowing. (**a**) Sagittal section: 1. Retrotraction of soft palate by levator veli contraction; 2. Rigidification of anterior third of soft palate by tensor veli contraction; 3. Tongue retro pulsion by mouth floor muscle contraction; 4. Epiglottis between the tongue vertical base and the arytenoid pad; 5. Up and down movements of larynx. (**b**) Posterior view: 1. Narrowing of isthmus of fauces; 2. Passive flexion of epiglottis deviating the digestive flow; 3. Lateralization of digestive flow; 4. Ascending arytenoid pad

8.4.1 Basic Notions on Sounds

To understand the intelligence of phonatory larynx, it is appropriate to treat some basics in acoustics, the physical science of sound; in phonetics, the science of phonatory organs; and linguistics, the study of language [8–12].

Sounds are produced by mechanical vibrations, which are oscillating air molecules without moving them. It is therefore an energy transport without material moving. The classical representation of the phenomenon is the stone thrown into water creating a series of waves gradually damped swinging a floating object without moving it.

The speed of sound depends on the density of the propagation medium and temperature (in air, 330 m/s at 0 °C and 340 m/s at 20 °C).

The frequency is the number of oscillations per time unit and is measured in Hertz: oscillation number/second.

The wavelength, λ, is the distance travelled by a wave during a period (full cycle) and uses metric units.

The variation of energy that decreases with distance from the source is embodied in a logarithmic function in decibel expressing a ratio of two lengths.

The intensity of the sound wave expressed in decibels can also be measured in Watts/cm^2.

The pitch defining treble and bass sounds depends on the frequency.

The human ear can hear sounds between 25 Hz and 20 kHz; below are infrasound and beyond ultrasound.

The Fletcher diagram defines an audible area (audiogram) reporting the frequencies, expressed in Hz, into dB with a perception threshold and pain threshold.

It is possible to study for sounds' various representations: temporal with shape, amplitude and period, spectral with the different frequencies and vectorial.

Joseph Fourier (1768–1830) has specifically developed a spectral analysis of sounds. We can thus identify a zone of greater sensitivity of the ear between 500 and 5,000 Hz. With age, the

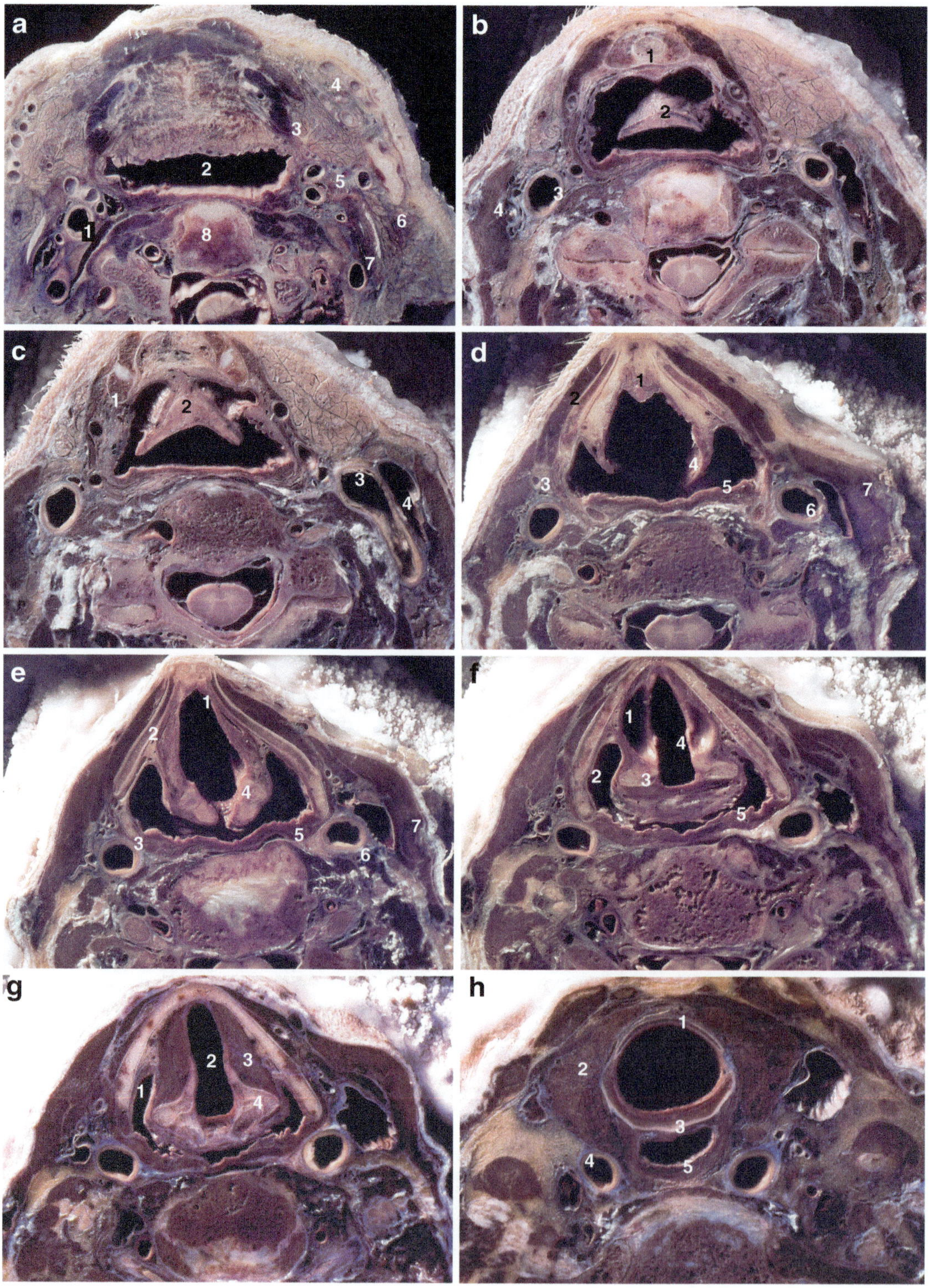

Fig. 8.8 Serial anatomical axial sections of larynx seen from above (up to down). (**a**) 1. Left carotid fork; 2. Oropharynx; 3. Intermediate tendon of the digastric muscle; 4. Submandibular gland; 5. External carotid artery; 6. Sternocleidomastoid muscle (SCM); 7. Internal carotid artery; 8. Cervical vertebra. (**b**) 1. Hyoid bone

presbyacousy characterises a loss of treble. An acoustic trauma in the same frequency band (e.g. use the jackhammer at 100 Hz without protection for years) can remove the corresponding frequency band of the audiogram.

We can distinguish pure tones that have single frequency, such as the one given by diapason, and complex or musical sounds that have fundamental frequency, the slower at spectral analysis which defines the note, mixed with other frequencies that are harmonics, integral multiples of the fundamental say pairs (asymmetric signals) and odd (symmetric signals). The noise is a more or less a structured combination of sounds.

The sound is thus defined with three parameters: height (high shrill or low), intensity (weak or strong) and timbre (harmonic mixed with the fundamental frequency for defining a musical instrument or the voice).

Producing a sound in a column of air was used since the beginning of humanity. It may be done by a vibration transmitted in the column of air (excitatory) contained in a space (resonator) that modifies it more or less with tightening, creating areas of compression optionally followed by decompression more or less explosive according to the aperture of the tube.

8.4.2 The Articulatory Phonetics

Phonetics studies how human speech sounds are produced.

The exciter is the centrepiece of the larynx: the glottic area with a posterior triangular space with a posterior base formed by the two vocal muscles surrounded by the mucosa (vocal fold) and fixed on the back onto the arytenoid cartilages.

They are topped with two ligaments, vestibular or ventricular bands, in the same orientation and circumscribing a mucous cavity, the ventricle described by Morgagni. The elastic cone of larynx contributes to the formation of sounds and is placed under the subglottic space.

Below is the subglottic wind supply with trachea, bronchial tubes and lungs activated by diaphragm bellows and the dynamics of abdominal wall muscles, so important for the control of the voice.

Above stands the voice channel going from glottis to lips and bordered by supraglottic resonators:

- pharynx,
- nasal fossa surrounded by aerial sinusal cavities (maxillary, frontal, sphenoid sinuses and ethmoidal cells),
- mouth with the tongue,
- bony palate forward with alveolar ridges and teeth,
- soft palate on the back with uvula and azygos muscles,
- upper and lower lips.

All these accidents that punctuate the voice channel will change the column of air that passes through it by generating "formants" which characterise the wealth of articulated speech.

The articulation point is the point where there is an obstacle to the passage of air which is used to define different types of articulation in relation with the place of articulation:

Fig. 8.8 (Continued) body; 2. Epiglottis; 3. Left common carotid artery; 4. SCM. (**c**) 1. Junction between hyoid body and great horn; 2. Epiglottis with its specific shape (roman tile) and hyoepiglottic ligament; 3. Right carotid fork; 4. Large internal jugular vein. (**d**) 1. Superior thyroid incisure; 2. Thyroid cartilage; 3. Superior horn; 4. Aryepiglottic fold; 5. Inferior pharyngeal constrictor muscle; 6. Right common carotid artery and internal jugular vein; 7. SCM. (**e**). 1. Piriform recessus; 2. Thyroarytenoid muscle; 3. Vestibule of larynx; 4. Corniculate cartilage (Santorini); 5. Inferior pharyngeal constrictor muscle. (**f**) 1. Arytenoid pad (oblique and transverse arytenoid muscles); 2. Piriformis sinus; 3. Ventricle of larynx; 4. Arytenoid cartilage with vocal process and ligament; 5. Glottis (rima glottidis). (**g**) 1. Piriformis sinus; 2. Glottis; 3. Vocal muscle; 4. Arytenoid cartilage base with vocal process. (**h**) 1. Trachea; 2. Lateral lobe of thyroid gland; 3. Cricoid cartilage base; 4. Common carotid artery; 5. Oesophagus

- lips (bilabial or labial);
- teeth (dental and interdental) or both (labiodental);
- gingival margin of the upper incisors (alveolar pre-mid-post);
- hard palate (pre-palatal, mid-palatal and post-palatal);
- soft palate (velar);
- uvula (uvular);
- root of the tongue (radical);
- pharynx (pharyngeal);
- glottis (glottal).

The articulation mode is also important: free passage, single oral way or nasal and oral, medial or lateral buccal way.

8.4.3 The Language

Languages spoken by different groups of humans inhabiting the planet Earth led to this present linguistic diversity: the 21 most spoken languages, the first being the Chinese Mandarin, concern 70 % of people of the world; the total number of languages and dialects spoken in the world is estimated at 6,000.

The speech, the main communication vehicle, imposed quickly the written conservation of logos and there were born writings of all civilisations. The scribe in Mesopotamia, Egypt and China belonged to the elite of society because he had the code and technical writing. Some civilisations use picto-ideographic writing which requires decryption keys prohibiting syllabic reading. This is still the case of kanji ideograms of Chinese and Japanese. Later on, appeared alphabetic writings accessible to all. The alphabets (from alpha (α) to beta (β) first Greek letters or primers) in each language provide a list of letters with corresponding phonetic values (syllables, words, sentences) for the spoken language.

We can therefore establish a series of fundamental distinctions, mainly between vowels and consonants (Fig. 8.9).

The vowels are characterised by sounds produced by the free passage of air in the supraglottal cavities. There are 6 written vowels (A, E, I, O, U, Y) and 16 phonetic (or vocoids). They are mostly resonant due to a vibration of the vocal cords. We distinguish different types according to the possible different resonators: nasal (an, on), rounded (labial cavity: ou, oh), anteriors (tongue forward: haie, ah), posteriors (back tongue: hou), closed (by minimum oral aperture) or open (by maximum oral aperture).

Consonants are defined as a result of the obstruction of the passage of air in the supraglottal resonators. They are phonetically classified according to their mode of articulation: occlusives (with complete blockage and "explosive" release), constrictive, sonorants and fricatives (mixture of laryngeal sounds and friction at the constriction point), or affricates (combination of an occlusive and a fricative). They can be oral or nasal depending on the interplay of the soft palate.

There is obviously a phonetic language that is specific to the specialists in the field of voice, or scientific and technical specialists (engineers, acousticians, researchers, song teachers, . . .) or therapeutic specialists (phoniatric specialists, speech therapists).

8.4.4 Phonatory Mechanics

Three structures play an important role: vocal cords, soft palate and tongue.

8.4.4.1 The Vocal Muscle

It cannot vibrate actively at the high frequency of sounds due to the inability of muscle fibres to contract at this frequency. However, their contraction can adjust their stiffness which alters their mode of vibration by the propelled column of air.

The voicing is the vibration of the vocal cords producing sounds whose characteristics depend on the degree of opening of the vocal cords and the frequency of vibration.

A more complex system allows to put them more or less in tension. In fact, they are attached to the inside corner of the thyroid cartilage which

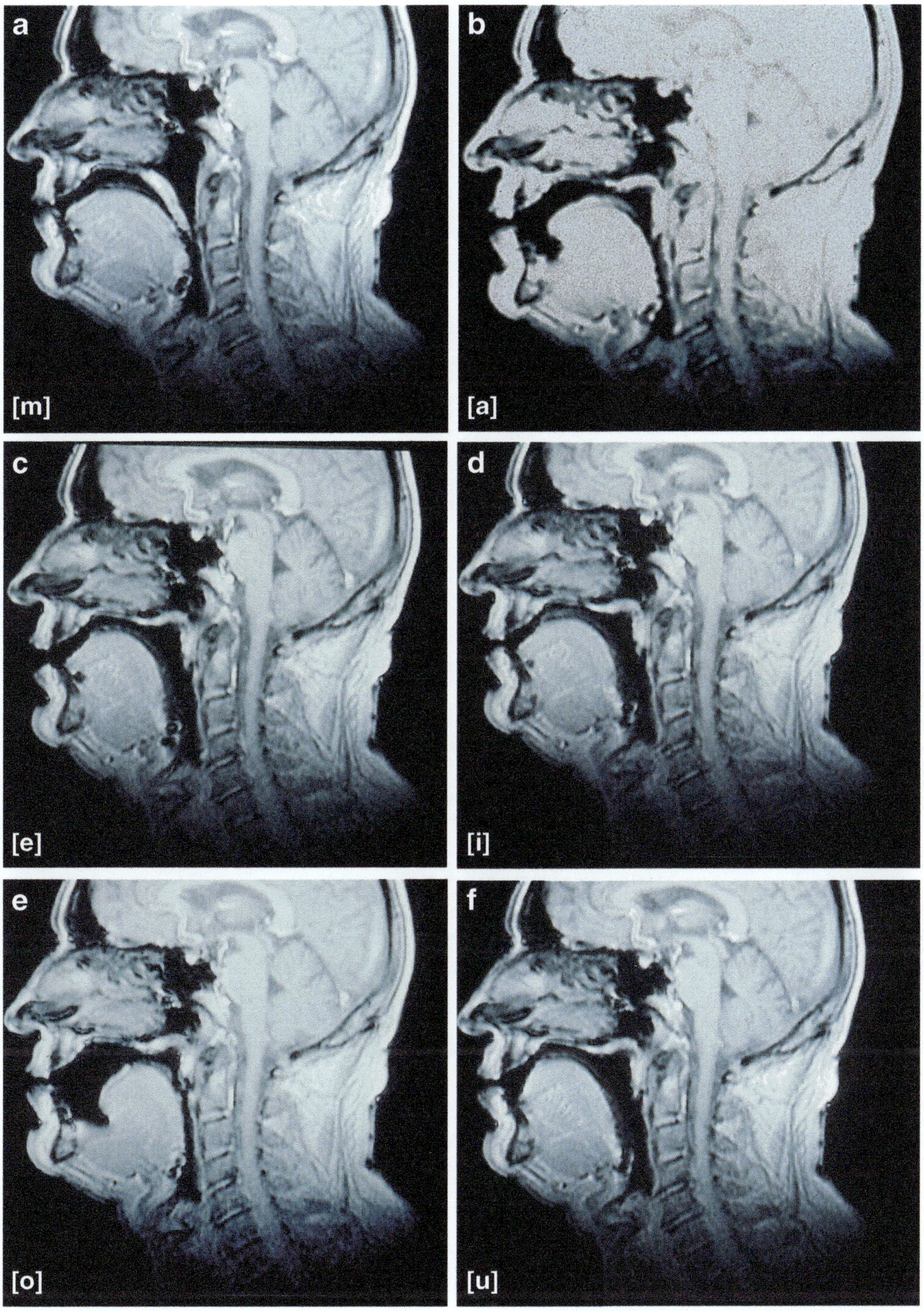

Fig. 8.9 Phonation study by MRI sagittal sections during sound production. (**a**) [m] Nasal way with relaxed soft palate. (**b**) [a] Levator veli contraction for nasopharynx closure and anterior buccal cavity. (**c**) [e] Levator veli contracted and posterior buccal cavity with vertical tongue slope. (**d**) [i] Palatal compression. (**e**) [o] Anterior buccal cavity. (**f**) [u] Posterior buccal cavity with tongue oblique slope

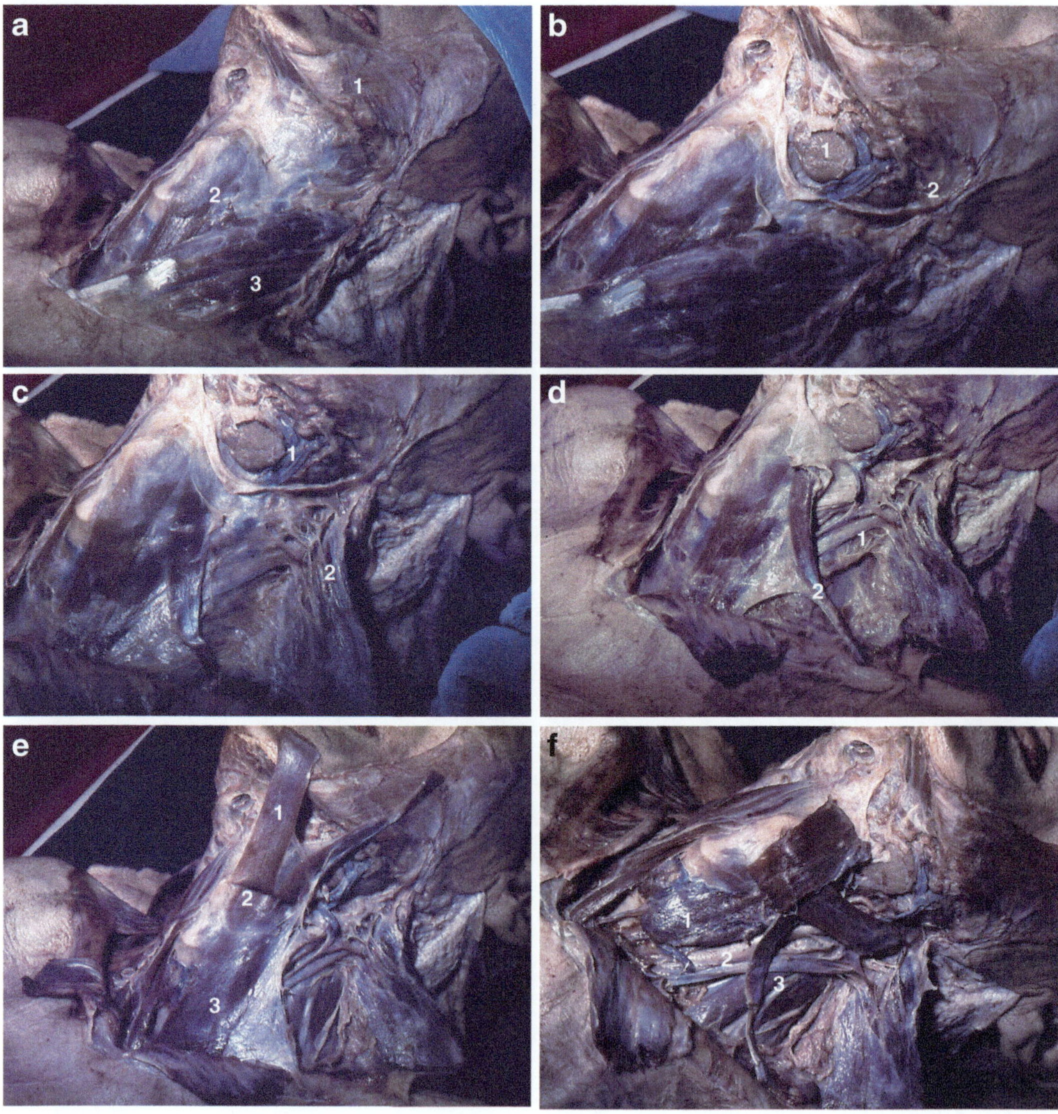

Fig. 8.10 Dissection of the anterior muscles of the neck. (**a**) Superficial plan: 1. Reclined platysma; 2. Larynx; 3. SCM. (**b**) Opening of the superficial cervical fascia: 1. Submaxillary gland; 2. Reclined aponeurosis. (**c**) Reclined SCM outside: 1. Venous thyrolinguofacial trunk; 2. Accessorius nerve (XI) innervating the SCM. (**d**) Vascular pedicle of the neck: 1. Common carotid artery and internal jugular vein; 2. Omohyoid muscle. (**e**) Section of sternohyoid and omohyoid muscles: 1. Reclined up sternohyoid muscle; 2. Thyrohyoid muscle; 3. Sternothyroid muscle; 4. Middle cervical aponeurosis. (**f**) Dissection of the thyroid gland: 1. Lateral lobe of the thyroid gland; 2. Vascular pedicle; 3. Phrenic nerve within the aponeurosis of anterior scalene muscle

can by its articulation, below with the cricoid cartilage, switch slightly forward through the action of cricothyroid muscle. The change in tension of the vocal cord will strongly influence the vibration behaviour by changing the frequency for the variations in height.

The detailed technical knowledge that we have concerning wind instruments (instruments with single or double reed like clarinet or oboe, or mouth like trumpet) allows us to understand that the change of notes must be made by a change in length of the column to vibrate. This

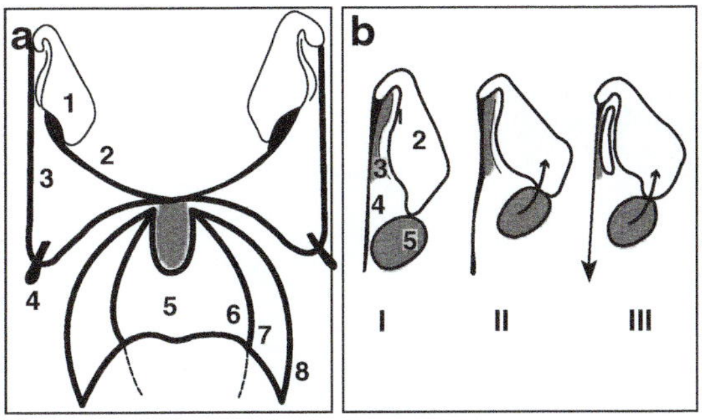

Fig. 8.11 Schematic representation of the velo-tubar dynamics. (**a**) "The suspended and arched bridge with four pillars": 1. Medial fibrocartilaginous valve of tuba auditiva; 2. Levator veli muscle (infratubar); 3. Tensor veli muscle (laterotubar); 4. Hamulus of medial lamina of pterygoid process; 5. Isthmus of throat; 6. Palato-pharyngeal muscle (posterior pillar); 7. Tonsils fossa; 8. Palatoglossal muscle (anterior pillar). (**b**) The three states of the pharyngeal ostium of tuba auditiva: 1. Ostium; 2. Medial valve; 3. Lateral fibrous tubar membrane fixed on the tensor veli; 4. Latero-tubar tensor veli; 5. Infra-tubar levator veli. The tree states of the tuba ostium: I. Relax: ostium closed. II. Phonation: contraction only of the levator veli without opening the tuba. III. Cocontraction of the both muscles with opening of the tuba (swallowing, yawning)

justifies the existence of holes or keys; Boehm, for the flute, was at the origin of an important technical innovation. It is obviously difficult to imagine drilling holes in the trachea to put fingers on.

Therefore, glottis, in order to cover a large range of acceptable scores and variable notes according to a large tessitura of singers (soprano, mezzo soprano, alto, tenor, baritone, bass) needs to move vertically and be fixed in different positions which is made possible by extrinsic muscles of the larynx (Fig. 8.10).

We can describe two functional systems to ensure:

- firstly, the mobility of the "hyoid trapeze": we have seen that the hyoid bone supports the larynx and the base of the tongue. It exists muscles firstly above the hyoid bone: digastric strap, stylohyoid, mylohyoid, geniohyoid; and secondly below the hyoid bone: medially, the sternohyoid muscle attached to the sternum and laterally the digastric omohyoid muscle fixed on to the scapular incisure of the superior border of the scapula. The latter is also the tensor of the middle cervical aponeurosis covering the whole subhyoid area;
- secondly, the vertical mobility of the thyroid cartilage which is attached, above, to the hyoid bone by the thyrohyoid muscle and, below, to the sternum and to the first rib by the sterno-costo-thyroid muscle. These two muscles are inserted on an oblique ridge of the thyroid cartilage which gives them an ideal distribution of muscle strength.

This extraordinary complex organisation responds apparently well to the need to lower the larynx to generate low sounds and raise it for treble. In addition, as the artist who plays a stringed instrument makes his notes by putting his finger at the right place on the rope and gives a particular resonance by vibrating his finger to create the classic "vibrato", a well-trained singer gives colour to his voice by vibrating the larynx with an alternate contraction of the two moorings muscles of the thyroid cartilage.

All of these extrinsic muscles are innervated by the cervical plexus (C2/C3) through a handle placed in front of the internal jugular vein and supported by hypoglossal nerve.

8.4.4.2 The Soft Palate

As Sébileau said, "it has the appearance of a suspended bridge and of an arch with four pillars" (Fig. 8.11). Indeed, it has above two muscles connected to the skull base and the tuba auditiva (Eustachian tube) as well as to the soft palate: levator veli and tensor veli. Below, it is tied to the tongue by the anterior pillar, glosso-staphylan muscle, and to the pharynx by the posterior pillar, pharyngo-staphylan muscle (Fig. 8.12). Between the two pillars is the palatine tonsil which is for ENT surgeons like appendix for visceral surgeons.

To understand the soft palate, it is mandatory to know the solution of the technical problem of the middle ear [13]. It is directly related to the tympanic equipressure and to the velum-tuba synergy.

The Tympanic Equipressure

The auditory function allows capture, process and identification of sound vibrations by measuring and controlling their intensity (ossicular

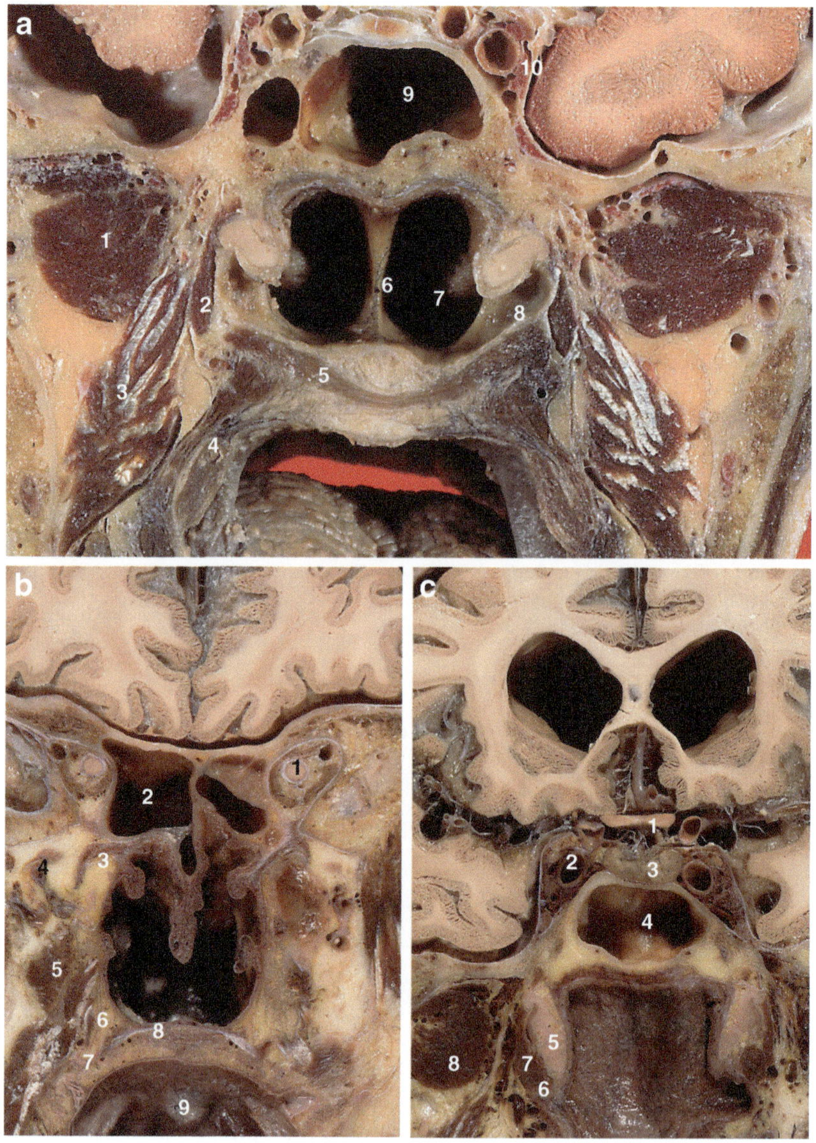

Fig. 8.12 The muscles of velum palatinum. (**a**) Coronal section of the rhinopharynx seen from behind: 1. Lateral pterygoid muscle (LPM); 2. Tensor veli muscle; 3. Medial pterygoid muscle (MPM); 4. Glosso-staphylian and palatopharyngeal muscles; 5. Levator veli muscles forming a muscular strap within the velum, with transversal fibres of the uvular muscle; 6. Septum nasi (vomer bone); 7. Choana (posterior opening of nasal fossa); 8. Pharyngeal ostium of tuba auditiva; 9. Sphenoid sinus; 10. Internal carotid artery. (**b**) Coronal section of the rhinopharynx seen from the front: 1. Apex of orbit with optic fascicle, ophthalmic artery and ophthalmic superior vein; 2. Sphenoid sinus; 3. Spheno-palatine artery; 4. Parasympathetic pterygopalatine ganglion; 5. MPM; 6. Tensor veli muscle; 7. Hamulus (hook) of the external lamina of pterygoid process; 8. Reflected tendon of the tensor veli on the velum surface; 9. Palatine uvula. (**c**) Coronal section of hypophysis: 1. Optic chiasm; 2. Internal carotid artery within the cavernous sinus; 3. Hypophysis; 4. Sphenoid sinus; 5. Tuba auditiva; 6. Levator veli; 7. Tensor veli; 8. LPM

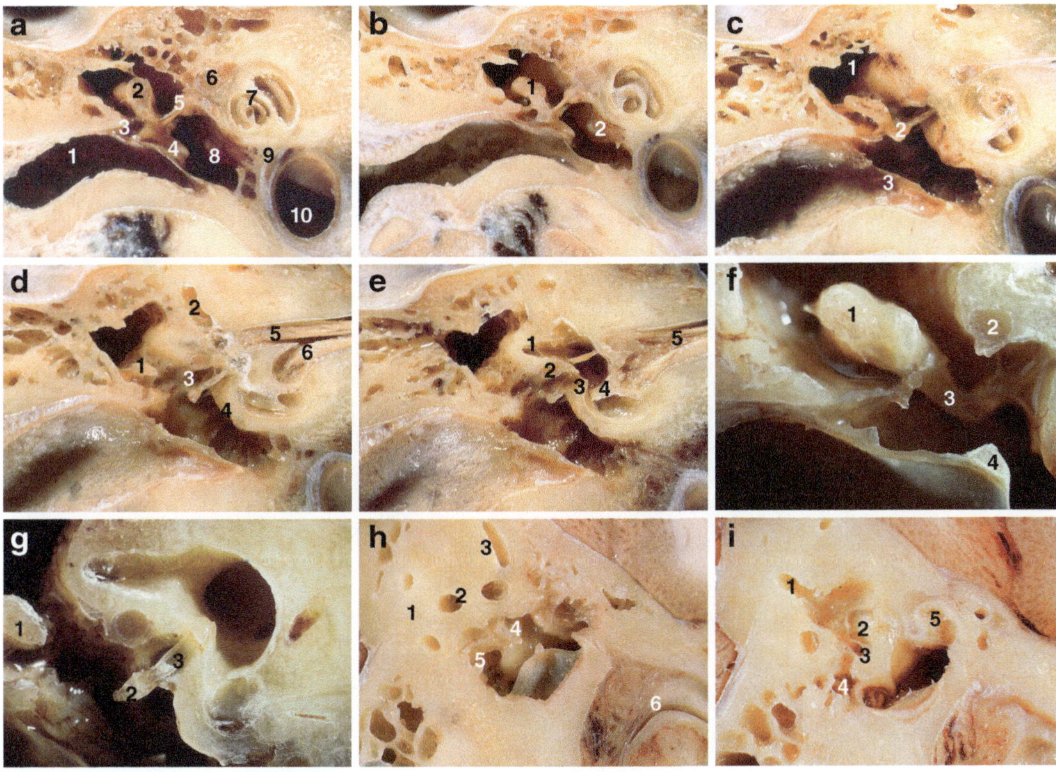

Fig. 8.13 Serial anatomy of the ossicular chain on coronal and sagittal sections of right ear a/b/c/d/e Successive coronal sections from front to back. (**a**) 1. Meatus acusticus externus; 2. Head of malleus with superior ligament; 3. Lateral process of malleus; 4. Hammer of malleus fixed on tympanic membrane; 5. Reflected tensor veli tendon on the cochleariform process; 6. Geniculate ganglion (VII); 7. Cochlea; 8. Tympanic cavum; 9. Sympathetic fibres around the internal carotid artery; 10. Internal carotid artery. (**b**) 1. Incudo-mallear joint; 2. Promontorium. (**c**) 1. Aditus ad antrum; 2. Incudo-stapedial joint; 3. Tympanic membrane. (**d**) 1. Short branch of incus and its ligament within fossa incudis; 2. Superior SCC; 3. Stapes and tendon of the stapedial muscle; 4. Tympanic nerve (IX); 5. Vestibular nerve (VIII); 6. Cochlear nerve (VIII). (**e**) 1. Lateral SCC; 2. Facial nerve (VII); 3. Stapes in oval window; 4. First roll of spiral lamina of cochlea; 5. Meatus acusticus internus. (**f**) Magnified view of incus and stapes: 1. Incus body; 2. Facial nerve in its tympanic portion; 3. Long branch of incus with lenticular process articulated with the stapes head; 4. Hammer of malleus on tympanic membrane. (**g**) Magnified view of stapes: 1. Short branch of incus; 2. Lenticular process of incudo-stapedial joint; 3. Posterior branch of stapes and base within vestibular window (oval window). (**h**) Sagittal section of right ear: 1. Posterior SCC; 2. Lateral SCC; 3. Superior (anterior) SCC; 4. Stapes; 5. Round window (cochlear); 6. TMJ. (**i**) Sagittal section inside the previous one: 1. Lateral SCC; 2. Utricular macula; 3. Saccular macula; 4. Round window closed by the secondary tympanic membrane; 5. Tensor tympani within its bony canal

chain) and frequency (cochlea) [14–18]. The long hearing chain is going from the ear (external, middle and inner) to brainstem (five relays from cochlear nuclei to inferior colliculus) and brain temporal cortex (transverse gyrus of first temporal lobe gyrus).

Among the various possibilities for picking up vibrations, the choice of a membrane (tympanic membrane) is technically optimal. But there is a major constraint which is the mandatory same pressure on both sides of the membrane in order to ensure the maximum efficiency with a maximum possible elastic deformation. Indeed, if the pressure is higher on one side of the membrane, it is then preloaded with a greatly reduced ability to deform.

Therefore, the eardrum is placed at the bottom of the external meatus acousticus (external auditory canal) set in the tympanic sulcus of the tympanic bone. Its outer face is relating to

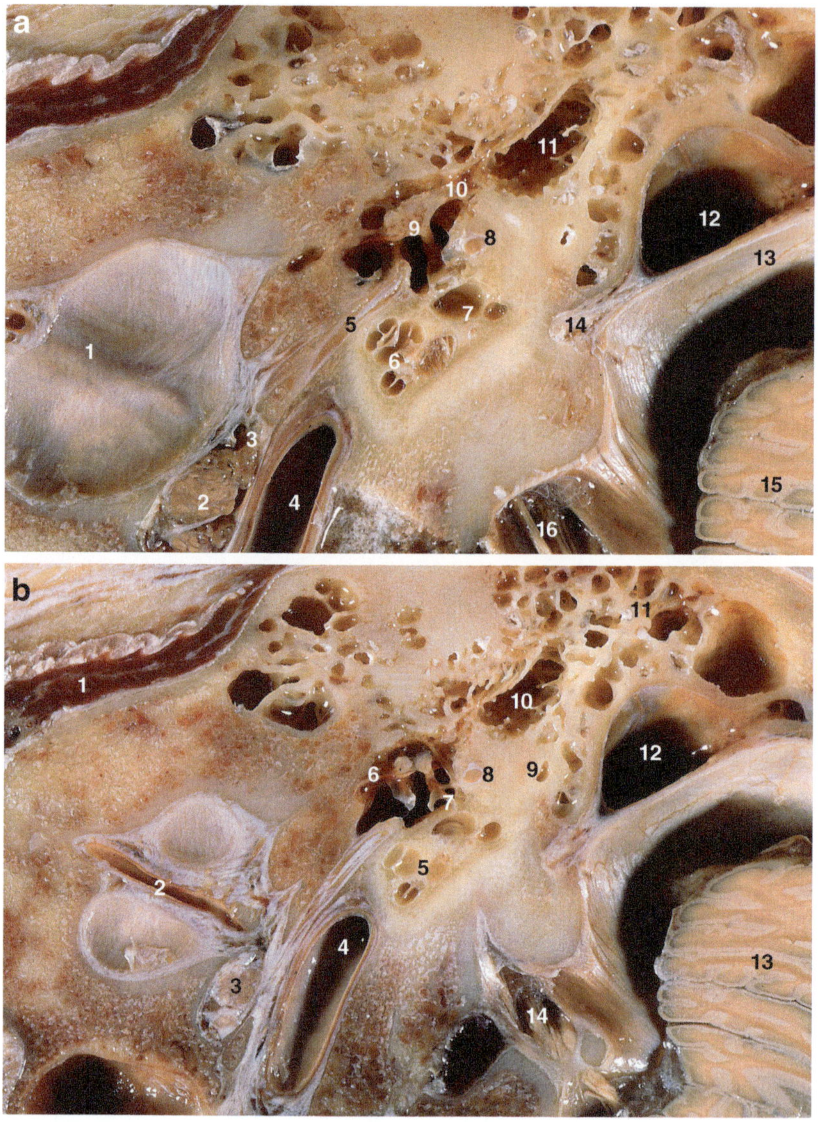

Fig. 8.14 Axial sections of a right ear seen from above. (**a**) Superior axial section: 1. Middle meningeal artery (MMA) seen through the dura of the middle cerebral fossa; 2. Mandibular nerve (V3) in foramen ovale; 3. Foramen spinosum (MMA) of ala major of sphenoid; 4. Internal carotid artery; 5. Tensor tympani muscle and processus cochleariformis for its tendon fixed on malleus; 6. Cochlea; 7. Vestibule; 8. Facial nerve (VII); 9. Incus (covering the head of malleus); 10. Short branch of incus in fossa incudis; 11. Aditus ad antrum; 12. Sigmoid sinus; 13. Dura of the posterior fossa; 14. Endo-lymphatic canal; 15. Cerebellum; 16. VII and VIII nerves. (**b**) Inferior axial section: 1. Temporal muscle in its fossa; 2. MMA; 3. Mandibular nerve (V3); 4. Internal carotid artery; 5. Cochlea; 6. Chorda tympani between malleus and incus; 7. Stapes within the vestibular (oval) window; 8. Facial nerve (VII); 9. Posterior semicircular canal; 10. Antrum; 11. Mastoid cells; 12. Sigmoid sinus; 13. Cerebellum; 14. Roof of the meatus acusticus internus

outside air. Its inside face is located in a bony cavity, the cavum tympani, and receives the hammer handle (malleus) to serve as a transmission lever of vibrations to the rest of the ossicular chain (malleus, incus, stapes) (Figs. 8.13 and 8.14). This is a closed cavity that extends back through the mastoid antrum (the access of which being the aditus ad antrum) in which there are

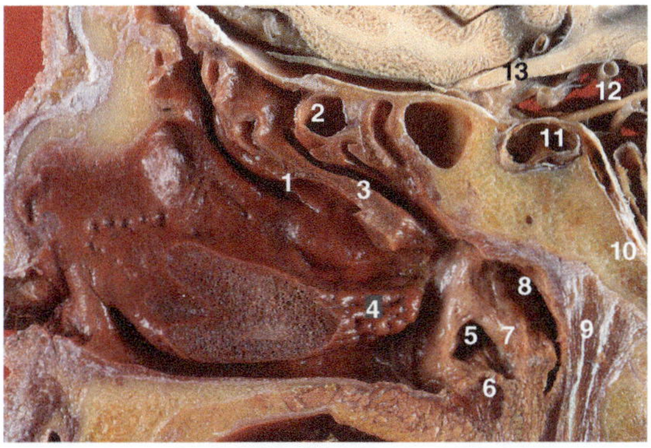

Fig. 8.15 The pharyngeal ostium of tuba auditiva on a parasagittal section of left hemiface. 1. Etmoidal infundibulum and frontonasal canal; 2. Ethmoidal cells; 3. Resected middle turbinate showing the middle meatus; 4. Posterior part of the inferior turbinate; 5. Ostium of the tuba with torus of levator veli; 6. Salpingo-pharyngeal fold (muscle); 7. Torus (tubar valve); 8. Pharyngeal recessus (Rosenmuller); 9. Prevertebral muscles; 10. Clivus; 11. Internal carotid artery; 12. Oculomotor nerve (III); 13. Optic chiasm

mastoid air cells comparable to the frame of the spongy bone tissue of vertebral bodies [19, 20].

It is in fact necessary to find a solution to connect the inside face of the tympanic membrane with the outside air to allow pressure equalisation. Trying to think about a possible solution, we could imagine putting from the cavum tympani an open tube to outside just below the external auditory meatus. But in this case, the vibration train analysis would hit at the same time both sides of the eardrum, cancelling one another out mechanically.

The technical riddle can stop there because we have the solution, which is the auditory tube. The nearest air cavity is the nasal fossa specifically made for conveying air from outside. The auditory tube is a channel connecting the cavum tympani to the lateral wall of the nasopharynx where is its clearly visible orifice [21–24] (Fig. 8.15). This way, any increase or decrease in the external pressure may be felt identically on both sides of the eardrum.

Then comes another technical problem, because putting a communication tube between the middle ear and the vocal tract creates two paths of unequal length for listening one's own voice. The long one goes out from the mouth to the external auditory meatus and the outer side of the eardrum, and the shorter one, from larynx to the inner face of the ear drum via the auditory tube (Fig. 8.16). Unequal length of the two channels can create a dephasing of acoustic signals. Patients with permanently open tube complain about disturbed listening to their own voice [25, 26].

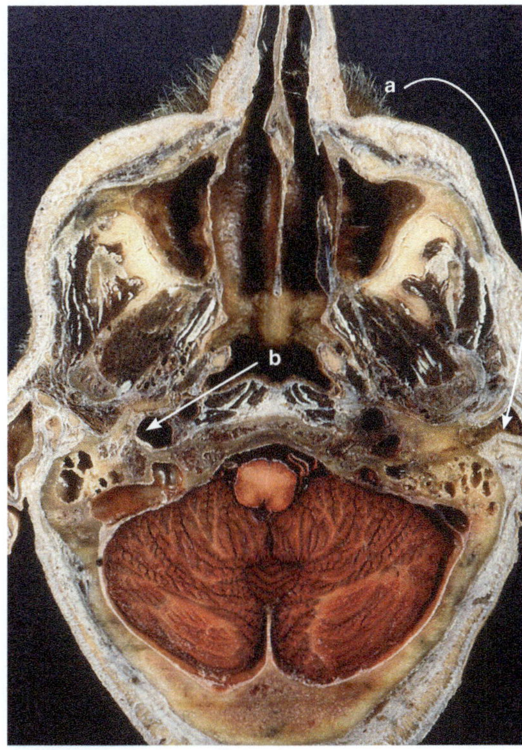

Fig. 8.16 Axial section of the head showing the two ways for the voice. 1. The long way from the mouth to the ear; 2. The short way from the rhinopharynx to the ear by the tuba auditiva. This justifies the closure of the tuba during phonation, avoiding the risk of dephasing the sounds

The solution is therefore not to let the auditory tube permanently open but only when the larynx does not work, during swallowing or yawning.

The Velum-Tubal Synergy

That is the reason why the two upper muscles of the soft palate are in relation with the auditory tube. It is formed by a medial fibro-cartilaginous valve which provides rigidity to the tube and a fibrous membrane on lateral side. It tapers from the pharynx to the protympanum, the bony canal at the

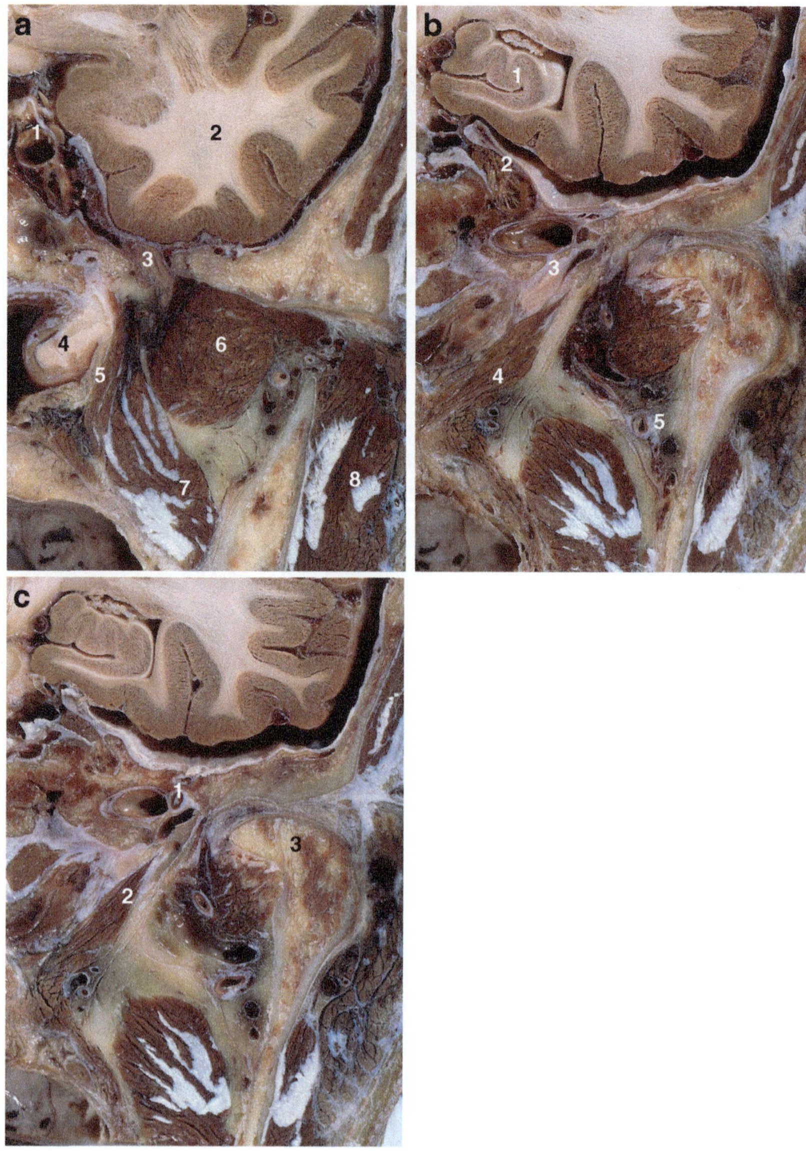

Fig. 8.17 Coronal sections, right hemipart seen from behind. (**a**) Posterior section: 1. Internal carotid artery; 2. Temporal lobe; 3. Mandibular nerve (V3); 4. Fibro-cartilaginous valve of the tuba auditiva; 5. Tensor veli muscle; 6. Lateral pterygoid; 7. Medial pterygoid; 8. Masseter muscle. (**b**) Middle section: 1. Hippocampus; 2. Triangular plexus of trigeminal nerve; 3. Tuba auditiva; 4. Levator veli muscle; 5. Maxillary artery. (**c**) Anterior section: 1. Tensor tympani muscle above the protympanum; 2. Levator veli muscle; 3. Mandibular condyle

origin of the tube where it binds in a groove of the skull base, the semi-channel of the auditory tube.

The levator veli muscle is oblique and passes below along the auditory tube (infra tubal) to end by the levator fold in the manner of a strap on the soft palate upper side [27]. Its contraction pulls the velum back and up and the uvula ensures sealing the nasopharynx. This is important not only for swallowing but also for producing nasal and oral sounds (vowels) as we have seen previously (Fig. 8.17). In contrast, although raising the fibro-cartilaginous valve of the tuba upward, it does not open the tuba. In addition, it has no muscle spindles (Winckler), but in fact only a motor innervation that does not require a proportional control but an on/off control of the nasopharynx closure.

The second muscle, the tensor veli, is laterotubal. It forms a triangular curtain with tip bottom ending on its tendon which is reflected around the hook (hamulus) of the medial lamina of pterygoid process of the sphenoid to spread on

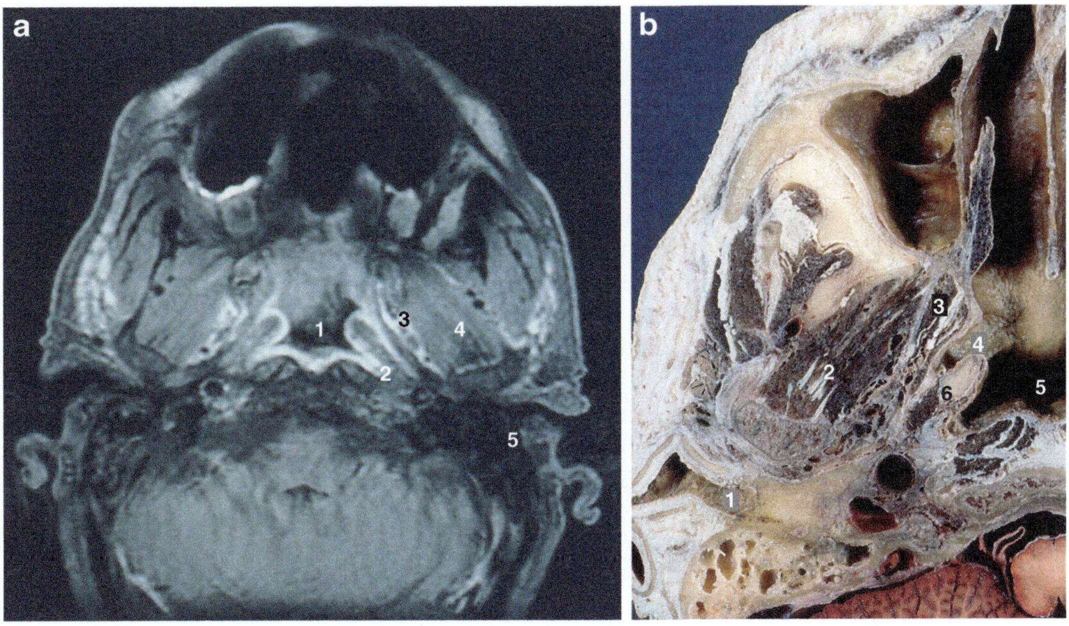

Fig. 8.18 Anatomical basis of the tuba auditiva opening. (**a**) Axial MRI section after spraying Gadolinium contrast medium in rhinopharynx (Dr Pharaboz): 1. Rhinopharynx; 2. Medial valve of the tuba auditiva; 3. Tensor veli muscle with curved shape in relaxing phase, becoming linear during contraction, generating a traction on fibrous part of the tuba and opening of the tuba for 40/100 s; 4. Lateral pterygoid muscle; 5. Meatus acusticus externus. (**b**) Corresponding anatomical axial section: 1. Meatus acusticus externus; 2. Lateral pterygoid muscle; 3. Curved shape of tensor veli muscle; 4. Pharyngeal ostium of tuba; 5. Rhinopharynx; 6. Medial valve of the tuba auditiva

tendon of upper face of the velum. The important detail is that this muscle takes fasteners on the lateral fibrous membrane of the tube.

Therefore, the isolated contraction of the levator allows it to play its phonetic modulator role. The cocontraction with the tensor veli during swallowing allows the opening of the tube for about 40/100th second by rocking up the valve by the levator and by fixing the lateral fibrous membrane by the tensor (Fig. 8.18). That is enough to equalise the pressures in the case of a tympanic pressure gradient on one side.

A straightforward verification of this phenomenon can be done either by climbing mountains or in air planes. Those are pressurised to a variable altitude of about 2,000 m. During takeoff or landing, the pressure varies and a particular sensation occurs with attenuation of the hearing, which may become painful, as it is often the case for children with rhinitis. To restore the pressure equilibrium, it suffices to make a series of swallows. This can be done by drinking a liquid that parents or hostesses should provide for children. It is not recommended to perform the Valsalva manoeuvre commonly used by divers: closing nostrils and mouth and pushing the air in the nasal cavity. Even efficient, this nasal hyper pressure pushes all the germs of the nasopharyngeal mucus into the middle ear. This is a good way to get an ear infection [28].

There remains the problem of the eardrum and malleus handle attached thereto. Because of pressure variations sometimes significant, they could be pushed outward. This justifies the existence of a special protection by a small muscle, the tensor tympani, lying in a small bony channel over the tuba auditiva. Its tendon is reflected at right angle to the cochleariform process (spout spoon) to get fixed to the base of the malleus handle. This muscle forms a digastric system with the tensor veli which has the same innervation by the trigeminal nerve. By contracting, these muscles maintain in position the malleus providing an effective protection of it. This muscle contains muscular spindles producing a myotatic reflex when stretched by the tensor veli during swallowing [29].

Particular emphasis should be given to another small muscle located at the other end of the ossicular chain, the stapedian muscle [30]. Located within a small canal in the posterior wall of the cavum tympani, its tendon, after reflection on the pyramidal eminence, is fixed on the head of the stapes. It plays a very important role as a regulator of the mechanical impedance of ossicular chain, limiting the amplification of acoustic signal and thus protecting the cochlea. Unlike the tensor tympani, it has no muscular spindles and only a motor innervation by the VII [29]. But there is an acoustic reflex by a connection in brainstem between the facial nerve nucleus, which innervates it and the nucleus of cochlear nerve and the superior olivary complex [31, 32]. This reflex is activated each time we talk or when there is a significant noise source. This is why young people, who use "players" at full power (100 dB or more) or who attend nightclubs—which are not resting places promoting peaceful conversation—undergo repeated acoustic trauma that are potential candidates for audiologists. Indeed, the stapedian muscle, a small muscle, gets tired and with overworking can no longer contract. The hunters who also suffer acoustic trauma could emit before shooting one discrete sound in order to launch the acoustic reflex to protect the cochlea. But it is not sure that this suggestion will have a great success.

To summarise the velum/tubal synergy, it can be remembered that there are three different muscles, operating either alone or in common: the levator veli which has a major role in phonation and therefore is innervated by the glossopharyngeal nerve (IX); the tensor veli and its digastric part, the tensor tympani which are rather digestive, innervated by the trigeminal nerve (V) and responsible by cocontraction with the levator for swallowing and auditory tube opening, and alone for the stiffening of the anterior third of the velum making compression of the bolus during passage of the isthmus and protecting of the malleus.

8.4.4.3 The Tongue

This is the third major operator of phonation. It is a highly mobile muscular mass, covered with different mucosa in the anterior two-thirds innervated by the sensory lingual nerve (V3) and the posterior third with taste buds of different forms: filiform, conical, fungiform, circumvalate, lenticular and foliated. They records taste sensations: sweet, salty, sour, bitter and glutamate by glossopharyngeal nerve (IX). At the union of the two parts is a V, the foramen caecum, which is the embryologic residue of the thyroglossal duct. In addition, the state of the lingual mucosa is very representative of the state of the digestive tract (white tongue, "roast aspect,"...).

The tongue has 17 muscles ($2 \times 8 + 1$) that give an extraordinary mobility combined to a discriminative sensibility making it a tactile organ for oral exploration. As said commonly, "the tongue is always where the tooth hurts".

To understand its organization, we can build it piece by piece:

- firstly, there are two sagittal muscle beams: the geniohyoid muscles fixed on the chin lower geni-apophyses forming the floor on which it rest. The floor of the mouth is a form of muscular "sandwich" with the mylohyoid diaphragm located between the anterior digastric below and geniohyoid above;
- secondly, in the middle of the lingual mass range a powerful sagittal muscle, the genioglossal muscle inserted above on the chin superior geni-apophyses and radiating from the tip to the tongue base. This muscle is one that is able, thanks to its lower horizontal fibres, to push the tongue outward aided by the contraction of the geniohyoid;
- thirdly, laterally, one on each side is a vertical muscle curtain inserted on the hyoid body and the great horn, the hyoglossal muscle that can flatten the tongue. It has also a high topographic importance having on its lateral side

two nerves: the lingual nerve (V3) on top and the hypoglossal nerve (XII) on bottom and, on its medial side, the winding lingual artery going to the tip of the tongue. There are in fact typically two surgical discovery triangles of this lingual artery through the vertical fibres in the hyoglosse: first, the Beclard triangle on the back limited by hyoid great horn, posterior edge of the hyoglossal, posterior belly of digastric and forward the Pirogoff triangle limited by hypoglossal nerve, intermediate tendon of the digastric, posterior edge of the mylohyoid muscle.

One can also easily dissect on the lateral side of the hyoglossal muscle the nerve anastomoses binding the lingual sensory nerve and the hypoglossal, a pure motor nerve. In fact, the muscles of the tongue are very rich in muscle spindles because of their required positioning accuracy in the oral cavity and thus have an important proprioceptive quota conveyed by the lingual nerve;

- fourthly, it can be added a suspension on the skull base by the styloglossal muscle. Under the mucosa is the superior longitudinal muscle, and at the bottom, the inferior longitudinal muscle from the base to the tip. Those two muscles can retract the tongue backward in the oral cavity. The odd transverse lingual muscle is able to bring together the two lateral tongue edges. Finally they are some interlaced vertical fibres in muscle mass.

The motor innervation of the muscles of the tongue is supplied by the hypoglossal nerve (XII) with a propioceptive contingent given by the lingual nerve. The floor of the mouth itself (mylohyoid and anterior belly of the digastric) is innervated by V3 and all the hyoid muscles without exception by the cervical plexus (C2/C3).

All the muscles of the tongue may be activated separately by the nervous system, which is essential for phonation, but as we have seen in the chapter on motor control, the human pilot of the machine does not know and does not need to know details of brain operations to make best use of this complex feature. He orders movements and not directly muscles.

8.4.5 Command and Control of Phonation

First of all, we must remember two essential notions: the spoken language mobilises 35 different muscles and the conductor of all these actors is hearing. Indeed, voice control is given by the ear. This explains why a child congenitally deaf is necessarily mute.

The most amazing thing is that the vocal apparatus we have described is ready at birth to put in place phonatory specific language used by the parents or the environment, which will gradually guide and structure phonetic motor patterns stored in the nervous system [33]. Learning another language is more easy if realised during this early period of plastic brain implementation.

The strong neural mandatory connection between the fair perception of syllables and words spoken and written must be well understood. In other terms, there is a close relationship between reading a text in a known language and its verbal articulation [34]. In addressing the technical problem, we can admit that it must exist, in the brain, a centre controlling the transmission of speech and another identifying the interconnected meaning of spoken and written language. It is also needed to have a huge central library for words, symbols, figures, notes and images, plus how to use them: syntax and grammar [35–41].

The methods that have allowed us to understand the complexity of written and spoken languages have been refined over time depending on the progress of our means of investigation in the nervous system. Initially, psychologists were limited in their access to the brain in an approach called "behaviourist" by varying the inputs to analyse the outputs.

Then was developed the anatomo-clinical method initiated by Broca and still in use today in some large chapels of neurology. The objective is to carefully observe clinical signs and symptoms in a patient with a neurological disorder and, if he dies, remove his brain for analysis on macroscopic sections or histological sections stained by different techniques. That is still the clinical major interest of the laboratories of neuropathology.

Thus, the anatomist and neurologist Paul Broca in 1861 studied the case of a man named Leborgne which was for twenty years at the Hospice de Bicêtre. He could not speak but understood everything that was said to him. An examination of his brain after his death showed a lesion of the left hemisphere at the lower part of the frontal lobe.

Some years later, Carl Wernicke isolated a region of arcuate shape including angular gyrus, supramarginal and perisylvian temporal as well as insula that would be responsible for speech understanding. Dejerine subsequently isolated in the rear area of understanding the written language location [42]. In that way, all the language areas were defined.

But the greatest progress came in the first place by the exploration through electrostimulation of the motor cortex during neurosurgical procedures on awake patients. Touching the brain is not painful except in certain deep zones. Thus, Wiler Penfield, a Canadian neurosurgeon, in 1920, has done this protocol on more than 1,000 patients in describing the motor "homunculus". It shows a large area for face and pharyngo-larynx corresponding to the area of language articulation.

This method was supplemented by the test made by Julin Wada in 1949 producing reversible anaesthesia of a cerebral hemisphere by intra-carotid injection of sodium amobarbytal.

Nowadays, the modern neuroimaging techniques allow us to "see" with functional MRI the brain areas that work thanks to the BOLD effect of the venous return of groups of activated neurons. In addition are PET scan and tractography by images of diffusion of water molecules along the nerve fascicles. Parallelly, neurosciences have developed and the work done in neuroanatomy and neurohistology gives a significant amount of knowledge that allows us to specify the location of cortical areas of verbal articulation with their verbal working memory, essential for speaking, reading, writing and memorising a text for theatre actors or a musical score for musicians [43].

Of course the cerebral cortex can store a huge amount of information in its vertical and horizontal matrix with interconnected six layers. It works with other neural structures, such as brainstem and cerebellum, central basal ganglia (striatum and pallidum). Finally, they are billions of neurons organised in cortical columns with a parallel architecture [44, 45]. This whole interactive system fits into what we may call the "cognitive brain" which is the subject of increasing research in specialised institutions.

In the end, our greatest difficulty with our educational reductionist tendency is to have a holistic perception of simultaneous operations in all these interconnected complex systems. We discuss the "thought" as a unifying element, but what is it exactly? We then have to return to the concept of "black box" in which we cannot enter through introspection. But finally it works very well.

It is therefore clear that the brain has a rigorous plan of construction and appropriate pre-wired circuits, so we can fill it with classified information and memorise by a code. Having two hemispheres and a single brain deserves some explanation. We have already discussed the technical problem of control of odd organs with two active halves like larynx, mouth, soft palate, … The motor control centres are double with one in each hemisphere or in each half of the brainstem then appears the notion of dominance or laterality that we discussed before. One centre initiates the activation and the other one is synchronised. Given the need to cross the nerve pathways to "integrate" the two halves, we see that laterality for hand and upper limb/foot and lower limb is a reality in humans: 64 % of pure right handed, 4 % of pure left and 32 % ambidextrous. The main area of spoken and written languages for articulation and understanding is in the left hemisphere of most humans. A stroke in left hemisphere will result in a right hemiplegia and a more or less important aphasia.

So, in this situation, what is the role of the right hemisphere? We currently have a large amount of clinical studies well documented in patients with right and left hemispheric lesions [46, 47]. It shows some specific differences right/left that are not fully determined at birth, since hemispherectomy in young children may allow a single hemisphere to regain a full normal function. But we can say, despite the predominant role of the left hemisphere in language, that the right hemisphere has not, as we thought, a minor role. It is responsible for the prosody, which means the intonation, accents of specific languages and nuances of an individual person, as well as hesitations or repetitions of parasitic words in speech. Right brain injury causes loss of melody of the language, emotional verbal expression and gesture with some indifference regarding the hemiplegia. Therefore, the emotional mimicry is better perceived by the right hemisphere.

Study of musical creation provides some confirmation [48, 49]. The melodies presented to the left ear and perceived by the right hemisphere are better identified than with the right ear. The specific clinical study made by Alajouanine on Maurice Ravel, who became apraxic, alexic and agraphic, showed that he could hear the melodies he had composed and can even detect errors of interpretation. He was also able to hear new music compositions in his head (right integrity) but could not transcribe it in notes and gestures for the piano (left lesion). He could still make scales at the piano, which can corresponds to motor automatisms stored in the motor library of basal ganglia.

The right hemisphere would be responsible for collecting melodic contour, timber and listening pleasure with a synthetic mode of thinking. The left hemisphere would be responsible for the musical technique and therefore a way of thinking rather analytical and logical. The existence of inter-hemispheric commissures and particularly the corpus callosum with 200 million fibres also explains the common work of both hemispheres and some possible substitutions in case of unilateral lesion [50, 51].

One final aspect that is particularly important remains: the "inner language". It is a human characteristic that allows us to speak or sing a tune or see a coloured image without verbal expression. While creating a sentence without speaking or writing, letters and words can be seen written in mind. It is also possible to talk internally with voluntarily changing intonations.

The question is whether we can think without internal language or not? In any case, the richness of the language determines the richness of the thoughts. A non-verbal communication exists with oneself evoking past events or reckoning on current issues. Of course, the transition to the dream is easy, with only one major difference which is the loss of awareness by real disconnection in the brainstem (locus coeruleus). There is room for daydream for these people staring without seeing and living inside a past or desired situation. The pathology may follow and thus explain some delirium lived with a colourful vision outside reality. Finally, the inner mental life of a musician can allow him to hear the music he made with all the nuances of the orchestra without using a music instrument. Ludwig van Beethoven or Gabriel Fauré, who were at the end of their lives completely deaf, nevertheless made great works including the famous IXth symphony.

Conclusion

In conclusion, every human has a specific biological identity card recognisable by his voice, his speech, his gesture, his step and his writing which allows him to sign official documents in an original and exclusive way. Language, by the complexity of the mechanisms mobilising the full neuronal potential, is the most typically human function.

References

1. Langman J (1972) Embryologie médicale. Masson et Cie, Paris
2. Aviv J (1999) The normal swallow. In: Carrau RL (ed) Comprehensive management of swallowing disorders. Singular Publishing Group, San Diego, CA, pp 23–29
3. Humbert IA, Robbins J (2007) Normal swallowing and functional magnetic resonance imaging: a systematic review. Dysphagia 22(3):266–275
4. Martin RE, Goodyear BG, Gati JS et al (2001) Cerebral cortical representation of automatic and volitional swallowing in humans. J Neurophysiol 85:938–950

5. Miller AJ (1982) Deglutition. Physiol Rev 62:129–184
6. Mosier KM, Liu WC, Maldjian JA et al (1999) Lateralization of cortical function in swallowing: a functional MR imaging study. AJNR Am J Neuroradiol 20 (8):1520–1526
7. Shaker R (1995) Airway protective mechanisms: current concept. Dysphagia 10:216–227
8. Titze IR (1994) Principles of voice production. Prentice-Hall Inc., New York
9. Fischetti A (2003) Initiation à l'acoustique. Editions Belin, Paris
10. Jessel M (1973) Acoustique théorique. Masson, Paris
11. Matras JJ (1990) Le son. PUF, Paris
12. Val M (2002) Acoustique appliquée. Dunod, Paris
13. Rabischong P (1994) Anatomia ragionata dell'orechio. In: Salvolini U, Chiesa A (eds) Otoradiologia. Guido Gnocchi, Napoli
14. von Bekesy G (1960) Experiments in hearing. McGraw Hill, New York
15. Fletcher H (1951) On the dynamics of the cochlea. J Acoust Soc Am 23:637–645
16. Kemp DT (1986) Otoacoustic emissions, travelling waves and cochlear mechanisms. Hear Res 22:95–104
17. Sterkers O (1984) La formation des liquides de l'oreille interne. J Fr Otorhinolaryngol 33:239–250
18. Aran JM (1973) Clinical measures of VIIIth nerve function. Adv Otorhinolaryngol 20:374–394
19. Mancuso AA, Hanafee WN (1982) Computed tomography and magnetic resonance of the head and neck. Williams and Wilkins, Baltimore
20. Veillon F (1991) Imagerie de l'oreille. Médecine-Sciences Flammarion, Paris
21. Riu R, Flottes L, Bouche J et al (1966) La physiologie de la trompe d'Eustache; Applications cliniques et thérapeutiques. Arnette, Paris
22. Aschan G (1955) The anatomy of the Eustachian tube with regard to its function. Acta Soc Med Ups 9:131
23. Buch NH, Jorgensen MB (1964) Eustachian tube and middle ear. Arch Otolaryngol 79:472
24. Tamari MJ (1953) Auditory tube and tympanic cavity during embryonal, fetal and prenatal life. Arch Otolaryngol 57:627
25. Miller JB (1961) Patulous Eustachian tube. Report on 30 cases. Arch Otolaryngol 73:310
26. Neveu M (1959) La béance permanente de la trompe d'eustache. Un trouble physiopathologique peu connu. Ann Otolaryngol 76:950
27. Basmadjian J, Dutta CR (1961) Electromyography of pharyngeal constrictors and levator palati in man. Anat Rec 139:561
28. Soros P, Lalone E, Smith R et al (2008) Functional MRI of oropharyngeal air-pulse stimulation. Neuroscience 153(4):1300–1308
29. Winckler G (1982) Particularités de l'innervation de certains muscles dans le domaine oto-rhino-laryngologique. Arch Anat Hist Embr Norm Et Exp 65:49–56
30. Lang J (1995) Skull Base and Related Structures. Schattauer, Stuttgart
31. Moller AR (1974) The acoustic middle ear muscle reflex. In: Keidel WD, Neff WD (eds) Handbook of sensory physiology. Springer, Berlin
32. Bratzlavsky M (1976) Human brain stem reflexes. Sahani Elsevion, Amsterdam
33. Mesulam M (2000) Principles of behavioural and cognitive neurology. Oxford University Press, Oxford
34. Philipose LE, Gottesman RF, Newhart M et al (2007) Neural regions essential for reading and spelling of words and pseudowords. Ann Neurol 62(5):481–492
35. Aboitiz F, Garcia RV (1997) The anatomy of language revisited. Biol Res 30(4):171–183
36. Bookheimer SY (2002) Functional MRI of language: new approaches to understanding the cortical organization of semantic processing. Annu Rev Neurosci 25:151–188
37. Damasio AR, Geschwind N (1984) The neural basis of language. Annu Rev Neurosci 7:127–147
38. Hickok G, Poeppel D (2007) The cortical organization of speech processing. Nat Rev Neurosci 8 (5):393–402
39. Jacquemot C, Scott SK (2006) What is the relationship between phonological short-term memory and speech processing? Trends Cogn Sci 10(11):480–486
40. Matsumoto R, Nair DR, LaPresto E et al (2004) Functional connectivity in the human language system: a cortico-cortical evoked potential study. Brain 127 (10):2316–2330
41. McCarthy R, Warrington E (1984) A two-route model of speech production. Brain 107(2):463–485
42. Dejerine J (1895) Anatomie des centres nerveux. Rueff et Cie, Paris
43. Schulze K, Zysset S, Mueller K et al (2011) Neuroarchitecture of verbal and tonal working memory in nonmusicians and musicians. Hum Brain Mapp 32(5):771–783
44. Duvernoy H (2005) The human hippocampus, 3rd edn. Springer, Berlin
45. Paxinos G, Mai JK (eds) (2004) The human nervous system. Elsevier, London
46. Ross ED (1981) The aprosodias: functional-anatomic organization of their affective components of language in the right hemisphere. Arch Neurol 38 (9):561–569
47. Powell HW, Parker GJ, Alexander DC et al (2006) Hemispheric asymmetries in language-related pathways: a combined functional MRI and tractography study. Neuroimage 32(1):388–399
48. Bigand E, Habib M, Brun V (2012) Musique et cerveau; nouveaux concepts. Nouvelles applications. EMPR Sauramps Médical, Montpellier
49. Garrel R, Amy de la Bretèque B, Brun B (2012) La Voix Parlée et la Voix Chantée. EMPR Sauramps Médical, Montpellier
50. Paul LK, Brown WS, Adolphs R (2007) Agenesis of the corpus callosum: genetic, developmental and functional aspects of connectivity. Nat Rev Neurosci 8(4):287–299
51. Schmahmann JD, Pandya DN (2006) Fiber pathways of the brain. Oxford University Press, New York

9 Bladder and Rectum Continence

Abstract

The bladder and anal sphincters are responsible for maintaining a safe continence of the pelvic reservoirs during the periods between evacuations (micturition and defecation), which have to be consciously and voluntarily controlled. These functions are dependent on the vegetative nervous system with two polarities: sympathetic and parasympathetic. The sensation of need is functionally important and different, depending on the two reservoirs. For bladder, kidneys produce urine permanently and the need is in relation with the stored volume and the pressure on the detrusor. For rectum, the need for defecation is linked with the beginning of the mandatory contraction of the rectum, which is a mental cortical conditioning explaining the frequent constipation, particularly in women. Two types of sphincters are acting: the smooth one guarantees the major part of the continence and the striated one provides a complementary action in case of urgent need. The two striated sphincters cannot be contracted separately in order, presumably, to avoid a too complex conscious control.

9.1 Introduction

This motor function is devoted to striated cylindrical muscular structures called sphincters, placed around the exhaust ducts of the pelvic reservoirs, marking the metabolic end of renal clearance and digestive tract.

9.2 Technical Problems and Solutions

The pelvic reservoirs occupy the bottom of the pelvic basin. The bladder collects the urinary vector produced continuously by the kidney filters. The rectum collects faeces, with a typical repulsive smell avoiding any temptation to reverse the normal progression of the oral/anal digestive tract.

The evacuation of the excreta being necessarily intermittent and controlled, the first problem is the storage in a deformable, elastic and waterproof reservoir in order to prevent any leakage out of washout periods. The tank must have an entry allowing the filling with a positive pressure against a check valve return. It must avoid reflux in the filling pipe during the phase of active evacuation.

This requires a reduction in volume of the reservoir by a contraction of its walls. Therefore,

P. Rabischong, *Comprehensive Anatomy of Motor Functions*, DOI 10.1007/978-3-319-04169-8_9,

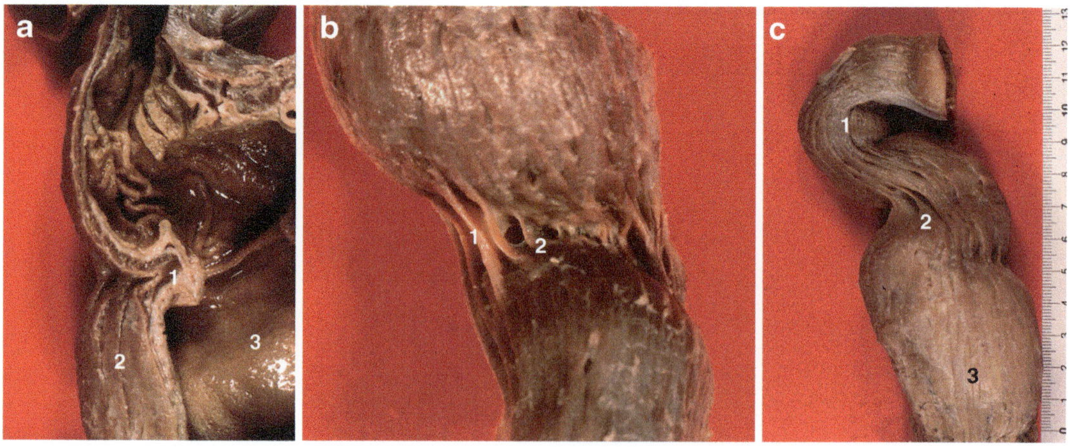

Fig. 9.1 Dissection of the rectum. (**a**) Opening of the ampulla showing the different layers: mucosa, deep circular and superficial longitudinal smooth fibre layers: 1. Permanent rectal fold with circular fibres playing the functional role of anti-reflux valve; 2. Longitudinal smooth fibres; 3. Mucosa with folds. (**b**) External view of a rectal permanent fold: 1. Fibrous tracts maintaining the fold; 2. Deep part of the fold. (**c**) General view of the rectal pouch: 1. Rectosigmoid junction; 2. Superior part of the rectum with antireflux fold; 3. Inferior part of the faeces container

both reservoirs need a contractile pocket. The bladder is a reservoir formed by a mucosa and a smooth muscle—the detrusor [1–4]—and the rectum is a pocket formed by smooth muscle fibres oriented in two directions, longitudinal and circular (Fig. 9.1).

9.2.1 The Mandatory Continence

The choice must be made between a central controlled striated muscle and a smooth muscle with vegetative autonomic control. This one has a better fatigue resistance and has obviously been chosen. The main problem is in fact the continence between evacuation periods.

There are, at the output channel, three different and complementary devices:

- Firstly, a smooth sphincter capable of maintaining permanent contraction providing the bulk of continence.
- Secondly, a striated sphincter quickly fatigable after 130 s in women and 60 s in men. But it can intervene to stop the flow of output under voluntary control or maintain continence in case of overpressure in the tank, for example, during coughing or in an imperious urination. For this purpose, inside the striated sphincter, there are neuromuscular spindles capable of inducing a clamping myotatic reflex.
- The third feature is based on the optimal use of geometrical resources derived from the shape of the output channels. For the bladder, it refers to the organisation of the vesical trigone showing a superficial muscle layer partly dependent on the outer longitudinal detrusor and a deep inextensible fibro-connective part. For the rectum, the ano-rectal angle of about 95° opened backwards is maintained by the puborectal beam of the levator ani muscle, creating a "back tie" fixed in front of the pelvis (Fig. 9.2). It also sends fibres to the skin of the anal margin, creating the "cutis ani muscle corrugator", well described by Milligan and Morgan [5].

9.2.2 The evacuation process

The sensation of need is essential to control evacuation. It comes from submucosal mechanoreceptors sensitive to distension during filling but especially to the active contraction of the reservoir. It works differently for the bladder and the rectum.

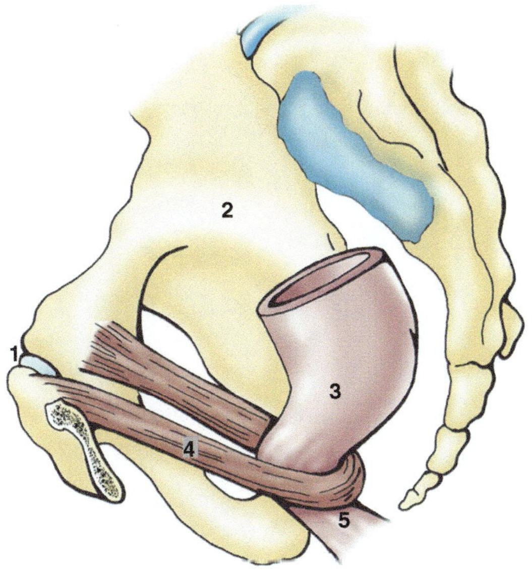

Fig. 9.2 Puborectal muscle strap on a lateral view of pelvis. 1. Pubic symphysis; 2. Iliac bone; 3. Rectal ampulla; 4. Puborectal muscle; 5. Anorectal angle of 80°

9.2.2.1 Miction

- For the bladder, the first sensation appears when the filling reaches between 175 and 200 cc, which can be easily inhibited, except in a spinal cord-injured patient where it can cause leaks. It becomes urgent to urinate for filling at 400 cc, which is accompanied by the beginning of detrusor contraction. Beyond, an imperious urination may arise. The conscious control of evacuation is made by sensory feedback from the mucosa of the urethra [6], the outlet canal, which facilitates the active contraction of the detrusor, but with a possibility of trailing blocking, sometimes imperfectly, by activation of the striated sphincter which inhibits the detrusor (Fig. 9.3).

9.2.2.2 Defecation

- For the rectum, the problem is different [7]. The sensation of need comes not from filling of the rectum, but from its active contraction. It operates not like a striated muscle controlled voluntarily, but by a cortical conditioning heavily dependent on education and individual habit for induction mechanism of the exemption: at sunrise, after breakfast or lunch, or after the first cigarette, or before bedtime. Constipation, so frequent in women, has always a psychological basis of demotivation and loss of the sensation of need, conditioned by the thing "kaka" (from the Greek word *kakos*: bad). The abdominal thrusts by contraction of the abdominal muscles and diaphragm blocked in inspiration are ineffective if the rectum does not actively shrink. This point has not always been well understood by gastroenterologists or paramedical staff involved in rehabilitation. A vaginal touch in a constipated woman always shows a rectum full of faeces.

In addition, the lining of the anal canal is rich in various tactile corpuscles, which explains its remarkable capacity to perception. It has also numerous mucosal folds lined with crypts, in connection with more or less important haemorrhoidal venous plexus, forming what Stelzner named "rectum corpus cavernosum", participating to some form of passive continence.

The important role of the anal mucosa to control output occurs specifically in the difficult problem of the expulsion of gas. These are formed by the fermentation of the colon contents and are pushed into the rectum with faeces. They distend the rectum and create a special feeling that some experts exploit for a more or less discrete resorption of these "basic breaths" by using a subtle striated sphincter relaxation to modulate the "cacophonous" noise. However, they cannot avoid an olfactory betrayal, fortunately most often anonymous among a group of persons.

The anorectal reflex [8] is used to trigger a contraction of the rectum by anal stimulation touch in paraplegics [9] or by inserting suppositories or laxative injections in people with constipation.

Pressure on the posterior perineum by the active contraction of the rectum associated with abdominal muscle contraction during the exemption fully justifies the anatomical configuration of the striated sphincter, well described by Winckler [10] and resumed at our request by Gagnard-Landra and Sambuc, and that of the levator ani muscle.

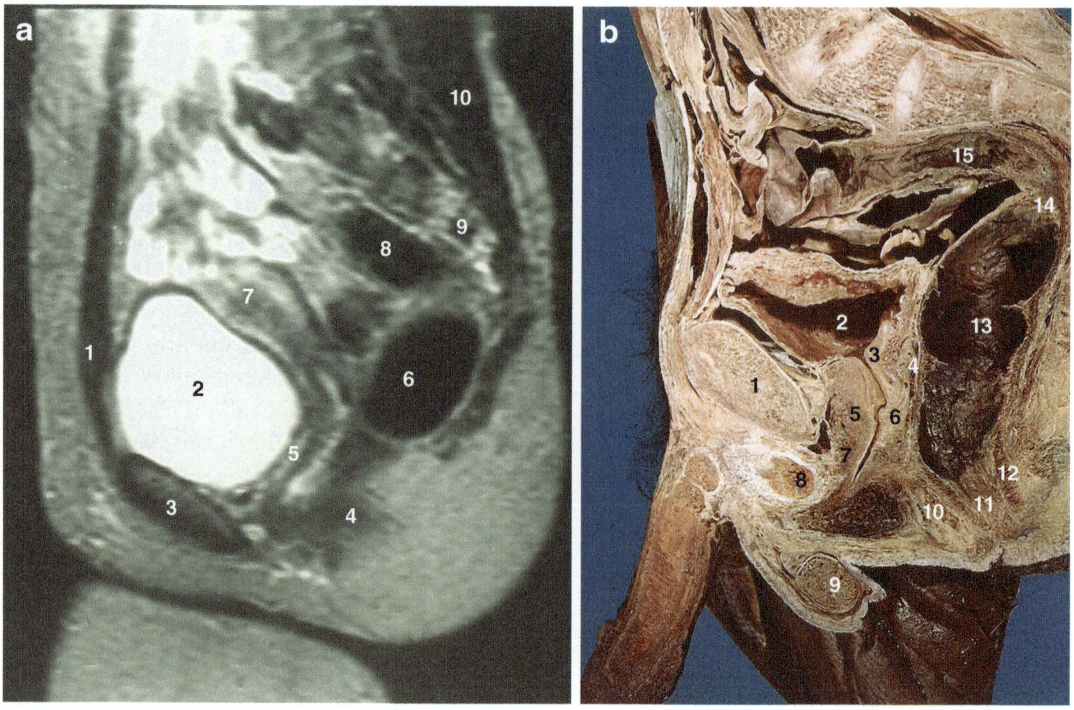

Fig. 9.3 General view of pelvis and sphincters. (**a**) MRI sagittal section of the pelvis of a young women: 1. Rectus abdominis muscle; 2. Full bladder; 3. Pubic symphysis; 4. Anus; 5. Vagina; 6. Rectum; 7. Uterus; 8. Sigmoid colon; 9. Sacrum; 10. Spinal muscles. (**b**) Sagittal section of pelvis in man (right half): 1. Pubic symphysis; 2. Bladder; 3. Bladder neck; 4. Seminal vesicles; 5. Prostate; 6. Veru montanum of prostatic urethra with ejaculator canal; 7. Urethral sphincters; 8. Cavernous body; 9. Testicle; 10. External striated sphincter; 11. Anal canal with anal columns (rich in venous plexus); 12. Internal smooth sphincter (in continuity with the smooth rectal muscles); 13. Rectum; 14. Rectosigmoid junction (anti-reflux valve); 15. Sigmoid colon

The striated sphincter has a navicular form with a deep part made of circular fibres surrounding the smooth anal sphincter and separated by a connective-adipose space and a superficial part formed by sagittal fibres curved in the manner of an arc (Fig. 9.4). These fibres attach on the back on coccyx by two parallel beams (rectococcygeal muscles). On the front, they attach in man by a median beam fixed to the central fibrous perineum node and bulbospongious muscle and in woman by a bundle surrounding the vaginal opening and ending in the ischiocavernous muscle attached to the clitoris. There is therefore a muscle continuity from coccyx to anterior perineum and pubic framework (the classical musculus pubococcygeus).

The levator ani muscle, so named by Holl in 1897 and Thomson in 1899, is a true diaphragm muscle going from the pubis and the medial side of the ilium, through the “arcus tendineus levatoris ani” attached to the internal fascia of internal obturator muscle, to the coccyx [10]. It closes the wide opening laterally limited by the sacrosciatic notches. It should be noted that modern imaging, in particular the MRI, gives a different image on the cadaver body, showing a concave dome form and not a convex one [11, 12]. Finally, as mentioned earlier, it is necessary to have an anti-reflux system made of mucosal folds with thick transverse circular musculature at the upper part of the rectum in the rectosigmoid junction. When the longitudinal and circular fibres of the rectum contract, the valves close and avoid a hunt to the sigmoid colon (anti-reflux) and at the same time the rectum height decreases, facilitating exemption. The fibres attached to the undersurface of the skin of the anal margin raise it at the same time

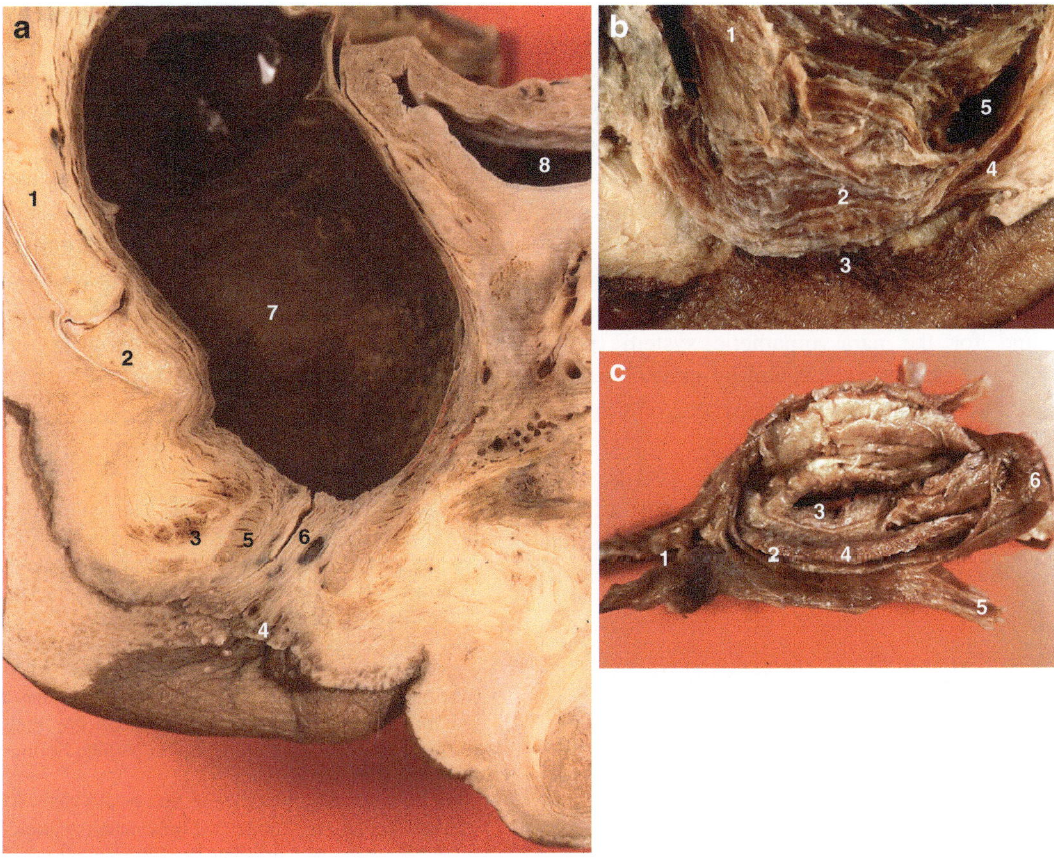

Fig. 9.4 Rectum and sphincters of the anal canal. (**a**) Sagittal section of rectum (left half): 1. Sacrum; 2. Coccyx; 3. External striated sphincter; 4. Anus; 5. Internal smooth sphincter in continuity with the smooth muscular fibres of the rectum; 6. Anal canal with venous hemorroidal plexus; 7. Rectal pouch; 8. Bladder. (**b**) Sagittal dissection of the anal canal (left half): 1. Levator ani muscle; 2. External striated sphincter; 3. Rectum; 4. Ano-coccygeal muscles; 5. Ischio-rectal fossa. (**c**) Dissection of anal canal: 1. The two posterior ano-coccygeal muscles; 2. External striated sphincter; 3. Anus; 4. Internal smooth sphincter; 5. Puborectal muscle creating a posterior anal tie (maintaining the anorectal angle); 6. Centrum tendineum perinei

and a final contraction of the striated sphincter creates a "cigar cutter" effect.

9.3 Command and Control

9.3.1 The Two Complementary Levels of Control

It is necessary to design, for command and control system of pelvic reservoirs [13–15], two distinct but complementary levels: one for conscious and voluntary decision necessarily involving the primary sensory parietal cortex for sensation of need and brainstem, responsible for the management of all the semi-automatic functions of biological maintenance and especially the micturition within the pontine centre of dorsal tegmentum; and another one, for *unconscious execution* of automated programmes, *filling with continence* and *controlled evacuation.*

Both vegetative nervous systems have the same organisation but with opposite functions. They have the following:

- intra-axial centres: C8/L2 in intermediate part of the grey matter of the spinal cord for sympathetic system and brainstem and sacral spinal cord S2, S3, S4 for parasympathetic;

- extra-axial ganglions: cervico-thoraco-lumbosacral sympathetic latero-vertebral chain plus visceral ganglia for sympathetic system and cranial and sacral ganglia for parasympathetic system;
- communication with the peripheral nerves: for the sympathetic system, white communicating branches (the preganglionic fibres) and grey communicating branches (the postganglionic fibres using all peripheral nerves as vectors) and for the parasympathetic system (post-ganglionic fibres conveyed by cranial nerves: III, V, VII, IX, X), vagus nerve going to the pelvis and by pelvic nerves;
- in the hypogastric visceral plexus, the sympathetic ensures releasing of detrusor and smooth muscle of the rectum with contraction of the smooth sphincter and the parasympathetic the reverse action [11].

9.3.2 The Somatic Innervation

For a fair analysis of pelvic–perineal pains, it is important to remember the nerves for the somatic innervation of the perineal region.

A first thoracolumbar contingent (T10/L1) comprises:

- the iliohypogastric nerve (T12/L1) with a genital branch to the skin of the pubis and labia and the supero-medial thigh region;
- the ilioinguinal nerve with the same origin and the same distribution;
- the genitofemoral nerve (L1/L2) terminating at labia.

The precise knowledge of these nerves is essential for an accurate diagnosis of radicular origin of pain in that territory. The obturator nerve has no sensitive territory but only muscular for the obturator externus muscle.

A second sacral contingent from sacral origin mainly includes the pudendal nerve (S2/S3/S4) which projects into the gluteal region going below the piriformis muscle and then passing in a doubling of the fascia of the obturator internus muscle. It ends with three branches: the sensory inferior rectal nerve for perineum and anal region; the perineal nerves for the anterior perineal muscular triangle and sensitive for labia majora and minora, urethra, vestibule of the vagina, and the dorsal nerve of the penis and clitoris. The cluneal nerve (S1/S2/S3) is a branch of the posterior cutaneous nerve of the thigh, which innervates the lateral region of the anus and labia.

A final important point is that the two striated sphincters for bladder and rectum cannot be activated voluntarily independently, probably for the same need for a simple command concerning two sphincters that in fact have the same functional goal.

References

1. Andersen J, Bradley WE (1976) The syndrome of detrusor sphincter dyssynergia. J Urol 116:493–495
2. Barrett DM, Wein A (1981) Flow evaluation and simultaneous external sphincter electromyography in clinical urodynamics. J Urol 125:548–551
3. Blaivas JG, Sinha HP, Zayed AH et al (1981) Detrusor external sphincter dyssynergia. J Urol 125:545–548
4. Bradling A (1987) Physiology of bladder smooth muscle. In: Torrens M, Morrisw JFB (eds) The physiology of the Lower Urinary tract. Springer-Verlag, Springer, New York, pp 161–191
5. Bartolo DCC, Jarratt JA, Read MG et al (1983) The role of partial denervation of the puborectalis muscle in idiopathic faecal incontinence. Br J Surg 70:664–667
6. Deleval J (1984) Contribution à l'étude du sphincter de l'urètre. Acta Urol Belg 52:147
7. Kerremans R (1969) Morphological and physiological aspects of anal continence and defecation. Ed. Arcsia, Brussels
8. Henry MM, Swash M (1978) Assessment of pelvic floor disorders and incontinence by electrophysiological recording of the anal reflex. Lancet 311:1290–1291
9. Girard R, Minaire P, Casteran JP et al (1978–1979) Anal and urethral sphincter electromyography in spinal cord injured patients. Paraplegia 16:244–247
10. Winckler G (1964) Manuel d'anatomie topographique et fonctionnelle. Masson et Cie Ed, Paris
11. Rouanet JP, Filhastre M, Maubon A et al (2011) Anatomie IRM du plancher pelvien. In: Bonneau D, Mares P (eds) Algies pelvi-périnéales et thérapies manuelles. Sauramps Medical, Paris, pp 21–34
12. Singh K, Reid WM, Berger LA et al (2002) Magnetic resonance imaging of normal levator ani: anatomy and function. Obstet Gynecol 99(3):433–438
13. Blok BF, Holstege G (1998) The central nervous system control of micturition in cats and humans. Behav Brain Res 92:119–125
14. El Badawi A, Chenk E (1971) A new theory of the innervation of bladder musculature. J Urol 105:372
15. Fowler CJ, Griffiths D, de Groat WC (2008) The neural control of micturition. Nat Rev Neurosci 9(6):453–466

10 Ejaculation and Orgasm

Abstract

The reproductive function is mandatory for species survival. This requires, to be efficient, an intelligent automation of the processes and a real motivation of the partners which is based on an important parameter: the pleasure, which is one of the main motor of reproduction. In order to make possible the union of two gametes (spermatozoon and ovule) within the genital tract of the female to create a new individual, it is needed to have some specific anatomical features: a variable stiffness penile injector, a bulbospongiosus urethral pump made by striated perineal muscles and an orgasmic ejaculation seizure generating expulsion of sperm and pleasure. This is provided by the firing of neurons of the ejaculation spinal centre in men, and the same orgasmic seizure can be recorded in women after appropriate stimulation of the clitoridian inductor. Most of the sexual problems are linked to a lack of knowledge of these neurophysiological precise processes.

10.1 Introduction

The reproductive function is essential for the maintenance of species by organising the union of male and female gametes to produce a new individual. It is also under the control of striated muscle functions that need a particular technical programme.

This "survival kit" represents the mechanisms of species survival, all of whom, except man, do not know the biology and are totally ignorant. That is why automation is an absolute necessity as well as a rigorous programming of the three survival processes: search for food [1] (individual survival), search for partner for reproduction (species survival) and marking a territory where to find food and partners.

These three aspects are dependent on the limbic system [2], which is a generator of survival behaviours, placed in the brain of all species including humans.

Some animals have only a limbic cortex and spend all their time eating, copulating and defending their territory. Man has an extensive surrounding limbic system with cingulate gyrus, hippocampus, fornix, thalamus, amygdala, septal area, hypothalamus, raphe nuclei but also a supralimbic cortex allowing him to control more or less well limbic impulses. For example, the typically limbic territoriality typically can be

P. Rabischong, *Comprehensive Anatomy of Motor Functions*, DOI 10.1007/978-3-319-04169-8_10,

very active in explaining human wars, crimes and jealousy in all its forms expressed in human couples, families and even in occupations, including universities.

This chapter is devoted to reproduction [3], a part of the limbic survival kit programme that requires examining a number of related technical problems answering the question: how to reproduce?

10.2 Technical Problems of Copulation

The objective is to put together in a receiving cavity two gametes, male and female, which are nanotechnological components. The spermatozoon has a head of 5 μm in length for a total length of 60 μm. The ovule has a diameter of 200 μm. Their union allows first fertilisation and the gradual development of an embryo, then a foetus and finally a newborn. Therefore several technical features are necessary to achieve this programme.

10.2.1 Variable Stiffness Injector

To penetrate into the genital tract of the female, the male organ should be stiffened on request. The solution for the human penis pendulus is to use the arterial blood flow injected into an extensible structure: the two corpora cavernosa that has an internal mesh wired cavity and below, the urethra [4] (Fig. 10.1).

But hydraulics has precise principles and stiffness of the penis can be obtained only by increasing the input flow and reducing the output flow. This is the classical problem of filling a bathtub that can be done by opening the taps and closing the outlet; otherwise the tub is half full.

A great deal of research was needed to understand precisely the mechanism of erection [5, 6]. In 1900 an Austrian histologist Viktor von Ebner (1842–1925) described pads on the wall of the cavernous arteries penetrating within the corpora cavernosa able to dilate the arteries to fill them up. But a French urologist G. Conti in 1952 defined active valves on the outside veins [7] responsible for blocking the venous flow and for generating erection. However the use of intracavernous injection of papaverine to create an erection as demonstrated particularly by R. Virag [8] allows to give the major role to the sinusoids which form the lattice of the corpora cavernosa very rich in smooth muscle cells. During erection they dilate by relaxation, which initiate the penile turgor [9].

The mandatory venous blocking was well explained in particular by Lue and Tanagho. The venous plexus under the tunica albuginea of the penis are compressed, and the venous return is reduced to maintain the stiffness of the erection [10].

There is in the retracted state, resulting in flaccid penis, a sympathetic activation of the smooth muscle of the cavernous walls and during erection, a collapse of them by a parasympathetic action.

At the distal end of the injector is a penile spongy glans that has its own vascular network and also its specific spinal cord centre activation [11]. It is therefore two erections: spongy with turgid glans induced by the spinal centre L1/L2 and cavernous by the spinal centre S2/S3/S4. The clinical study of paraplegic patients clarified these functional notions. These spinal erection centres are under brain control with a "conditional" command related to mental factors, which explains the frequent erection inhibitions of central origin [12]. We must also remember the importance of venous factors in erection [13] which condemns the surgical methods using an adjacent artery like epigastric artery to cover deficiency of cavernosal arteries.

10.2.2 Bulbospongiosus Urethral Pump Activating the Injector

Once introduced, the penis injector needs to propel the sperm out of the urethra which can only be done by a muscle action. This acts on the urethral spongy bulb placed below the prostatic portion of the urethra. Inside is the verumontanum, a micro-canal connected with the seminal vesicles—the storage sperm tanks—placed behind and above the prostate.

Thanks to a striated fibre layer from the striated bulbospongiosus muscle in the manner of a

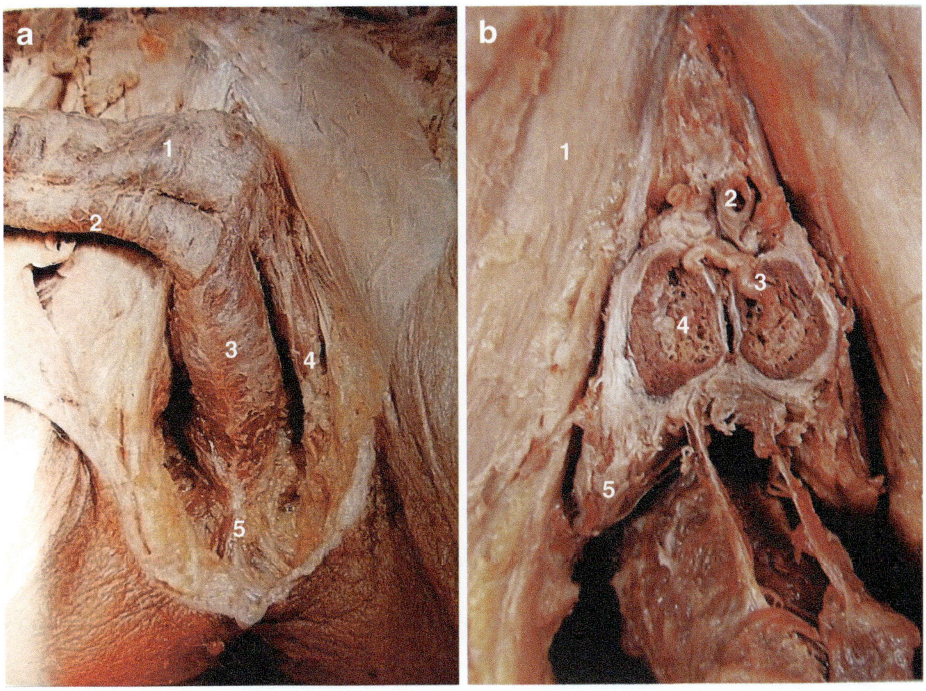

Fig. 10.1 Dissection of the anterior perineum in man. (**a**) Seen from the front: 1. Cavernous body of the penis; 2. Urethra; 3. Bulbospongious muscle; 4. Ischiocavernous muscle; 5. Centrum tendineum perinei. (**b**) After resection of cavernous bodies: 1. Inferior branch of pubis; 2. Dorsal vein of penis; 3. Penetration of the cavernous artery into the cavernous bodies; 4. Section of cavernous body; 5. Pillar of the cavernous body

surrounding sleeve around the spongy bulb, a pump effect is obtained by crushing rhythmically the urethral tube. That is a similar work as peristaltic pumps used for extracorporeal circulation (Fig. 10.2). The real problem, however, is to find the solution to order this rhythmic pump during ejaculation.

10.2.3 The Perineal Paroxystic Crisis of Ejaculation

We have performed in our research centre ejaculation records seeking the pump effect [14]. Ejaculation leads to sperm jerky propulsion in a process that is impossible to stop when started, which excludes the inclinations of the ejaculatory. The fact that the striated muscles of the anterior and posterior perineum (anal sphincter) cannot be voluntarily ordered separately but synchronously has enabled us to develop a non-invasive method using a pneumatic sensor connected to a measuring chain and placed in the anal canal (Fig. 10.3). Due to the synchronisation of the entire perineal striated musculature, the contraction of the bulbospongiosus pump can be measured by synchronous contractions of the striated anal sphincter.

We found and recorded a succession of four phases: precritical stage, tonic contraction, clonic contractions and post-critical phase. However, this neurophysiological phenomenon is strongly reminiscent of the clinical course of the epileptic seizure, which cannot stop when it has started and which is formed of a similar succession of three phases: first, tonic contraction of the muscles including respiratory generating face cyanosis; second, clonic phase with rhythmic contractions of the muscles of the face and limbs; and finally, a post-critical phase during which the patient who has lost consciousness progressively regains it with a complete retrograde amnesia. This allows even nowadays to use in psychiatry a method of induction of seizures, called sismotherapy or electrochoc, for therapeutic purposes.

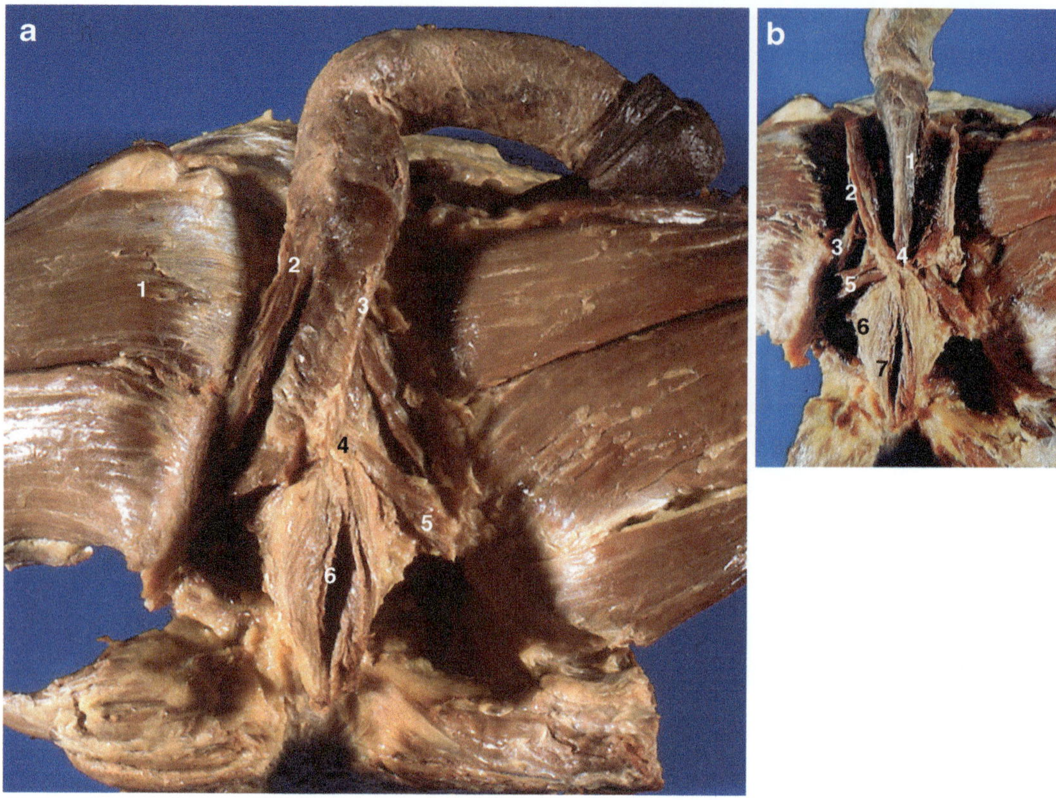

Fig. 10.2 Dissection of the muscles of the anterior perineum and of the bulbospongious pump. (**a**) Superficial plan: 1. Adductor muscles; 2. Ischiocavernous muscle; 3. Bulbospongious muscle surrounding the spongious urethra and bulb; 4. Centrum tendineum perinei; 5. Transverse perineal muscle; 6. Anus. (**b**) After section of the bulbospongious muscle: 1. Spongious urethra; 2. Bulbospongious muscle; 3. Ischiocavernous muscle; 4. Centrum tendineum perinei; 5. Transverse perineal muscles; 6. External striated anal sphincter; 7. Anus

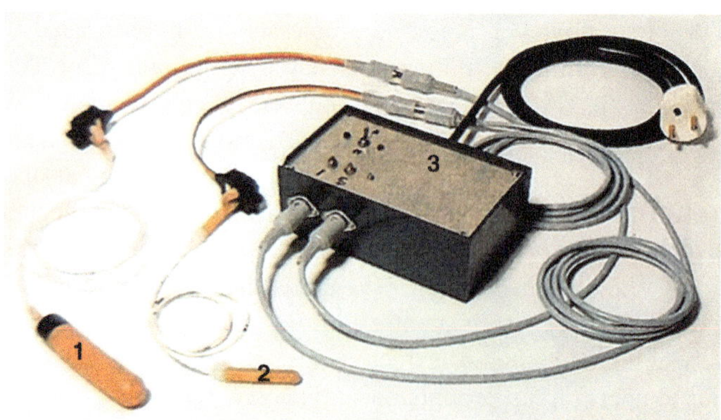

Fig. 10.3 Measuring device of the anal and vaginal contractions. 1. Pneumatic intra-vaginal transducer; 2. Pneumatic intra-anal transducer; 3. Electronic measuring and recording system

Of course, ejaculation is not a seizure as described, but the neurophysiological process is of same nature. It is a synchronous firing of all the neurons of a centre, which is for ejaculation, the spinal centre of ejaculation, and for epilepsy, the whole cortex for a great seizure and/or a part of temporal cortex for temporal lobe epilepsy. When we presented these results in a Congress of Sexology, some sexologists, who are rarely physiologists or neurologists, have expressed aggressive criticism, which is not surprising in a population so polymorphic and sometimes fanciful.

But the reality is there and the intelligence of the system is to use effectively a neurophysiological phenomenon of synchronous "firing" of all the neurons of the spinal ejaculation centre. That creates a paroxystic crisis allowing an ejaculatory pump effect with rhythmic contractions of bulbocavernosus and ischiocavernosus muscles [15, 16], opening the verumontanum with contraction of seminal vesicles to let out sperm and contraction of the urethral striated sphincter to avoid intrabladder sperm reflux. This phenomenon can be seen in some paraplegic patients who have always sphincter disturbances.

In the end, this physiological process of ejaculation is normally triggered by a stimulation of tactile receptors of sexual areas which lead to a sensitive critical activation of the ejaculation centre neurons up to the outbreak of a synchronous ejaculatory sequence. As it is impossible to stop, it largely guarantees a successful sperm injection.

10.3 Motor of Inference: The Orgasm

It is important to understand the data of the problem of reproduction, which is the necessary condition for the maintenance of life on earth. The question is to know how to organise this function to make it widely available and used by all animals, including humans.

A careful analysis of the reproductive biology allows to understand a mandatory basic principle, also similar for the other component of survival kit, the search for food which is to create a need whose satisfaction leads to a fundamental parameter: pleasure, which has a very high motivating power. Therefore biology is functioning with pleasure. The need is metabolic for the search for food and endocrine for copulation.

The female animal who knows nothing about the biology of reproduction, however, does not accept the male if she is not fertilisable during the periovulatory period, called the heats. During that time, it is pheromone secretion attracting and triggering the male in rut and facilitating her acceptance of copulation with perineal trappings like secretions, turgor, ...

In mankind the situation is different. Even if a woman also often ignores the biological constraints of the ovarian cycle, she accepts "in principle" the male without restriction in relation to its fertility, only possible in the periovulatory period about a week around the ovular laying normally on the 14th day of the cycle. Sexual activity is no longer a preprogrammed stereotype to ensure reproductive success, as is the case with most animals copulating "a posteriori". But it becomes a form of culture with multiple variations, not infinite. Seduction strategies depend largely on the partners' mental abilities.

The orgasmic pleasure (orgasm from the Greek *orgao*: bubble of ardour) is in the ejaculation for males and humans, which is mandatory for reproduction. It is clear that this type of emotion is experienced very differently depending on the variable configuration of the brain in different species. However, the orgasmic seizure described before does not exist in female animals with often rapid coitus, and above all, it is not essential for reproduction.

This raises the problem of the woman orgasm. A non-negligible percentage of them may have several children without experiencing any orgasm. However, the orgasmic process recorded in women with the same method described above has exactly the same patterns as in man [14]: tonic and clonic phases with contractions of the perineal muscles including the anal sphincter, a privileged witness of orgasm (Figs. 10.4 and 10.5).

In addition, a woman's sexual organ [17, 18] is seen above the entrance of the vaginal tube, one specialised little body designed only for delivering pleasure: the clitoris (etymologically the key) with, as in men, a visible and palpable erection, and a particular sensitivity in connection with a large number of tactile receptors (Fig. 10.6). It should be also remembered that talking about orgasm in women means only describing the precise perineal crisis, but with different possible inductions, mostly by clitoral

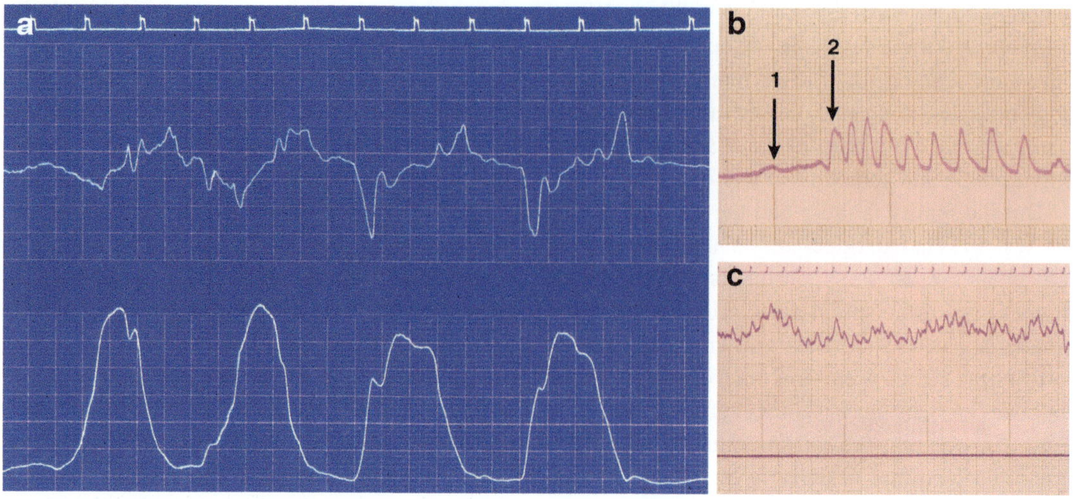

Fig. 10.4 Recording of the orgasmic seizure. (**a**) Voluntary contractions: anal transducer (*superior line*) and vaginal transducer (*inferior line*). (**b**) Recording of an ejaculation by anal transducer: 1. Tonic phase; 2. Clonic phase. (**c**) Preorgasmic excitation phase: recording of the heart pulses and preorgasmic contractions

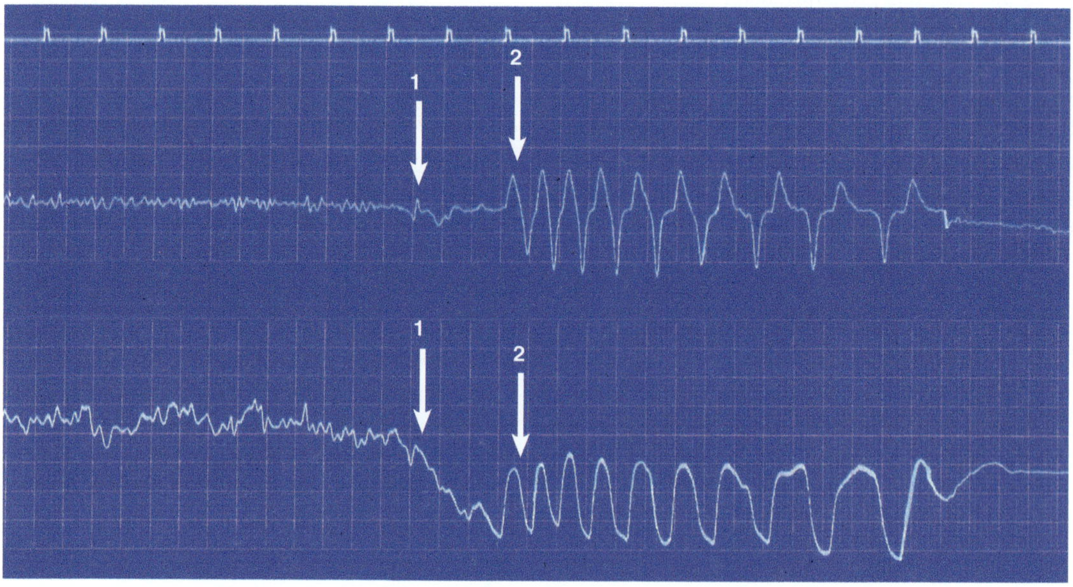

Fig. 10.5 Orgasmic seizure recorded in women by anal transducer (*superior line*) and vaginal transducer (*inferior line*). 1. Tonic phase more visible with the vaginal transducer; 2. Clonic phase: synchronized contractions of the external striated anal sphincter and the vaginal striated constrictor

stimulation by means of different techniques, hopefully not painful, with additional vaginal sensations including the famous G-spot, recently histologically demonstrated as a concentration of nerve fibres in the anterior wall of the vagina, around the urethra.

Finally, it is interesting to know that an orgasmic woman's typology exists from rapid to very slow induction and variable clonic contractions in terms of number, sexual rhythms... A frequent lack of knowledge causes disharmonies which feed sexologists activity [19, 20], but could be greatly improved by discussion without taboos between partners and especially by acquiring more information about the precise mechanisms we have described.

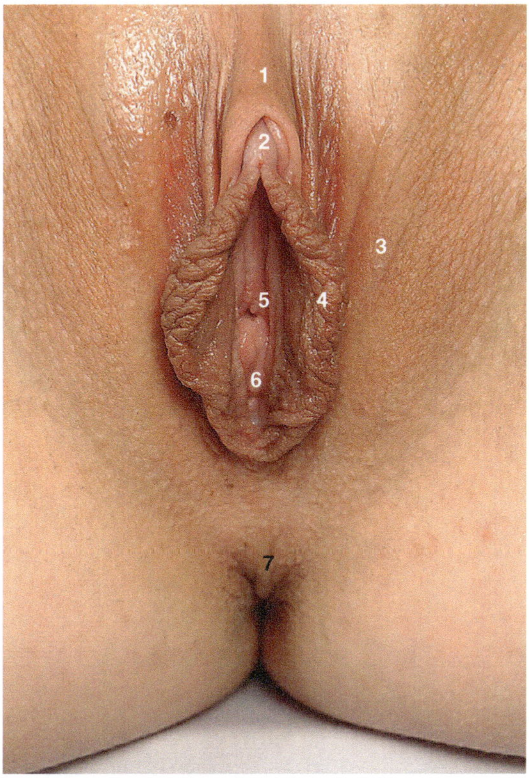

Fig. 10.6 External aspect of the vulva. 1. Prepuce of the clitoris; 2. Glans of the clitoris; 3. Labia majora; 4. Labia minora (nymphae); 5. Opening of the urethra; 6. Opening of the vagina; 7. Anus

References

1. Saper CB, Chou TC, Elmquist JK (2002) The need to feed: homeostatic and hedonic control of eating. Neuron 36(2):199–2111
2. Hockman CH (1972) Limbic system mechanics and autonomic function. Charles C Thomas, Springfield
3. Zwang G (1972) La fonction érotique. Robert Laffont, Paris
4. Fitzpatrick TJ (1975) The corpus cavernosum intercommunicating venous drainage system. J Urol 13:494–496
5. Newman JF, Northup JD (1981) Mechanisms of human penile erection: an overview. Urology 17(5): 399–408
6. Wespes E, Schulmann CC (1984) Parameters of erection. Br J Urol 56:416–417
7. Conti G (1952) L'érection du pénis humain et ses bases morphologico-vasculaires. Acta Anat 14:217
8. Virag R (1982) Intracavernous injection of papaverine for erectile failure. Lancet 2:938
9. Friedman D (2001) Mind of its own a cultural history of penis. Free Press, New York
10. Lue TF, Schmidt RA, Tanagho EA (1985) Erection cavernosography. J Urol 133:189
11. Bors E, Blinn KA (1959) Bulbo-cavernous reflex. J Urol 82:128
12. Schiavi RC (1981) Male erectile disorders. Annu Rev Med 32:309–320
13. Delcour C, Wespes E, Brion JP et al (1983) Explorations artérielle et veineuse de l'impuissance d'origine vasculaire. Ann Urol 17(6):356–359
14. Rabischong P (1977) Approche physiométrique de l'acte sexuel. Actualités gynécologiques, huitiéme série, Masson, pp 37–44
15. Kolberg S, Petersen I, Stener I (1962) Preliminary results of an electromyographic study of ejaculation. Acta Chir Scand 123:478–483
16. Kegel AH (1952) Sexual function of pubococcygeal muscle. West J Surg 60:521–524
17. Perry JD, Whipple B (1982) Multiple comportments of the female orgasm. In: Graber B (ed) Circumvaginal musculature and sexual function. Karger, New York
18. Zwang G (1996) Le sexe de la femme. La Musardine, Paris
19. Zwang G (1976) Abrégé de sexologie. Masson, Paris
20. Zwang G (1990) Pathologie sexuelle. Maloine, Paris

Conclusions

Trying to understand the anatomical structures in their technical complexity is the main objective of this book. Of course, a choice was made in the abundance of descriptive details accumulated by anatomists over time in order to foster the clearest possible perception of the logical constructive principles of human motor functions.

Once understood, they always appear extraordinarily cleverly done, with a biological engineering using the best selected original and reliable materials, even if normal wear gradually degrades the harmony of the whole, which is one of the basic rule of the biological game as difficult to accept as it can be.

The obvious concern for making as simple as possible the control of this complicated human machine is an interesting feature of its design, which is rather surprising when one has some technical knowledge in mechanics, computer science and control systems.

Man, creator of machines, found, thanks to advances in his own biological knowledge, the technical means to realise complex systems searching more and more for very small or micro- and nanotechnologies, which progressively are approaching the miniaturisation of biological elements that the microscope in all its forms is able to describe. The great complexity of the living structures is not perceived by the pilot of the human machine who, most of the time, has no knowledge of biology: he only needs little information to perfectly run difficult motor programmes. This is probably the most amazing point that we tried to demonstrate in this book.

Whatever conclusion we can draw from the detailed analysis of the functions with this comprehensive approach we propose, we cannot fail to be stroked by the real need for preprogramming of the wiring connections between organs. Putting them in place during ontogeny with multiple cell specialisations and neural integration starting and ending at the central nervous system is finally calculated very widely with billion of neurons processing and storing information.

We can quickly get lost in the labyrinth of pathways and nervous centres that our reductionist learning tendency cannot understand in its entirety. The holistic view of the whole machine, where various nervous areas richly interconnected, corresponding to completely different functions, are working at the same time, still eludes us. But we must permanently be astonished and amazed by so many wonderful systems to discover. This will push us to always stubbornly try to understand. Every living organism is an original programme with its specific characteristics, best suited to different technical problems, which gives to comparative anatomy all its educational value.

We can finally conclude that in such an organisation, there cannot be room for chance or improvisation, but rather think that a great rigorous method is required. Life, as we know it through more than three million different species, cannot be the result of a mutational unpredictable and non-programmable game, as unfortunately many still believe. If this book was able to convince those who will use it, it would have fully achieved one of its main objectives.

P. Rabischong, *Comprehensive Anatomy of Motor Functions*, DOI 10.1007/978-3-319-04169-8,

Index

P. Rabischong, *Comprehensive Anatomy of Motor Functions*, DOI 10.1007/978-3-319-04169-8,

GPSR Compliance

The European Union's (EU) General Product Safety Regulation (GPSR) is a set of rules that requires consumer products to be safe and our obligations to ensure this.

If you have any concerns about our products, you can contact us on ProductSafety@springernature.com

In case Publisher is established outside the EU, the EU authorized representative is:

Springer Nature Customer Service Center GmbH
Europaplatz 3
69115 Heidelberg, Germany

Batch number: 10683152

Printed by Printforce, the Netherlands